Tobacco

Tobacco

by

Anil Kumar Singh

2010

SATISH SERIAL PUBLISHING HOUSE

403, Express Tower, Commercial Complex, Azadpur, Delhi-110033 (India)

Phone : 011-27672852, Fax : 91-11-27672046

E-mail :info@satishserial.com, hkjain1975@yahoo.com

Website : www.satishserial.com

Published by :

SATISH SERIAL PUBLISHING HOUSE
403, Express Tower, Commercial Complex, Azadpur, Delhi-110033 (INDIA)
Phone : 011-27672852 Fax : 91-11-27672046
E-mail : info@satishserial.com, hkjain1975@yahoo.com

ISBN 81-89304-83-6

Typeset at: **Harminder** *for* **Laxmi Art Creations** # 98 11 48 23 28
Printed at: Salasar Imaging Systems, Delhi-35

Dedicated
To
Tobacco Nature, Nurtures
&
My Guru Ji
Sri H.P. Singh Ji

Prof. Panjab Singh
FNA AAS, FIAS, FNIE, FBRSI, FSEE
Ex Secretary, DARE, GOI & DG, ICAR &
Ex Vice Chancellor, Banaras Hindu University Varanasi - 221 005 (U.P.)

Foreword

Tobacco is an important commercial crop, could be grown under wide range of climatic and soil condition which play a crucial role in world as well as Indian economy. The cultivation practices and post harvest procedure are very considerably because of the widely varying conditions under which the crop is grown in India. Due to globalization and implication of international treaties it is high time to make Indian tobacco more competitive in world market, therefore, there is urgent need to develop suitable tobacco genotypes and equip them with latest production and protection technologies. To compete in International market our golden leafs should be raised under good agronomic management (GAP) condition to match with International standard, as trader can pay any thing if the produce is tagged with good physical condition and excellent quality (texture and aroma).

Tobacco is an excellent source of phytochemicals *viz.* nicotine, solanesol, organic acids, proteins, seed oil, vaccines and enzymes which are beneficial to the mankind. These secondary metabolites can be utilized more profitably for alternate uses such as protein from green leaf, lipids from seeds and nicotine sulphate from leaf waste. There is urgent need to intensify research for process developments and enhancing the yield of these value added product. Inclusion of seed production technique, tips to farmers to produce own seeds are worth appreciating. Addition of new topics like alternative uses of tobacco & alternates to tobacco crop and providing information about green tobacco sickness (GTS) for the first time is worth parsing. Additional information on tobacco uses, abuses, ethnobotany, tobacco alkaloids and other many things is worth to acknowledge. Special attention has been given to tobacco improvement and brief account on tobacco varietals development is also a sincere attempt to make book more useful. Tobacco economics and trade is worthy feature added in this presentation. Every aspect is dealt delicately in concentrated form to enhance the apatite to know more about this miracle crop.

I am glad that Dr AK Singh has brought almost all aspects in one volume. I am confident enough that timely publications of this book will surly help to scientist / researcher working in tobacco and tobacco growers for enhancing tobacco production to meet domestic and global needs and equally beneficial to the personnel engaged in pharmacy and other alternate medicines as well as social worker of tobacco abuse. In nutshell I hope this book must be useful to understand tobacco in better way without any preconceived ideas in healthy manner.

August, 2009
Varanasi

(Panjab Singh)

Dr. M. A. Khan
Director, ICAR-RCER
Patna 800 014

Message

Tobacco (*Nicotiana* sp.) is a native of America, introduced in to India in early 17th century. The narcotic quality of tobacco was discovered by Christopher Columbus during his course of voyage from Persia to Spain some time in 1492. The scientific name and the word nicotine are derived from French ambassador in Portugal *Jean Nicota.* Though tobacco fetches good income to the farmers but due several reasons discussed in detail in appropriate place is the major cause of fluctuations in its coverage area. Cast of production, plant protection challenges and marketing are main deciding factor of tobacco productions. With the improve crop management practices and availability of high yielding, short duration, photo & thermo insensitive varieties of range of crops provides ample scope for experimenting with several crops combinations to evolves tobacco based cropping systems to scale up the economic status of tobacco growers of this state.

To make Indian tobacco more competitive in world market therefore, there is urgent need to develop suitable tobacco genotypes and equip them with production and protection technology. There is no other remunerative and alternative crop to tobacco in many tobacco growing tracks. Tobacco growers are major beneficiary with agro technology of tobacco recommended from time to time.

To bring the gap between existing old practices and new research to improve the quantity and quality of tobacco, this publication has been brought out. I hope the tobacco farmers will follow the techniques of production in cropping systems. Other aspect related to tobacco post harvest management, processing, product manufctring and its trade, as well as inclusion of social and health aspect of producer and consumer are welcome addition. The book is well written off and having all most all aspect of tobacco and worthwhile to go through for any one wish to grow tobacco and know tobacco.

Dated: August, 2009

(M.A. KHAN)
Patna

Preface

Man survival has always been depends on the plant kingdom, as a source of food raw material and energy. This is simply because of the ability of green plant to transform light energy in to chemical energy through the noble process called photosynthesis. A plant produces a wide variety of chemical compounds, which can be divided primary and secondary metabolites (often constitute less than 1% of the total dry weigh). Nicotine is the most abundant alkaloid (Secondary metabolites) in cultivated tobacco. Secondary metabolites are responsible for execution of several important functions for plants survival in the environment *e.g.* in protection against ultraviolet light, as well as in various defiance related reactions.

The tobacco is believed to be originated from America. In India it was introduced by Portuguese during 1605 and used as *Hookah* tobacco. Initially tobacco was grown in Khera and Mehsana district of Gujarat. Importance of tobacco is much more in Indian context, being still primarily an agrarian country and nearly 75% population still lives in rural area. Despite occupying less than fraction of the country arable land (0.4 million hectare) tobacco is a profitable commercial crop producing 700 million kg of leaf. It provides livelihoods to nearly 38 million people, almost 80% of whom are in agricultural sector. Tobacco contributes significantly towards Indian economy being fourth largest exporter and provides 9% of total excise collection. This is to the tune of 3% of total Indian agricultural export. Though it fetches good income to the farmers but due several reasons discussed in detail in appropriate place is the major cause of fluctuations in its coverage area. Cast of production, plant protection challenges and marketing are main deciding factor of tobacco productions. As such there is no other remunerative and alternative crop to tobacco in many tobacco growing tracks, therefore, efforts are continuing to develop improved tobacco varieties of high yield and quality from time to time. Tobacco growers are major beneficiary with agro technology of tobacco recommended by the various agencies time to time. To bring the gap between existing old practices and new research to improve the quantity and quality of tobacco, publication has been brought out.

The purpose of writing of this book is two fold, *firstly* is the immediate need to educate people in the basic principles of tobacco cultivation so that tobacco growers may able to produce quality tobacco scientifically with modern tools and techniques and equips them with desired post harvest management handling for quality concern tobacco trade under WTO regime and earn smart profit and *secondly* to update their knowledge about the beneficial uses and abuses arises due to indiscriminate and unethical uses of tobacco and tobacco products and to create awareness and sense of understanding, responsibility, and ability to improve tobacco utilization as whole. This book presents wide spectrums of topic which cut across the boundaries of various scientific disciplines including tobacco trade and social aspect. I am optimistic enough to believe that such broad treatment will provide better understanding about "**tobacco**", although I am fully aware of the scope and purpose for which it is to be used. Material included in this book has been freely drawn from various sources, while writing this manuscript, I have drawn information's from various published books, reports, research papers, news letters, symposium proceedings, technical literature including information available on electronic media (web sites of respective agencies/organizations engaged in research and development on various aspects related to tobacco). I wish to acknowledge our special gratitude to the author (s) of various scientific and technical publications which forms the basis of this outcome. Various reports of All India Network Research Project on Tobacco were consulted and formulated our own method of presentation are duly acknowledge. I also wish to acknowledge on record, contributes/ authors/researcher whose name has been left from citation list in hurry. In presenting the information on each component the write up is mostly proportional to its importance and based on availability of literature. I hope the farmers will follow the production techniques mentioned in this book scrupulously and earn better profits through quality tobacco production as well as through efficient and competitive marketing. A brief account of tobacco economice, trade, tobacco products and companies involves in import and export, during pre and post WTO regime. Issues and concern about tobacco production and other related aspects like judicious and wisely uses to avoid tobacco abuses, health related aspects, Indian laws governing tobaccos is a useful addition to know tobacco in totality though it is in a capsulated form.

No work is perfect; there is always scope for its improvement, being first edition, scope exist, my sincere desire to convey my feeling to the academic community that I will welcome any suggestion at any time for improvement of this manuscript. All such suggestion will be gratefully acknowledge and incorporated in the further edition of this book. It is fervently hope that publication of this book will serve usefully as a basic text book to the entire spokesman to the tobacco development. If this book is able to create interest even in the fraction of tobacco grower, academic people and researcher who directly or indirectly concern to tobacco, I will consider my efforts will be rewarded.

August 05th 2009

Anil Kumar Singh
Patna

Acknowledgement

I gratefully acknowledge Dr. Punjab Singh (Ex. Secretary Department of Agricultural Research and Education, Govt. of India) for writing foreword to the book and for his keen interest in the publication of the book and positive criticisms throughout my professional life. I remain thankful to Dr. Mathura Rai, Director Indian Institute of Vegetable Research, Varanasi for sincere advice and encouragement. Special acknowledgement is made to Dr. MA Khan, Director ICAR-Research Complex for Eastern Region, Patna, for sparing time for compiling this manuscript and rendering useful suggestions for shaping up manuscript in the present format.

I would also like to express my particular thanks to Dr. RK Bata Head and Principal Scientist, Division of Land and Water Management for their stimulants comment and valuable suggestions. I would like to express my gratitude to Dr. B.K. Krishnamurthy, Director, Central Tobacco Research Institute and Dr. P.R.S. Reddy Project Coordinators All India Network Research Project on Tobacco, CTRI, Rajahmundry for giving spark for the task. Thanks are also due to Drs KA Singh, Director Indian Grassland and Fodder Research Institute, Jhansi. I gratefully acknowledge the following for reading the part of manuscript and offering suggestions to make the content technically sound. Drs. RD Singh, Principal Scientist (Agronomy), SS Singh Principal Scientist, (Agronomy), SK Singh, Principal Scientist (Agronomy) Naresh Chandra, Senior Scientist (Agricultural Economics) and Sanjeev Kumar, Senior Scientist (Agronomy).

I am deeply indebted to my colleagues Dr. Brijesh Singh (Senior Lecturer, Agricultural Botany) and Dr. Lal Singh (Assistant Professor, Agronomy, SKUAS&T, Srinagar) for their valuable suggestions and critical advices during manuscript preparation. I am equally thankful to my professional friend Drs. A. Rahman, LK Prasad, Manibhushan for moral support and other needful help rendered during the course of manuscript preparation. I am grateful to my colleague and seniors from National Bureau of Plant Genetic Resource, New Delhi, especially Drs Vandana Tyagi, Deep Chand, Nidhi Verma, Arjun Lal, and SP Singh for providing need full literature as and when required during the process of manuscript preparation. My friend from IARI, New Delhi, Dr. VB Patel, his help is utmost important during this manuscript preparation. I deeply appreciate to Shri AK Pandey, Smt. Meera Pandey, CD Rai, AS Mahapatra and Kamlesh Mahto for their pains taking assistance for manuscript preparation. My thank is also due to Shri Satish Kumar (Library section) for providing needful literature during preparation of manuscript.

I must thank for constant encouragement and assistance given by my family members especially my betterhalf Smt. Manjulata Singh and my little children *viz.*, daughter Gargi Singh and son Divyanshu Singh for taking their valuable time to compile this book as they ungrudgingly endured partial loss of my company when I was busy with the manuscript. I am equally grateful to the **"Society for Upliftment of Rural Economy" (SURE)** for making consistance request for the tobacco update and the book for the benefit of tobacco culture. Thank is also due to **"Satish Serial Publishing House"** for publishing the book in present shape.

Anil Kumar Singh

List of Reviewer

Dr. Pravesh Kumar
Division of Agronomy
BHU, Barkachha Campus
Mirzapur (Uttar Pradesh)

Dr. R.K. Batta
ICAR Research Complex for Eastern Region
Patna -800 014 (Bihar)

Dr. Deepchand
National Bureau of Plant Genetic Resources
New Delhi -110012

Dr. Nidhi Verma
National Bureau of Plant Genetic Resources
New Delhi -110012

Dr. Surendra Singh Gangawar
Plant Protection Specialist
Pilibheet (Uttar Pradesh)

Dr. Vandana Tyagi
National Bureau of Plant Genetic Resources
New Delhi- 110012

Dr. Lal Singh
Division of Agronomy
Sher - E- Kashmir University of Agricultural Science and Technology,
Shalimar, Kashmir, (J&K)

Dr. V. B. Patel
Division of Horticulture & Fruit Technology
IARI, New Delhi- 110012

Dr. Naresh Chandra
ICAR Research Complex for Eastern Region
Patna -800 014 (Bihar)

Dr. S.K. Dimree
CSAzad University of Agriculture and Technology
Kanpur -208 002 (Uttar Pradesh)

Dr. Brijesh Singh
Purvanchal University
Balia (Uttar Pradesh)

Dr. Surendra Singh
Institute of Agicultural Sciences
BHU, Varanasi (Uttar Pradesh)

Contents

Part-I : Crop Production

Part-II : Crop Protection (Stress Management)

Part-III : Crop Improvement

Part-IV : Tobacco Trade

Part-V : Issue and Concern

Part-VI : Others

PART-I
Crop Production

1

Introduction

Tobacco is the only commercial non-food crop that enters the world trade as a leaf. It is priced for its aroma, taste and flavour. Tobacco is an important commercial crop in view of revenue generation, export earnings and employment potential. It is aptly called as the golden leaf of India.

Tobacco is claimed to be miracle crop because of its nature and properties, which is used for range of purposes right from narcotics, pesticide, stimulants, and medicinal uses. Tobacco was one of the first material to be used as an insecticide as early as 1763. The use of finely powdered tobacco mix with water was recommended to destroy aphids without injury to the foliage. Tobacco is better known as a cause of cancer rather than a potential cure. In a recent research development in Arizona and elsewhere, believe that tobacco plants may hold the key to developing a personalized cancer vaccine as well as treatments for other diseases. America is said to be the native place of today's cultivated tobacco, from America it was spread all over the world, Portuguese people introduced in India during 1605 and used as *Hookah* tobacco. Gujarat is the first state where it was cultivated; initially

Hookah tobacco was grown in Khera and Mehsana district. India is still an agro based country and around 3/4 population is directly or indirectly earning its livelihood from agriculture. During its early days, tobacco was primarily used for hookah, cigarette and pipe purposes. Since then tobacco is used in various forms other than hookah and cigarette, now in India tobacco is consumed by human beings in seven most common forms which is known as type of tobacco mentioned in table 1.1.

Table 1.1 Type of Tobacco grown in various States of India

S. No.	Type of Tobacco	States cultivating
1	Flue Cured Virginia	Andhra Pradesh, Karnataka
2	Bidi	Gujarat and Nipani area of Karnataka
3	Cigar & Cheroot	Tamil Nadu & West Bengal
4	Hookah	Assam, West Bengal, Bihar, Uttar Pradesh Punjab
5	Chewing & Snuff	Tamil Nadu, West Bengal, Bihar, Gujarat, Assam, Uttar Pradesh & Punjab
6	Natu, Lanka & HDBRG (Burley hybrid)	Andhra Pradesh
7	Pikka	Orissa

Source: Adopted from Anonymous (2006a)

Historical Background

Tobacco is one of the important commercial crops in India owing to its significant role in Indian economy on account of its considerable contribution to the agriculture and commerce including export. Currently India holds second position in terms of tobacco production in the world arena just after China and before USA. Other major producing countries are Brazil, Turkey, Zimbabwe and Malawi.

Table 1.2 Global leaders in tobacco production

S. No.	Country	Production (in tones)		
		1998-99	2002-2003 (p)	2003-2004 (f)
1	China	2,010,250	2,365,988	2,224,481
2	United States	604,131	358,363	339,241

Contd...

3	India	572,200	592,000	595,000
4	Brazil	373,150	551,250	515,720
5	Turkey	217,570	133,812	142,190
6	Indonesia	123,653	144,700	135,000
7	Malawi	95,996	124,301	122,580
8	Greece	127,000	120,000	121,000
9	Italy	112,225	108,460	106,250
10	Argentina	98,100	106,000	97,700
11	Pakistan	84,636	84,721	86,389
12	Others	1,815,666	1,335,094	1,187,016
13	World total	6,234,577	6,024,689	5,672,567

P: preliminary; f: forecast,
Source: United States Department of Agriculture (USDA) 2007

Tobacco is the crop of temperate origin well adopted all over the world, it has unique adaptability and can be grown as rainfed and under assured water supply system, it is mostly transplanted but at some places it is still a direct sown crop. It requires heavy nutrition and weed free cultivation, its growing period varies from 70 days (Virginia tobacco) to 150 days (bidi tobacco). Time of tobacco raising varies in different stage depicted in table 1.3.

Table 1.3 Growing season of tobacco in different states in India

S. No	States	Growing period
1	Andhra Pradesh	December to February
2	Karnataka	July to September
3	Gujarat	September to November
4	Uttar Pradesh	September to November
5	Bihar	September to February- March

Tobacco Type

Indian tobacco is grouped in two major categories i.e. *Nicotiana tabacum* and *N. rustica. Nicotiana tabacum* occupies 10 per cent of total cultivated area but produces about 90 % volume of this crop, *N. rustica* is major share holder in terms of acreage but is poor yielder and contribute only around 10 per cent of whole. This type of tobacco is mainly confined to northern and north eastern state of India. Tobacco is further classified in to two major groups on the basis of consumption known as cigarette tobacco and non cigarette tobacco. Different kinds of tobacco under both groups are as under (Table 1.4).

Table 1.4 Cigarette and non cigarette tobacco type

Cigarette tobacco	Non cigarette tobacco
FCV	Bidi
Dark air/sun-cured Burley	Chewing tobacco
HDBRG	Cigar
Oriental	Hookah
DWFC	Snuff

Source: www.indiantobacco.com

Tobacco is further classified based on their type of product/ consumption in India

Bidi Tobacco

This occupies 30%-35% of the total area under tobacco cultivation and is grown in Gujarat, Karnataka and Maharashtra. Nearly 85% of the world's bidi tobacco is grown in India. The average yield varies between 1000 and 1700 kg/hectare in Karnataka and Gujarat, respectively.

FCV Tobacco

It is grown in Andhra Pradesh and Karnataka. A small quantity of this tobacco is also grown in Orissa (Rayagada) and Maharashtra (Gadchiroli). It is the most remunerative crop due to the demand from domestic cigarette manufacturers and importers. In 2000-2001, the average yield of FCV tobacco was 1710 kg/hectare. Fifty per cent of the FCV grown in India is used by the domestic cigarette industry while the remaining is exported.

Burley Tobacco

This air-cured form of tobacco is used for cigarette blends and is grown in Andhra Pradesh.

Hookah Tobacco (*Nicotiana rustica*) Variety

It is used for smoking and is grown in Uttar Pradesh, Bihar, West Bengal and Orissa.

Chewing Tobacco

This is used for gutka, snuff and pipe tobacco. It is grown in Tamil Nadu, Uttar Pradesh, Bihar, West Bengal and Orissa.

Natu Tobacco

It is sun-cured and grown in Andhra Pradesh.

Cigar Tobacco

This type of tobacco is grown in West Bengal and Tamil Nadu.

Cheroot Tobacco

It is grown in Tamil Nadu and Andhra Pradesh.

Pikka Tobacco

This type of tobacco is grown in Orissa.

History of Tobacco Development in India

The foundation of tobacco use and its cultivation was laid by portages during sixteen century but concrete efforts toward its land mark development was started by East India Company at flag end of eighteenth century by establishing botanical garden. Chronologically the development in tobacco production and its commercial exploitation is given here.

Year	Milestone Development
1787	Establishment of the Botanical Gardens at Sibpur, Calcutta (trials to grow tobacco were conducted).
1829	The government decided to promote cultivation of superior tobacco. Imported seeds were made available to the Agri-horticulture Society of Calcutta and trials on an improved variety continued for several years.
1875	Attempts were made to produce Virginia tobacco at Ghazipur in Uttar Pradesh and growing and curing of tobacco continued at the Pusa farm in Bihar.
1901	The British and American Tobacco Company expanded their trade into India and set up three companies, which later together became the Imperial Tobacco Company India, i.e. the present Indian Tobacco Company (ITC) Ltd.

1903 The establishment of the Imperial Agricultural Research Institute and College at the Pusa farm initiated the cultivation of a new variety of tobacco.

1920 The Indian Leaf Tobacco Division (ILTD) of ITC experimented on the black soils of Guntur, Andhra Pradesh.

1928 successfully cultivated Virginia tobacco in Guntur, Andhra Pradesh.

1929 Commercial and large-scale production of tobacco was initiated by the ILTD. The company established demonstration barns, provided technical guidance to them and encouraged local farmers to grow tobacco by providing financial assistance to construct barns, purchase fertilizers, wood fuel, etc. Slowly, tobacco cultivation spread to all the coastal districts of Andhra Pradesh.

1933 The ILTD introduced flue-cured Virginia (FCV) tobacco into the international market.

1936 A cigarette tobacco research station was established in Guntur to study the effect of soil and manure on the flavour of tobacco.

1937 Tobacco cultivation was introduced in Karnataka, (Mysore State) by the Mysore Tobacco Company Ltd.

1938 India produced 499 million kg of tobacco and ranked second in production next to the USA (628.7 million kg) China was the third largest producer (446.8 million kg).

1940s Cultivation of FCV tobacco was initiated in north Bihar (1940), Uttar Pradesh (1940) and Gujarat (1945- 1946). In the first year (1943-1944), excise revenue from tobacco was Rs. 9.65 Crore. This led the government to explore other types of tax on tobacco as well as initiate measures to promote tobacco cultivation.

1943 The government set apart an annual, non-lapsable grant of Rs. 10 lakh from the proceeds of excise duty imposed to extend the cultivation of high-quality leaf and improve the production of tobacco.

1945	The Tobacco Grading Inspectorate was established at Guntur to ensure the quality control of tobacco for exports, and the Indian Central Tobacco Committee (ICTC) was set up to look after the cultivation, technical and economic aspects of tobacco cultivation in India.
1947-1950	The Indian Central Tobacco Committee (ICTC) established the Central Tobacco Research Institute for undertaking research on cigarettes and the Lanka type of tobacco. Later, four research stations were established in Tamil Nadu (in 1948 for cigarette, cheroot and chewing tobacco), Bihar (in 1950 for hookah and chewing tobacco), West Bengal (in 1952 for wrapper and hookah tobacco) and Karnataka (in 1957 for flue cured Virginia [FCV] tobacco) to study the different types of tobacco.
1956	The Tobacco Export Promotion Council (TEPC) was established to support, protect and promote the export of tobacco.
1965	The ICTC was abolished.
1966	The Directorate of Tobacco Development was established to act as an information agency on tobacco production, trade, marketing, export, consumption, etc.
1975	The Tobacco Board was constituted under the Tobacco Act, 1975, replacing the TEPC. The Tobacco Board is responsible for regulating the cultivation, production, marketing and export of FCV tobacco.
1980-81	The Agricultural Price Commission recommended a minimum support price for FCV tobacco grown in light and black soils.
1983	The National Cooperative Tobacco Growers Federation Ltd. (TOBACCOFED) was established by the Ministry of Agriculture and Rural Development to promote the production and marketing of Non-FCV tobacco in the country.
1984	Auction sale of FCV tobacco was introduced for the first time by the Tobacco Board in Karnataka and in Andhra Pradesh in 1985.

(Adapted from Tobacco Board 2002; Directorate of Tobacco Development 1997)

Tobacco Growing Tracts of India

1. Guntur Area

It is famous for cigarette and Natu types. The area consists of Guntur, Krishna, East-west Godavari and Telangana districts of A.P.

2. Charottar Area of Gujarat

It is famous for the cultivation of Bidi tobacco and is limited to khera district.

3. Nipani Area

This is bidi tobacco growing area which consists of Belgaon, Satara, Kolhapur, Songali etc. districts.

4. North Bihar and Bengal Area

Here mainly hookah tobacco but in limited area chewing and snuff types are also grown. This area consists of Muzaffarpur, Darbhanga, Poornia, Cooch Behar, Jalpaiguri, Berhampur and Dinajpur districts.

5. U.P. and Punjab Area

The area is famous for Hookah and Chewing type of tobacco consisting of Furrukhabad, Mainpuri, Sitapur, Badaun, Muradabad, Jullundhar, and Firozpur districts.

6. South Tamil Nadu Area

It includes Coimbatore and Tamil Nadu where chewing, cheroot, Sunff and wrapper types of tobacco are grown.

Table 1.5 Estimate of area, production and yield of tobacco in India from 1949-50 to 2005-06

Year	Area (Lakh ha.)	Production (Lakh tones)	Yield (Kg/ha)
1950-51	3.6	2.6	731
1955-56	4.1	3.0	739
1960-61	4.0	3.1	766
1965-66	3.8	2.9	778
1970-71	4.5	3.6	810
1975-76	3.7	3.5	950

Contd...

1980-81	4.5	4.8	1065
1985-86	4.0	4.4	1111
1990-91	4.1	5.6	1353
1995-96	3.9	5.4	1356
1998-99	4.6	7.0	1522
2000-01	2.62	3.45	1316
2002-03	3.26	4.91	1505
2005-06	372.8	552.2	1481

(*Source:* Directorate of Economics & Statistics, Government of India, New Delhi 110 001)

Due to globalization and implication of international treaties & Conventions viz. TRIPS agreement under WTO, CBD, ITPGRFA it is high time to make Indian tobacco more competitive in world market therefore, there is urgent need to develop suitable tobacco genotypes and equip them with production and protection technology, and integrated pest management (IPM) strategies. Despite occupying fraction of the country arable land (4 lakh hectare), tobacco is a profitable commercial crop producing 7000 lakh kg of leaf. It provides employment for about 6.5 Million people, including growers and curers. In addition, cigarette industry provides employment for 5 Million people and bidi industry for about 30 Million people, almost 80% of whom are in agricultural sector.

The profitability of tobacco cultivation has led to a mismatch between demand and supply resulting in a glut situation and crashing prices. Excess production and carry-over with the trade led to crop holiday in Andhra Pradesh during 2000-2001. It is expected that tobacco production will further plunge due to efforts in different countries to liquidate stocks and lower volumes of cigarette use in selected markets. There is a down word trend in FCV exports largely due to poor quality of grading and other issues like pesticide residues. Thus, excess production, declining consumption on health related problems, trade barriers and increasing anti tobacco campaign world over are the major factors affecting Indian tobacco prospects. Tobacco accounts for 3.5 per cent (by volume) of India's agri exports to the exchequer, with cigarettes contributing to 80 per cent of this amount. The area under tobacco crop in India is given in table 1.8.

Table 1.6 All India estimates of state-wise area and production of tobacco

State	Area	% of Total Area	Production	% of Total Production	Cumulative % of Total Production	Yield	Area	% of Total Area	Production	% of Total Production	Cumulative % of Total Production	Yield	Under Irrigation During
	2005-06						2003-04						
Andhra Pradesh	134.0	35.94	198.0	35.86	35.86	1478	132.5	36.15	195.5	35.60	35.60	1475	28.5
Uttar Pradesh	20.2	5.42	121.5	22.00	57.86	6015	22.0	6.00	119.0	21.67	57.28	5409	99.8
Gujarat	71.3	19.13	113.9	20.63	78.49	1597	71.3	19.45	113.9	20.74	78.02	1597	76.8
Karnataka	101.0	27.09	62.0	11.23	89.71	614	91.0	24.83	62.0	11.29	89.31	681	6.5
Bihar	14.9	4.00	17.3	3.13	92.85	1161	16.1	4.39	18.7	3.41	92.72	1161	79.8
Tamil Nadu	5.1	1.37	7.8	1.41	94.26	1435	4.6	1.26	6.9	1.26	93.97	1435	95.4
Maharashtra	6.0	1.61	7.0	1.27	95.53	1167	6.0	1.64	7.0	1.27	95.25	1167	9.4
Others	20.3	5.45	24.7	4.47	100.00	@	23.0	6.28	26.1	4.75	100.00		-
All India	372.8	100.00	552.2	100.00		1481	366.5	100.00	549.1	100.00		1498	46.2

Source: Directorate of Economics & Statistics, Government of India, New Delhi 110 001

Table1.7 Mandated type of tobacco research at different centres

S. No.	Name of the Centre	Type of tobacco
A. Main Centre		
1.	Rajahmundry, C.T.RI. Rajahmundry - 533105	FCV
2	Shimogatrom 1976-77 RARS, Naveli, Shimoga-577 204 Bangalore till 1976 (UAS, Bangalore)	FCV in light soils Burley and FCV in light soils
3.	Anand ,BTRS, Anand - 388 110 (AAU, Anand)	Bidi, chewing and Rustica
4.	Pusa, CRP, Pusa - 848 126	Chewing
B. Sub Centre		
1.	Dinhata CTRI RS, Dinhata - 736 135	Cigar tiller and wrapper as well as chewing (Jati and Motihari)
2.	Hunsur C.TRI RS, Hunsur - 571 105	FCV
3.	Guntur CTRI RS, Guntur - 522 004	FCV and Natu (rainfed)
4.	Nandyal trom 1.4.1992 RARS, Nandyal- 518 503 Venkataramannagudem (From 1.6.1978 to 31.3.1992) Kavali (trom Qct. 1970 to 31.5.1978) (ANGRAU, Hyderabad)	Cigarette Natu and Bidi Natu (irrigated) FCV
5.	Nipani ARS, Nipani - 591 237 (UAS, Dharwad)	Bidi
6.	Berhampur trom 1987-88 Pulse Research Centre Berhampur - 761 001 (OUAT, Bhubaneswar)	*Pikka* (Natu)
7.	Araul trom 1.4.2004 Araul, Kanpur Nagar district, U.P. Gursahaiganj from 1.7.2001 Saraimiran from 1988-89 to 30.6.2001 (CSAUAT, Kanpur)	*Rustica* (Hookah), Bidi *Rustica* (Hookah), Bidi *Rustica* (Hookah)
C. Voluntary centre		
8.	Ladol ,(GAU, Anand)	*Rustica*
9.	Kandukur, CTRI RS, Kandukur- 523 105	FCV
14	Jeelugumilli, CTRI RS, Jeelugumilli- 534 456	FCV, Irrigated natu
15	Vedasandur, CTRI RS, Vedasandur- 624 710	Chewing

Source: Adopted form Anonymous, 2006a

Table 1.8 Area, production and yield of tobacco leaves

Particulars	2001-02	2002-03	2003-04	2004-05	2005-06
Area (M ha)	0.35	0.33	0.37	0.37	0.37
Production (M T)	0.55	0.49	0.55	0.55	0.55
Productivity (Kg/ha)	1565	1506	1486	1498	1481

(Source: Adopted form Anonymous, 2006a)

Nicotine Content in Indian Tobacco

Tobacco is crop mainly cultivated for its secondary metabolite known as nicotine. The sheet of nicotine synthesis is root and translocated to leaf via stem through phloem tissues. In India several types of tobacco is grown for various purposes as listed in table 1.1, the nicotine content of particular type of tobacco differ, the type of tobacco and nicotine content in them is presented in table 1.9.

Number of factors affects the nicotine content in any tobacco, to name few, agronomic management practices, plant nutrition prevailing environmental conditions, genotype, method of harvesting and curing etc. Nicotine content in Indian tobacco is reported as low as 0.5% in oriental tobacco produced from Gujarat and Andhra Pradesh and as high as 8.0% in bidi tobacco in general. Details of nicotine contains in various tobacco is given in table 1.9.

Table 1.9 Nicotine Contents of Indian tobacco

S. No.	Tobacco type	Growing area	Nicotine content (%)
1	Flue Cured Virginia	Traditional tobacco area	1.0 to 2.50
2	Flue Cured Virginia	Mysore	1.5 to 1.80
3	Flue Cured Virginia	Northern Light Soil (NLS)	1.5 to 3.00
4	Flue Cured Virginia	Southern Light Soil (SLS)	0.0 to 1.10
5	Sun cured country	Kurnool and Telangna	1.0 to 3.25
6	Sun cured natu	Eluru	3.0 to 5.50
7	Kentucky fire cured	Andhra Pradesh	2.5 to 3.75
8	Bidi	Gujarat and Karnataka	5.0 to 8.00
9	Cigar wrapper	West Bengal	3.0 to 4.90
10	Cigar filler	West Bengal type	0.9 to 1.90

Contd...

11	Cigar filler	Tamil Nadu type	2.0 to 4.00
12	Lanka	Andhra Pradesh	2.5 to 4.50
13	Cheroot	Andhra Pradesh	1.8 to 3.00
14	Chewing *rustica*	Calcutta	3.0 to 6.00
15	Chewing tobacco	Bihar	3.0 to 4.00
16	Chewing (Motihari and Jati)	West Bengal	3.0 to 6.00
17	Sun cured	Tamil Nadu	3.5 to 4.50
18	Smoke cured	Tamil Nadu	2.5 to 3.50
19	Pictured	Tamil Nadu	3.5 to 4.50
20	Indian Budey (Monsoon)	Karnataka	1.25 to 3.50
21	Burley (Traditional)	Gujarat, Andhra Pradesh	2.5 to 4.50
22	Oriental	Gujarat, Andhra Pradesh	05 to 0.80
23	Bright cigarette air-cured	Andhra Pradesh, Karnataka	1.5 to 2.70

(*Source:* www.indiantobacco.com)

❑❑❑

2

Climate and Soil

Climate and soil is first and for most important factor greatly influence the successful completion of life cycle of any living things, though tobacco is flourished under divers agroclimatic conditions, however the different tobacco type require specific climate and soil. In this chapter emphasis has been given to soil and climate for various tobacco type grown in various part of the country.

Tobacco is grown under a wide range of climates, basically being tropical crop (in origin), but it is now grown from subtropical to temperate region of world. It's only because of the cultivated tobacco is behave largely as a day-neutral in its response to flowering, except for some short-day varieties. Tobacco plant is very sensitive to the environment. Ideal conditions required for successful production of high quality leaves needs liberal and well-distributed rainfall during active vegetative growth stage, long day lengths, a mean temperature of 80°F (26°C) during growing season, a high relative humidity of 70-80%, with a frost-free period of 90 to 120 days from transplanting to last harvest of leaves. Well drained, relatively poor in humus content but should contain a good amount of potash, phosphoric acid, and iron, slightly acidic to neutral soil with good nutrient supplying capacity with optimum pH ranges from 5.5 to 7.0 is considered ideal soil for tobacco cultivation.

Tobacco being a short duration crop, it can be grown at any latitude and altitude provided that required optimum temperature and other parameters meet out satisfactorily. The most important factor being temperature, the optimum temperature for better germination is around 28°C but it may germinate at a temperature ranging 15°C to 35°C. However optimum mean daily temperature for growth is between 20 and 30°C. A dry period is required for ripening and harvesting of the leaves. The lower temperatures resulted in small, relatively wide leaves at the base of the plant and elongated leaves at the top. Higher temperatures tended to reverse the normal shape and size, ordering sequence of leaves; at the highest temperature condition the largest, as well as the broadest, leaves were developed at the uppermost stalk positions. The specific leaf weight was increased by an increase in the duration of the high intensity light and by decreases in temperature. The increased light duration also enhance the relative width of leaves. Lower temperature coupled with bright sunshine and low humidity during early growth stage of crop plants help in build up of thick and heavy bodied leaf, thus producing higher leaf yield with desirable physical and chemical quality.

Tobacco needs 100 cm rainfall well distributed throughout growing period. When grown as rainfed crop, it requires at least about 500 mm of well distributed rainfall during crop growing season. It is not usually grown where rainfall exceeds 1,200 mm during the season. Rainfall less than 75cm may results leathery leaves with reduced leaf size. Prolonged rainfall and late onset of winter may result in more incidence of leaf curl disease. Excess rainfall results in thin, light weight leaves. In nutshell cloudy weather with drizzling sun after transplanting is good for seedling establishment. Intermittent winter rains from October to January are good for tobacco growth and development as well as expansion of leaves, but continuous rains and high humidity particularly after 15th December may result in severe incidence of leaf spot and black shank disease. Leaf area, shape, and specific leaf weight (dry weight per unit leaf surface area) were determined at six stalk positions. Light rainfall at early growth stage and at grand growth stage bright sunshine with occasional moderate rain are conductive for better growth and development of quality leaves. Rainfall during maturity and harvesting period washes away the gummy substances from the leaf surface, spoils the leaf colour, as well as quality and reduces the yield. If rain is received just before harvesting, harvesting is postponed for 8-10 days in order to allow the gummy substances to recoup on leaf surface and to attain to normal maturity *i.e.* desirable yellowing stage.

In India, all types of tobaccos except cigar, filler and binder tobaccos and chewing tobacco of south India receive very little rain during growing season. The later tobaccos of south India require irrigation since the rainfall received is only around 400 mm during the crop season. The hookah and chewing tobacco types grown in Bihar and Bengal are given two irrigations while the hookah tobacco of North West India is grown only with single irrigation. The FCV tobacco of Andhra Pradesh grown in post rainy season is, generally, not irrigated since it is grown on deep black soils with high soil moisture retentive capacity.

Relative humidity also play crucial role in determining the quality of produce. Optimum relative humidity for tobacco leaves production as well as for better curing is 80-90 %. Relative humidity below 70% cause serious harm and leaves become unworkable due to very brittle in nature. Wind velocity and its direction is one of important weather parameter for successful tobacco cultivation. Speedy wind at any stages of crop cycle after seedling establishment is serious concern which caused harmful effect as in general tobacco suffers severe injury from strong winds and hail due to large size and tender nature of its leaf. Wind storms at later stage when plan has full grown canopy, are detrimental and affects quality and production adversely because the leaves splits and reduce the quality. In order to avoid development of mould or blackening of leaves, the harvested crop should be dried in dry cloddy field and provided with shade in the night to protect it from dewfall. On the other hand crop harvested during low humidity and intense sun heat in February-March, the harvested crop should be dried and cured on grassy field as it helps check fast drying and loss of leaf colour. During fermentation good sunshine and low humidity with westerly wind is desirable for development of good colour, desirable aroma, good taste and white incrustation on the leaf surface which is sign of good quality leaf. Sun-cured or oriental tobacco requires a relatively dry climate to develop its full aroma. Tobacco has unique adaptability to be grown as rainfed as well as under assured water supply system, it is mostly transplanted but some place it is still a direct sown crop. It requires heavy nutrition and weed free cultivation, to reap best yield and quality tobacco leaves, the ideal weather condition and rainfall distribution is as follows in table 2.1 to produce best quality tobacco leaves.

Table 2.1 Ideal weather condition and rainfall required during growth and development of tobacco

Growth and developmental stage	Ideal weather condition and rainfall
Seedling establishment (soon after planting)	Cloudy weather with drizzling
Early growth stage	Light rainfall
Peak growth stage	Bright sun shine with occasional moderate rain
Maturity (Repining of leaves)	No rains at all because the rain will washout the gums from the leaves surface resulting in poor curing

Note: Heavy rains may split the leaves and speedy wind may leads to crop lodging, both impair the quality of produce.

Growing Season

Tobacco growing season in India is greatly varies in different state even in one state its also depends on kind of crop (rainfed or irrigated) and type of tobacco. Length of growing period of tobacco crop is as short as 70 days in case of Virginia tobacco to 150 days for bidi tobacco. A generalized time of tobacco cultivation practiced in different state depicted in table 2.2.

Table 2.2 Growing season of tobacco in different states in India

S. No	States	Growing period
1	Andhra Pradesh	December to February
2	Karnataka	July to September
3	Gujarat	September to November
4	Uttar Pradesh	September to November
5	Bihar	September to February- March

Agro climatic conditions of important tobacco growing states in India are as under.

Andhra Pradesh

Andhra Pradesh is leading tobacco producer of India engaged mainly in FCV production. The crop is exclusively grown as rainfed on conserved soil moisture. *Rabi* is main tobacco cultivating season extending from October-November to February-March. Rainfall is received mainly from South-West monsoon is around 800-1000 mm starting from June to October. Unprecedented rain by the North- East monsoon, though it's occasionally through cyclonic storms, becomes very useful to this crop and boost yields. The FCV crop needed in general good sun light, low evaporation, high relative humidity, sustained soil moisture in black soils and free from wet footing and cyclonic storms. The another important tobacco type is "Lanka tobacco" needs 80 to 90% relative humidity during plant growth period that is from mid October to November ending with a mean day temperature of about 26.7°C. The peculiarity of the Lanka tobacco is its requirements of gradual incremental temperature with duration till its maturity. The crop needs 30 to 32°C temperature so that the crop will come to proper maturity.

Bihar

Quality chewing tobacco is produced in "**Saharsha Belt of Bihar**" famous in all over India. Climate and weather prevailing in North Bihar during *rabi* season is quite favorable for quality leaf production of tobacco especially chewing and hookah type as well as some other types. The warm and humid weather of 2nd fortnight of October is considered congenial for establishment of transplants with desirable initial growth. Low temperature, bright sunshine and low humidity during early growth stage of crop plants help, in build-up of thick and heavy bodied leaf, thus producing higher leaf yield with desirable physical and chemical quality parameters. Intermittent winter rains from October to January are good for tobacco growth and development as well as expansion of leaves. December second fortnight onwards continuous rains and high humidity may result in severe incidence of leaf spot and black shank diseases. Long lasting rains beyond November along with delay in onset of winter may cause heavy leaf curl attack.

Karnataka

Mysore produces FCV tobacco famous for its taste and aroma. The transitional zone of Mysore tract is endowed with ideal climatic conditions like lower diurnal temperatures, higher relative humidity and monsoonic rains during the crop growth period from May to September which are highly conducive for production of high quality flue cured tobacco. Mysore district is under the influence of tropical monsoonic climate with an average annual rainfall varying from 650 mm to 1000 mm of which over 80% is received during the south-west monsoon period from May to October which coincides with the FCV tobacco crop season. February to April is the dry summer months. The mean minimum temperature in the cold winter month of January is around 15 °C while the mean minimum temperature in the hot summer month of April is 34 °C.

West Bengal

West Bengal is known for its quality "**Motihari tobacco**" basically used for hookah and chewing purposes. Agro-climatologically, this zone is characterized by humid and damp weather throughout the year. The day temperature varies from as low as 5 °C during December-January (winter) to as high as 35 °C during April-May (summer). The rains of North Bengal are long and persistent and the average annual precipitation exceeds 3000 mm, major part of which is received during the month of June, July and August. Relative humidity of the atmosphere is about 85% during April and May facilitating easy handling and conditioning of cured leaf.

Soils

Quality of tobacco is a combined result of taste and aroma; greatly influenced by the soil conditions and its properties. As a general rule, heavy and naturally fertile soils are preferred for cigar, fillers, pipe, hookah and chewing type tobaccos. Light sandy and somewhat less fertile soils are the best suited for the bright curing thinner and high priced leaves such as cigarettes, bidi, cigar wrappers and high grade pipe tobacco. The crop is very sensitive to water logging and demands for well aerated and drained soils. Tobacco is adapted to moderately acidic soils, optimum pH ranges from 5 to 6.5. Quality of the leaves is affected by soil salinity.

Well drained, neutral to slightly alkaline soil with good nutrient supplying capacity is considered ideal for chewing and hookah tobacco cultivation. Saline and alkaline soils (usher soils) are not considered suitable for tobacco cultivation and should be avoided. However, after proper reclamation by applying 2-3 tones of pyrites per ha before tobacco cultivation, such type of soils can also be used for tobacco cultivation with suitable agronomical management and addition of sufficient organic manures including green manuring, paddy cultivation etc. Non saline, non-calcareous alluvial soil with neutral pH generally produces tobacco leaf with low nicotine and light colour.

Well aerated soils of tobacco field due to high organic matter content are most favourable for microbial activities and mineralization of unavailable or bound nutrients. Biosynthetic substances produced by fast multiplying microorganisms helps growth and development of plants as well as maintain good soil health. Tobacco, in general, needs an open aerated, well drained light textured, rich in P_2O_5 and K_2O and Fe, but poor in organic matter contents. Sodic soils are unfit for tobacco production because the plants absorb a lot of chloride ions which result in a poor burning quality of leaves. Mild acidic soil reaction is always better for the production of superior quality leaves. Tobacco from low-lying lands has been found to be very poor in burning quality. It absorbs moisture and deteriorates in colour during storage. Physical and chemical properties of soil influence the quality of tobacco leaves, therefore, soil should be selected according to the purpose for which crop is being grown.

The suitability of soil, as mentioned in table 2.3 differs according to the types of tobacco to be grown.

Table 2.3 Tobacco finished product and suitable soil type

S. No	Finished product	Suitable soil type
1	Cigarette	Light sandy soils
2	Filler (Cigar)	Light sandy soils
3	Binder (Cigar)	Heavy soils
4	Wrapper (Cigar)	Distinctly heavy soils
5	Bidi	Light to medium loams of old' alluvium origin
6	Hookah. Chewing & Snuff tobacco	Sandy to silt loam alluvial soils

In India mainly two species *N. rustica* and *N. tabacum* are widely cultivated both the species differ in their soil requirement drastically. ***N. rustica*** thrives best in heavy, low lying soils and contrary to this the *N. tabacum* require light and high lands. Though the ideal tobacco soils greatly vary with is species to be grown but in general a good tobacco soil is one which has following characteristics:

(i) Low native fertility status

(ii) Good water holding capacity with

(iii) Excellent drainage facility

(iv) Medium sandy surface soil

(v) Friable sandy loam to loamy sub-surface soil.

(vi) Low humus or organic matter content.

(vii) Neutral reaction (pH 5.5-7.0 preferably).

Importance of soil in relation to tobacco leaf quality has been widely recognized. Apparently, apart from soil physicochemical properties, certain other factors such as aeration, drainage, micro relief and such similar factors, which cannot be easily measured and controlled, profoundly influence the yield and quality of tobacco. Owing to this, in specific tobacco producing area, certain fields are reputed to produce tobacco of excellent quality while the produce from other fields very close to them are relatively poor.

FCV Tobacco Soils of Andhra Pradesh and Karnataka

In India, FCV tobacco is grown under different conditions due to monsoon dominated rainfall, which is in general terminate after mid-November, there are practically no rains and the crop has to depend on stored soil moisture. In Andhra Pradesh, FCV tobacco is grown on a variety of soils; heavy black clays containing 50-80 per cent clay throughout the soil profile of Andhra Pradesh considered most suitable soils for the purpose. Karnataka FCV tobacco grown on red loams soils with one to two light irrigations to supplement rainfall, has a better texture and is more aromatic than that grown on heavy soils of Andhra Pradesh. Tobacco grown on saline soils absorbs moisture quickly before and after redrying; changes colour rapidly during storage and have poor burning quality due to high chlorine content. Tobacco from low lying areas and from wetlands loose colour rapidly while that from high level as has lemon yellow colour but is thin and brittle and burns very fast.

Cigar Tobacco Soils of Tamil Nadu

Three kinds of tobaccos are needed to serve as filler, binder and wrapper in the manufacturing of cigars. These three types of tobacco are produced on a variety of soils to meet the three different requirements. Sandy to loamy red soils of Coimbatore, Tiruchirapalli and Madurai districts of Tamil Nadu are best suited soils for cigar and binder tobaccos In India. This crop needs larger quantities of nitrogen, potassium, calcium and magnesium. Indian cigar tobacco contains 2-4 per cent chlorine attributed by high levels of chlorine in irrigation waters, since the cigar tobacco receive around 20-25 irrigations. As per International Standard maximum permissible limit is of 1.0 per cent chlorine in American cigar tobacco.

Bidi Tobacco Soils Gujarat and Karnataka

Bidi tobacco is mainly cultivated in Charotar zone in north Gujarat and Nipani in Karnataka. Bidi tobacco requires light to medium loams of old alluvium origin because the spangle development, which is of vital characteristics in bidi tobacco, is found to be more pronounce under light loams soils conditions . In the Charotar area, it is grown on sandy to sandy loam soils called garadu soils. In Nipani area, bidi tobacco is grown on silt loams. The chemical composition of irrigation water greatly influences the quality of produce, if the bidi tobacco gets irrigated.

North India Soils of Hookah and Chewing Tobacco

Hookah and chewing tobacco are grown on a wide range of soils in north India. Major producer of this type of tobacco is Uttar Pradesh, Bihar, West Bengal and Punjab. In Uttar Pradesh this tobacco type is mainly grown on saline alluvial soils and in Punjab this crop is basically cultivated on sandy loam soils. In Bihar, these are grown on sandy to silt loam alluvial soils of northern tract containing free calcium carbonate with alkaline reaction. In Bengal, they are grown on sandy to loamy soils.

Chewing Tobacco Soils of Tamil Nadu

Though the chewing and hookah tobacco are principally cultivated in Northern States in India but chewing tobacco is also grown in Tamil Nadu, on well drained red soils of districts Coimbatore, South Arcot, Tanjore, Salem and Madurai . As chewing tobacco is a heavy remover of nitrogen and potassium, it responds well to higher doses of nitrogen and potassium. Chlorine enriches the chewing tobacco to retain the sauces and essences added to it during processing , hence chewing tobacco is preferably irrigated with waters rich in soluble salts, particularly, chlorides.

❑❑❑

3

Agro-Techniques for Growing Tobacco

Raising healthy tobacco is art based on scientific input as we know that tobacco is basically produced for its quality leaves; agrotechniques involves right from choosing appropriate site for raising tobacco seedling to harvesting of fully mature crop. A series of activities bound to happen having its own importance. Timing to perform any activities with precision is root to successful tobacco cultivation. Tobacco seed are very minute and delicate too, needs care full expert handling to grow healthy seedlimg, topping desuckering etc to name few are very much importance in quality conscious tobacco leaves production. This chapter is dedicated to various agrotechniques of different tobacco type.

Selection of Site and Preparation of the Nursery Beds

The primary objective is to raise healthy nursery with proper management for planting of the required tobacco area. The nursery bed should be made always on higher elevation so that they do not get flooded any time but their must be assured irrigation facility. The soil of seedbed preferably sandy texture so they are open and well aerated. It is advisable that to take healthy seedling, one should change the place of nursery every year because it reduces the chances of soil borne diseases and pests. Always prepare raised bed for quick drain out of excess water. The seedbeds should be prepared 1.2 m wide (1.0 m net) and of convenient length. There 0.5 m wide channel should be provided between the beds and beds should be 15 cm higher than the side channel. Seeds are very small; therefore, they required very fine tilth, hence bed must be free from stubbles and weeds. To avoid soil borne diseases nursery bed must be sterilized before sowing. Preparation of tobacco fields in Andhra Pradesh begins in June-July with the onset of south west monsoon. Land is ploughed 5-8 times. There is also practice of digging the field with crowbars soon after the harvest of preceding crop. During the active monsoon period, hand removal of weeds is resorted to for keeping the field weed free since ploughing of heavy soils is impracticable. In Kamataka, land preparation commences in January-February. A wooden plank is worked to smoothen the soil prior to planting. In general, preparatory tillage is almost common for all tobaccos. However, the number of ploughing varies from soil to soil depending on soil physical properties.

Nursery Bed Sterilizations

Sterilization of nursery beds take care of weeds, seeds of other undesired crops, seed borne diseases and pests, thus it provide protection to young tiny and tender tobacco seedlings. This can be done by rabbing and chemical treatments.

(1) By rabbing. Rabbing is the process of burning of trashes, weeds or any organic refuse on the soil surface. For rabbing 15-20 cm thick layer of leaves, weeds or paddy straw is uniformly spread over the soil surface and then it is burnt. The rabbing burns established weeds, stubbles, weeds-seeds and kills the insect-pests found on the surface-soil and also from certain depth of soil, which is besides improving physical properties of the soil.

(2) Sterilization by chemical treatment. This treatment is done by applying fungicides and herbicides on the soil surface. The treatment is done after application of organic manure in soil. Formalin solution of 2.0 per cent concentration in water or formaldehyde takes care of damping off. Sometimes Bordeaux mixture, Chloropicrin, Di Chloropropanc and Methyl Bromide can also be used as sterilizing agent.

Raising Tobacco Seedlings

The seeds being so small that can not be sown directly in the field therefore, raising healthy seedlings are an important aspect of tobacco cultivation. Tobacco seed germinates in 5-6 days under ideal conditions and remains viable for 10-15 years if stored in a dry state. Seed will germinate between 10° and 32°C and most rapidly between 27° and 32°C. Freshly harvested seed of tobacco may include a high proportion of seeds, which can only be germinated in the light. However; this light requirement either disappears or gradually decreases with age. The light requirements can be removed by priming the seeds in dark (soaking and drying).

Seed and Sowing

A seed rate of 3-5 kg/ha or 3-5 g/ per 10m^2 bed (750 seed m-2) should be sufficient and higher seed rate should be avoided as it results in over crowding of seedlings which in turn lowers the quality of seedlings and induces diseases like damping off. Under hot weather conditions a high seed rate of 4 to 5 kg is recommended otherwise usually 2 to 3 kg seed/ha is sufficient. The seed should be of high purity and germination for economic and healthy seedlings production. The seeds of tobacco are so minute that fine seed must be prepared and before sowing the seeds should treated with 2.5 % formalin solution or 0.25% solution of Dithen Z-78 or Dithen M -45. Pre sowing seed soaking in 50 ppm gibberelic acid solution for 48 hours increases and hastens the seed germination. Prior to sowing the seed should be mixed with fine sand and are sown uniformly by an experienced person and then leveled in order to maintain the optimum depth of seeding at 0.25 cm. After seeding and covering compact the seed beds by planking.

Tiny seeds of chewing tobacco

Table 3.1 Days taken to germination (>85%) of tobacco seed under different environments

Environment	Days taken to germinate		
	Pusa Tobacco -76	Vaishali Special	DP-401
Germination paper	3-4	3-4	3-4
Pot culture	4-5	3-5	4-5
Nursery	5-7	4-5	4-5

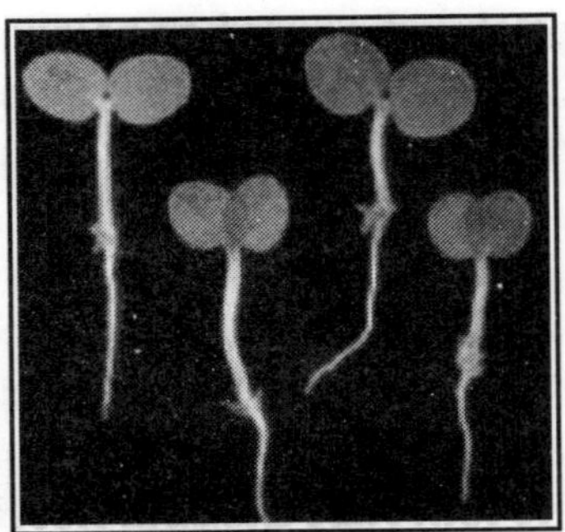

Tobacco seedling at very early stage

Sowing Time

The sowing time of seed in nursery vary from type to type which may be mentioned below:

Table 3.2 sowing time of tobacco seedling in different states in India

State	Sowing Time	Tobacco type
Andhra Pradesh	August -September	Cigarette, Cheroot Bidi (Nipani Area)
	July -October	Chewing and cigar types
Karnataka	April-May	Cigarette, Cheroot
Gujarat	May-June	Bidi
Tamil Nadu	April-May	Snuff, Cheroot, Chewing Wrapper
U.P.	June-July and September-December	Chewing Hookah
Bihar	August -September	Hookah and Chewing

Method of Nursery Sowing

Usually the treated or water soaked seeds are sown broadcast on the surface of the beds and later they are mixed in the soil up to a depth of 1.5 to 2.0 cm with the help of a rake. Before sowing the seeds are mixed with ash or fine sand in 1: 15 or 1: 20 ratio for their proper or uniform distribution on the beds. Sometimes the seeds are also sown in shallow furrows of 2.0 cm depth prepared at 5.0 cm distance from each other. To protect the tender seedlings from desiccation on a sunny day and from washing off during heavy rains, the seedbed is covered with different mulches. They are tobacco midribs in Andhra Pradesh, paddy straw in Karnataka, dry hariyali (*Cynodon doctylon*) in Tamil Nadu, mango twigs in Gujarat and thatched mat covers in Bihar. The mulch should be removed as soon as germination is noticed; otherwise, the seedlings become unduly elongated and dry out.

Nutrient Management in Nursery

For steady and uniform growth of seedlings in the nursery beds it is essential to apply organic manures in the soil. Addition of organic manures improves physico-chemical properties of the soil besides checking the cracking of the soil surface. For this purpose addition of six to eight tones of well rotten F.Y.M., compost, leaf-mould or 3 to 5 quintals of well powdered neem cake or mustard cake to the soil is recommended which will supply 25 to 50 kg N to the soil. Apart from the organic manures addition of urea, calcium ammonium nitrate or ammonium sulphate to supply 25 kg N is added in cases of very light or poor soils for maintaining the seedling vigour. This dose of N should be split into 4-5 equal amounts and should be top dressed at uniform intervals. The requirement of N, P and K by seedlings differs from type to type of tobacco as mentioned below.

Table 3.3 Nutrient requirement by seedling of different tobacco type

Type of tobacco	N	P_2O_5	K_2O
Cigarette tobacco	80	60	30
Bidi tobacco	160	80	30
Hookah tobacco	150-250	80	60

The nursery manuring schedules recommmended (ICAR 1997) for different tobacco growing areas are given below:

Andhra Pradesh

Farm yard manure 25 t / ha or filter press cake as a layer, 100 kg P_2O_5 as basal application and 100 kg N/ ha as top dressing in splits.

Tamil Nadu

Farm yard manure at 45 t / ha as a layer or compost at the same rate.

Karnataka

Farm yard manure 100 t / ha or 90 kg N / ha either in the form of ammonium sulphate or half as ammonium sulphate and half as groundnut cake with 56 kg P_2O_5 / ha.

Gujarat

Application of 180 kg N / ha half in the form of castor cake or groundnut cake as basal application a remaining half as top dressing.

Bihar

Application of 25 kg P_2O_5 / ha as basal dose and top dressing of 50 kg N / ha as ammonium sulphate.

West Bengal

Farm yard manure 10 t / ha and 56-112 kg P_2O_5 / ha as basal application and 22 kg N / ha as ammonium sulphate for top dressing.

Nursery Care

For raising pure and healthy seedlings which are the prerequisite for good crop production. The surface of the beds must be broken and the soil be pulverized with a pointed bamboo peg at the stage when second pair of leaves develops in the plant. This helps in a better root development apart from gathering little soil around the growing plants which are displaced by rain water.

Healthy Tobacco Seedlings

To minimized the damage of young seedling against scorching sun or from beating from heavy rains cover the seed beds with paddy straw or thatch grass. The thatch should be removed soon after the seeds have emerged and seedlings have two leaves. If possible thatch should be prepared which prevent from above mentioned and also minimized the water losses form the bed. Seed bed should be kept moist but not wet, hence watering should be done as per requirement but not more than once or twice per day till germination. The seedlings must be thinned out 10 to 15 days after sowing in order to protect them from damping-off. Resetting of 3 to 4 week old seedlings at a distance of 5 x 5 cm in new beds may be done for hardening of the seedlings. Top dressing with urea or ammonium sulphate should be done @ 10 g urea or 25 g ammonium sulphate per 10 m2 bed, if seedling growth is not up to the mark. After application of fertilizer, the seed beds should be copiously watered to wash off the fertilizer from the leaves. Over-growth of seedlings can be checked to some extent by clipping off some leaves, stop or cut down watering 3-4 days before pulling of seedlings in order to harden the seedlings. Spraying of Bordeaux mixture or Dithane Z-78 at the rate of 11 to 12 liters of 0.03 per cent solution for every 40 sq.m. area prevents spread of fungal diseases. For control of pests a spraying of Carbonyl 50 w.p. (Sevin 50 g in 22 liters of water) is useful. Transplanting age of tobacco seedlings is 7 to 9 weeks for *tabacum* and 5 to 6 weeks for *rustica*. Transplanting of smaller or younger seedlings causes a higher percentage of gaps while aged or woody seedlings result in a poor growth of plants and thus both the types have a tendency to lower the yield of cured leaves. Irrigation of the beds should be stopped about 10 to 15 days before transplanting so that they may become drought resistant and may resist the setback. However in the evening of the preceding day of pulling the seedlings, the beds must be watered copiously to facilitate the pulling of the seedlings without injury to the root systems. After each pulling, seed beds should be top dressed with 100g ammonium sulphate or 50g urea per 10 m bed to boost the growth of tiny tobacco plants.

Preparation of the Field for Transplanting

For quick establishment, better growth and development final field of tobacco should be well ploughed and nicely pulverized. Preplanting tillage operation should be done in manner that fine tilth may be produced and field should have no weeds and leftover stubbles. Tillage operation includes one or two ploughing with moldboard or disc plough followed by two cross ploughing with cultivator followed by planking (running patella) for leveling of the field. Use of Singh patella helps in collection and removal of weeds and stubbles from the fields. The tillage operation must be completed before planting time. Though the field preparation for tobacco planting starts with summer deep ploughing (twice) in the month of April-May. The practice is found highly beneficial in minimizing weeds and *Orobanche menace*, reducing insect pests and diseases problems and improves water and nutrient conserving capacity of the soil.

Time of Planting

Transplanting of Seedlings

Seedling of tobacco become fit to be transplanted when it attain the age of 7-9 weeks in case of N. tabacum and 5-6 week for N. rustica. The nursery beds should be given light irrigation about 7 to 8 hours before uprooting the seedlings from the beds so that the roots are not damaged during their uprooting process. Healthy seedlings with intact roots are selected for transplanting and they should be transplanted soon after they are uprooted. Transplanting is done in rows which are made by help of a rope in the finally prepared field. If possible, the transplanting should be done in the evening and the field should be given a light irrigation for quick establishment of seedlings. The field should be watered every after 3-4 days so that the seedlings do not desiccate in -the sun. The gaps created because of death of seedlings or because of their non-establishment should be filled 8-10 days after transplanting by planting freshly uprooted ones. Time of planting varies from place to place depending on the soil, weather and type of the tobacco. The time of transplanting plys vital role in quality tobacco leaves production. Early planting results in production of papery leaves accompanied with severe leaf spot infestation, while delay in planting results in production of small, leathery and brittle leaves. Optimum planting time for different tobaccos in different parts of the country is given in table 3.4.

Table 3.4 Time of transplanting and spacing of tobacco seedlings for different types

State	Tobacco type	Time of planting
Andhra Pradesh	FCV Tobacco	Mid-October to mid-November
	Burley	Mid-July to Mid-August
	Natu	Mid-October to mid-November
Karnataka	FCV tobacco	Mid-October to mid-November
	Bidi tobacco	Mid-August
Tamil Nadu	Cigar and cheroot	October to first week of November
West Bengal	Jati tobacco	Mid-February
	Motihari	Early March
Bihar	Desi tobacco (Chewing)	Mid-September to mid-October
	Rustica types	October
Punjab	Hookah tobacco (*N .rustica*)	Second fortnight of February
	N. tabacum	Mid-February to first week of March
Maharashtra	Bidi tobacco	Last week of August to first week of September

Method of Transplanting

In India, planting is done by hand for all tobaccos. The plants are set in holes (8-12 cm deep). If the soil is not thoroughly wet, the holes are pot-watered before transplantation. The position of plants is marked either by running a marker in cross directions or with marking ropes. The success of transplanting and establishment of seedlings depends much on the skill of the transplanter. Undue bending of roots or loose planting of seedlings leads to poor establishment. In the case of FCV tobacco in Andhra Pradesh and bidi tobacco in Nipani, plant holes are pot watered just prior to setting the seedlings. Planting of bidi and hookah tobacco in Bihar is generally, arranged on a rainy day. For cigar and chewing tobaccos in Tamil Nadu, field is irrigated prior to planting and the seedlings set in wet soil. In Uttar Pradesh and Punjab, planting is done in dry soil followed by irrigation. Planting on flat beds is the practice in Andhra Pradesh, Bihar, West Bengal and Uttar Pradesh. In Punjab, hookah tobacco is planted both on flat beds and ridges. In Tamil Nadu, cigar, cheroot and chewing tobaccos are planted on ridges. Generally, ridge planting is ideal for irrigated tobaccos. Under dry conditions, where moisture conservation is an important consideration and in sandy soils, flat bed system is preferable.

1. General Method

After seedlings are uprooted from the nursery beds they become ready for transplanting. While planting, a shallow whole of 2.5 cm depth is made with finger at each planting position which is marked by a marker with desired spacing both ways after soil surface is completely leveled. One seedling is placed at each place and the soil is pressed all-round the seedling to provide a firm foot-hold for the plant. Gaps are filled within 10 days of planting.

2. Furrow or Salia Method

This Method is practiced when the weather remains dry during the normal season for transplanting. In this method the markings in the prepared field are made in only one direction at first and shallow furrows called "Salia" are opened at desired spacing with country plough. A wooden plough is then lightly moved over the furrows and the cross markings are made with the marker. Before planting, water is let in the furrow and a seedling is set out at every crossing along the side of the furrows.

3. Flat Ridge or Ghora Method

In parts of Baroda district where tobacco is transplanted in heavier soils, this method is adopted. In this method furrows are made with three tined seed drill or a wooden plough followed by planking. Cross marking is then carried out with a marker. The furrows are then irrigated and the seedlings are transplanted on alternate ridges on each crossing. A light irrigation should be given to delayed nursery bed 6-7 hours before prior to uprooting of seedling to facilitate minimum damage to the seedling roots. If possible transplanting should be done in the evening with the healthy seedlings having intact roots. A light irrigation should be provided for quick and better establishment. The field should be watered every 3-4 days so that seedlings do not desiccate in the sun. Gap filling operation should be done after 8-10 days of transplanting. The transplanting should be done in line and spacing varies from variety to variety, time and place.

Plant Spacing and Optimum Plant Population

Tobacco seedlings are always planted in rows. Distance between rows and between plants within the row varies with type of tobacco and soil type. Based on the research findings, the following spacing is recommended (ICAR, 1997) for different tobacco growing states (table 3.5).

Table 3.5 Tobacco type and spacing (cm) indifferent states

State	Tobacco type	Plant spacing (cm) (Row X Plant)
Andhra Pradesh	FCV- Black soils	70 x 50
	FCV- Red soils	100 x 60 cm
	Burley	80 x 45
	Natu	90 x 90
Karnataka	FCV -Black soil	70 x 50
	FCV - Light soils	100 x 60
	Bidi	100 x 75
Tamil Nadu	Cigar and cheroot	75 x 50
	Chewing	75 x 75
Gujarat	Bidi	105 x 45
Bihar	Chewing	90 x 60
West Bengal*N*	*. rustica*	60 x 45
	N. tabacum	90 x 90
Uttar Pradesh	*N. tabacum*	90 x 60
	N. rustica	60 x 60

Relatively low plant population per unit area will produce close to the maximum yield where growing season is long. In less favoured environments, where the rate of growth in smaller and the duration is shorter, a larger population per unit area is required to attain near maximum yields. A change in plant population affects quality as well as yield. Increasing plant population from 7,500 to 25,000 plants ha^{-1} decreased total alkaloid content of cured leaves from 13.57 to 2.73 per cent of the whole plant. On the other hand, the content of reducing sugar increased from 13.1 to 17.2 per cent. An increase in plant population also decreases the proportion of orange and darker coloured grades in FCV tobacco crop. The optimum population for a particular locality, therefore, depends on the growth potential of the plants, the type of the crop required (large or small nicotine content, heavy or light leaf etc) and the economic factors, which dictate the relative importance of yield and leaf quality. In general, plant population for FCV tobacco is relatively higher (around 27,000 plants/ ha) than that (18,000 plants/ ha) for non-FCV tobaccos. Light soils require relatively higher plant population than heavy soils.

Growth and Development

During the expansion of a leaf, its rate of net photosynthesis has been shown to increase from 5 mg CO_2 dm^2/hr during cell division phase to a maximum of 20 to 25 mg CO_2 dm^2/hr at the time the leaf is 75-85 per cent expanded. Thereafter, it decreases fairly rapidly as the leaf ages and as the shade from expanding leaves above it increases. High dry matter accumulation occurs during 6-8 weeks after transplanting. The peak dry matter production in FCV tobacco was 64 kg/ ha / day^{-1} while for burley tobacco it was 140-168 kg/ ha / day. Dry matter production in burley tobacco could be as high as 5,670 to 6999 kg ha^{-1} with an average around 4,287 kg/ ha. In the maturation phase, individual leaves continue to increase in dry weight but change little in leaf area. Ripening begins in the bottom leaves while top leaves are still expanding rapidly. The remaining leaves mature in sequence from bottom to top of the stem, while the stalk and root dry weights continue to increase. By the time the last leaves have been harvested, the stem and roots accounts for about 20 per cent of total dry matter produced.

Topping

Topping is the process of removal of flower heads, either or alone with some upper leaves from plants. When flower heads begin to show, the plants are topped by removing off the top of the plant. Topping consists in removal of the terminal bud with or without some of the small top leaves just before or after the emergence of the flower head. This practice stimulates the development of the remaining leaves. Only as many leaves as can be brought into fullest development are left on the plant. The main objective of topping is to improve the size, body texture and quality of leaves. It gives a uniform quality product and prevents excessive coarseness in the leaves. It is observed that the photo-synthates and nutrient were diverted towards flower heads rather than to leaves, thus topping helps in full developments of top leaves or otherwise they will remain relatively shorter when the lower leaves are going to mature. The effect of topping is more pronounce in younger leaves than older. Topping increase the leaf size, thickness and dry weight of leaves to the range of 80-85, 20-26 and 45-50 pre cent respectively. Under North Bihar conditions the optimum level of topping in variety Pusa Tobacco -76 is at 16 leaves and in vaishali special it was at 14 leaves. The stage of topping for both the newly released varieties was found between 35 and 55 days after planting. The number of leaves in the plants to be topped is as mentioned in table 3.6.

Topping operation in tobacco
Courtesy by AINRP on Tobacco

Table 3.6 Tobacco type and leaves left on the plant at topping

S. No	Type of tobacco	No. of leaves left on the plant
1	Cigarette & cheroot	8-9
2	Wrapper cigar	10-12
3	Hookah, Chewing & Bidi	10-14

De-suckering

After the tops have been removed the buds in the axils of the leaves which otherwise remain dormant become active and suckers soon develop in the axis of the leaves. These suckers or lateral branches should be removed before they become large enough to

Chemical de-suckering
Courtesy by AINRP on Tobacco

retard the development of the leaves. Removal of these suckers is called de-suckering.

These active lateral branches/ side shooting / suckers should be removed by hand or by Chemical de-suckering before they become large enough to retard the development of the leaves.

To reap maximum benefit of topping, suckering should be avoided and suckers should be removed periodically. The process of removal these suckers commence after 6-7 days of topping operations and it will continue for 2-3 times for effective result. The main aim of topping and de-suckering operations is to divert the energy and nutrients of the plant from flower head to leaves which influence the yield and quality of tobacco. Desuckering can be done either manually or with the help of chemical.

Manual De-suckering

De suckering can be done manually by piercing of the stem after topping with 20-25 cm long needle, which also improve the productivity and quality of tobacco leaves.

Chemical De-suckering

To avoid time consuming laborious and more expensive process of manual de suckering, it can also be performed by using chemicals. Application of coconut oil to the top six axils soon after topping suppresses the emergence of suckers in flue cure virginia (FCV) tobacco. In case of cheroot tobacco of Tamil Nadu application of 2 per cent Naphthalene, Acetic Acid in triethenolemin to the topped portion suppressed all the suckers. Application of suckericide i.e. Decanal or Royalten @ 4 to 5% enhanced leaf yield by 9 per cent over the hand suckering and cut down the labour cost by 31 man days per ha. Maleic hydrazide (MH) is the only true systemic sukericide, because it is absorbed by leaves and translocated through the plant to small sucker buds. Rate of Maleic hydrazide (MH) application is 100 ppm. Indole Butyric Acid @ 0.2 % is also recommended for an effective de-suckering. Piercing of the stem after topping with 20 to 25 cm long needle is found to be beneficial in improving yield and quality of tobacco leaves. De-suckering also improved the leaf weight per unit area and physical quality parameters such as spangling, puckering, maturity etc. considerably and additional net profit realized.

Harvesting

Economic produce of tobacco is its quality leaves which largely depends upon the stage of the leaf at which they are harvested. The sign of maturity and method of harvesting differ from one type to another type of tobacco. Maturity could be identified by visual observation of leaf when its turned colour from normal green to yellowish, thickening due to starch accumulation in them; accumulation of gummy substances on them. Starting of flowering bitterness of the leaves is some important characteristic feature of maturity of tobacco leaves. Thus the leaves possessing all above sign are the matured ones which may be harvested but all of them do not mature at one time because of which the method of harvesting differ from one type to another type and their prerequisite for quality. In general there are two methods of tobacco harvesting, namely priming and stalk cut methods.

Priming

Incase of priming the leaves are harvested as and when they are fully developed and matured. Priming is supposed to be the best method of harvesting as every leaf gets sufficient time and nutrients for its full development but if the plants are cut the younger leaves get lesser time to grow and as a result they remain underdeveloped and immature which ultimately affect the quality. Thus for the production of quality leaves priming is more appropriate method of harvesting, though it is more time consuming and expensive. Cigarette and wrapper tobaccos are harvested by priming. In cigarette tobacco, maturity of leaf is judged by the development of greenish yellow colour of the leaf. Generally, it takes about 70 to 75 days from planting for the crop to be ready for the first priming. Leaves are primed by hand and at each priming 2-4 leaves are harvested. Subsequent priming is done at intervals of about one week. Entire harvest is completed in 5-6 priming. About 15-22 leaves fit for curing are obtained from an untopped plant. Harvesting is, generally, completed before 8 AM. Priming of FCV tobacco in Guntur district (AP) commences in January and continues up to March. Primed leaves are taken to a shaded place for stringing and finally loaded in the barn for curing. Yellowish brown mottling indicates maturity of hookah tobacco in West Bengal. First priming commences about 90 to 105 days after transplanting of motihari (*Nrustica*) and about 120 to 130 days for jati (*N. tabacum*). Harvesting of motihari tobacco is completed in about 3 priming and that of jati in about 4 priming. After priming, the leaves are spread in the field in morning hrs for partial drying. In the evening, the leaves are tied into bunches and carried to the shade. Next morning they are kept in the barn or other shady place for air curing.

Stalk Cut

In case of stalk cut the plants consisting of all the leaves intact are cut from close to the ground. It is essential to see that the maximum number of leaves is matured so that the quality is assured. The leaves are not separated until 3-4 days. Usually following method of harvesting is adopted for different types of tobaccos. Chewing and hookah tobacco are generally harvested by stock cut method with the help of a sickle or khurpi close to the ground level. In this method of harvesting leaves are not separated but intact for few days. It worth to mention here that harvesting is to be done when maximum leave are matured. Cigar, cheroot, chewing, bidi and hookah tobaccos are harvested by stalk cut method. Subsequent handling varies with the method of curing. Cigar and cheroot types of Tamil Nadu will' be ready for harvest 20-25 days after topping as indicated by yellowing of tip and margins, slight puckering and brittleness of leaf which breaks on folding. Generally, chewing tobacco comes to harvest in 35-45 days after topping. Harvesting is done in the evening. In case of cigar tobacco, harvested plants are wilted in the field for 36 hrs, first 12 hrs in the open and next 24 hrs in covered heaps. Plants are then strung on scaffolds for sun curing. Maturity of hookah tobacco is indicated by formation of yellowish brown spots of puckering. Harvesting of bidi tobacco is done at considerably advanced stage of maturity. Maturity is judged by pronounced development of brown spots called spangles. Normally, the crop is left in the field for 2-3 month after topping to ensure proper maturity. Entire field is harvested in 2 or 3 stages selecting each time only mature plants. After the harvest, the plants are ground cured.

Harvesting Method based on Tobacco Types

Table 3.7 Harvesting method of tobacco

Finished products	Method of harvesting
Cigarette & cigar (wrapper)	Priming
Cigar (Binder & filler), Cheroot, Chewing, Bidi and Hookah	Stalk cut method

Normally harvesting is done 4-5 weeks after topping but it again depends upon the type of finished products.

Table 3.8 Type of tobacco and time of harvesting after topping

Finished products	Time of harvesting in weeks after topping
Cigarette & cigar (wrapper)	3-4
Cigar (Binder & filler), Cheroot	4-5
Chewing, Bidi and Hookah	4-7

□□□

4

Post Harvest Processing

Freshly harvested leaves are not fit for direct consumptions, hense by post harvest treatment/processing leaves are prepared for consumption in various form. Post harvest processing is most important component which has great impact on its quality. Curing, aging and fermentation of tobacco leaves are some major events which all together make leaves suitable for consumption.

Curing

Curing is a process by which the harvested leaves are made ready for marketing. In more technical terms it may be stated that "Curing is a process of drying, decomposition of chlorophyll until the green colour disappears from the harvested leaves, hydrolysis of starch into sugars and respiration of fermentation of sugar thereafter". In curing the mineral salts also crystallize out producing the grain of the leaves. The sole object of curing is to dry the green leaf in such a way that the final product had required colour, texture and aroma. Thus curing is probably the most important operation in the production of tobacco and the method and efficiency of curing determines, to a large extent, the quality of the product. The methods of curing adopted in India are governed by several factors viz. tradition, convenience, market value of the finished product, economic considerations, etc.

Table 4.1 Curing method of different tobacco finished products

Name of the finished products	Method of curing
High quality cigarette	Flue curing
Inferior quality cigarette	Rack curing
Cigar, cheroot, Chewing, snuff and Bidi tobaccos	Rack curing
Very low grade cigarette, Bidi	Ground or sun curing
Chewing, hookah and snuff tobaccos	Ground or sun curing
Chewing and hookah tobaccos	Pit curing
Bidi, snuff, chewing and Hookah tobaccos	Fire curing

The most evident event of quantitative change in the leaves, harvested by any method, is the loss of water from the leaves. Thus it is observed that the concentration of the solute in the cell-sap of the leaves under curing was doubled after four days and increased to five times after eight days. Another significant visible change during curing process is the change in colour of the leaves. The matured leaves change their colour form green to yellow and finally to brown. Connecticut shade tobacco leaves contained about 0.5 to 4.0 per cent chlorophyll and most of it disappeared in 12 days during air-curing. The yellow colouration in leaves is due to carotene and Xanthophylls and carotene was found to decrease up to 60.0 per cent in stalk -cured leaves. With the advancement of curing the yellow colour is over powered by brown pigments which are formed partly due to the oxidation of phenols, polyphenols and tannins. The green leaves contain about 1 per cent rutin (phenol glucosides) that is destroyed during curing process, the dark brown coloured product of oxidized rutin probably makes an important fraction of the brown compounds formed during curing. There is also loss of organic solids from the leaves due to respiratory activities during curing. It was observed that about 16 per cent loss of organic solids during air-curing in the first few days when the leaves were yellowing.

Freshly cured leaves are not fit for direct consumptions as it gives pungent and irritating smoke, until it goes for aging commonly known as fermentation. Fermentation process is usually applied to cigar, wrapper, binder and filler tobaccos. On the other hand, bidi tobacco never undergoes fermentative process. Fermentation is exothermic process with evolution of CO_2, NH_3 and uptake of O_2 with a substantial loss of organic matter. Besides improvement in aroma and taste, there is uniform formation of dark colour. By this process colour are fixed the rawness and bitter taste of the freshly cured tobacco leaves are removed and a fine taste and aroma are developed. Chlorophyll consists of a substantial part of these compounds which are lost probably because of the decomposition of chlorophyll. Of the inorganic substances nitrogen and its derivatives are the most important substances that undergo various changes. Hydrolysis of protein results in the formation of soluble forms of amino acids, amides and ammonia. These reactions result in a loss of about 6.0 per cent of total nitrogen during curing that occurs particularly after the leaves turn brown. It was reported that the inorganic substances like ash constituents (phosphate, potassium, sulphur and chlorine), oxalic acid and crude fibre undergo a little change except some may migrate from leaf to stalk. Watering is done to enhance the hydrolysis of various organic and inorganic compounds and fermentation is process to convert an chemical compound in to different kinds desired in this case, hence Watering is most important process in curing process, because excess watering may spoil the colour and aroma of the leaf and less watering may depress the formation of white incrustation (deposition of sulphate and chlorides of Na, K Mg, Ca etc.) on the leaf. However, watering and fermentation process should be handled by experienced person in order to ensure good fermentation and preparation of high quality leaf. Quality of tobacco is matter of subjective evaluation. Sugars are considered to have positive influence on quality while nitrogenous compounds have a negative once. However best quality leaves must posses more sugar content and very less quantity of nitrogenous substances.

Factors Affecting the Attainment of Good Cured Qualities of Tobacco

There are several factors which are responsible for the quality of the finished product. The important ones of these are mentioned below.

1. Temperature

Curing is a vital process in production of quality tobacco, and it is expected to proceed normally within the prescribed limit of temperature. A temperature below or above the normal limit will inhibit or slow down the curing process but the drying will not necessarily be stopped and in fact it may be accelerated. Temperature should be within 60 to 180° F depending upon the stage of curing.

2. Relative Humidity

Relative humidity is highly important factor influencing the rate of drying. It is observed that within normal curing temperature the actual rate of drying in the air curing process is relatively slow, requiring several week for completion whereas, in the flue curing, the whole process is completed within 3 to 4 days. On an average the relative humidity should be 85 per cent at beginning but after the leaves begin to turn yellow a lower humidity that permits a rapid drying is advised.

3. Air Supply

Curing is largely a process of oxidation in which an adequate air supply is required. Normally the air supply is met with when ventilators are opened to regulate the humidity of the chamber. Opening of ventilators also governs the expose of leaves to the light. Total darkness is not a necessary condition for successful curing whereas bright light may be injurious in some types; especially it affects the colour of the cured products.

4. Condition of the Leaf

The capacity of tobacco to cure and the exiled of the process depend on the type of leaf involved and the condition. It was produced and the amount of tobacco mass, its degree of matunity and other physical characteristics. Aging of tobacco is a milder process than fermentation although fermented tobacco may continue to age. The initial step is to redry the cured leaf for uniform moisture content of 9 to 10 per cent. Since aging requires 18 months or more, forced ageing or artificial ageing for about two months can be done under controlled environmental conditions. However, depending upon type of tobacco, the method of harvesting and maturity of leaves at harvesting, there are four distinguishes methods of curing which are prevalent in India namely (i) air-curing (ii) fire-curing (iii) sun-curing and (iv) flue curing. Major curing methods are summarized as under.

Curing of FCV Tobacco

Flue-curing

In the curing process, moisture in the tobacco leaf is drained out with the help of radiated heat in barns. Usually, brick barns are used in which heat generated from wood, oil or gas is supplied in closed flues. Absolute control of temperature and humidity till the barn is essential. Coal is generally used as a fuel in the barns. Flue-curing is improved method of curing. The curing consists of drying green leaves under artificial atmospheric conditions by adopting a process which does not allow the green leaf to come in direct contact with smoke or flames of the fuel and which permits the regulation of temperature and humidity. The main feature of flue curing is the drying of leaf under controlled conditions, where the starch gets converted into sugars and the green tobacco becomes bright, aromatic and fine textured. This method is adopted in almost entire Virginia cigarette tobacco growing areas like Guntur, Godavari, Mysore and other areas at Hyderabad, Madras, U.P., M.P. and Orissa. The flue-cured tobacco was first used for chewing purpose but later, its mildness and low nicotine content made it popular as a smoking tobacco. The cigarette smoking came into existence since 1880-1890 and it became increasingly popular with its manufacture from flue-cured Virginia tobacco. Now every priced cigarette is prepared with flue-cured tobacco having a certain proportion of other types of tobacco got from sun-curing like Natu tobacco of Andhra Pradesh. The production technique of flue-cured tobacco is little different from other types viz. it is raised with low nitrogen supply and harvested at an advanced stage of maturity so that the tobacco might become rich in starch and poor in nitrogenous substances like nicotine at the time of harvesting. The leaves are always harvested by priming 3-4 leaves at a time. The harvested leaves are strung on sticks and then stacked into a specially constructed flue cured bam which is artificially heated with gradual increase in temperature until the leaves dry. There are three distinct stages in flue curing; how ever, there is no line of demarcation between different stages. There are three stages in flue curing i.e. yellowing of leaf, fixing of colour, drying of leaf, are briefly discussed here.

(I) Yellowing of Leaves

In the yellowing process of leaves, which usually requires 24-36 hours, or slightly longer, only gentle heating is done with a temperature ranging between 90 to 100°F and relative humidity between 85 to 90 per cent. It is a slow starvation process and the object of slow heating is to give optimum humidity and retain as much moisture in the leaves as is possible to keep them alive for 30-36 hours. At the end of this stage the leaves attain a bright golden yellow colour and the starchy matter gets converted in to sugar.

(II) Fixing of Colour

The fixing of colour takes another 24 to 30 hours for completion. The stage, being critical, should be attended with utmost care because slight change in temperature deteriorates quality which lowers the market price. The temperature of the barn should be gradually increased when yellowing is over with a decrease in the humidity by opening the ventilators. Any delay in drying of leaves, after yellowing is over, results in a development of a brownish tinge on the yellow leaves due to oxidation of tannin-like substances that is called "sponging". But, on the other hand, a sudden rise in the temperature, when the leaves are still wet, results in a bluish back discoloration called "scalding". These defects reduce the market value of the finished products. The fixation of colour merely involves drying of the tissues to the point where oxidation of colour producing matter cannot take place. Ample ventilation is provided so that relative humidity of the barn is effectively reduced and the temperature is advanced at the rate of about 4-6 degrees per hour until it reaches to 140 °F which is maintained until the leaf blade becomes practically dry. At this stage about 80 per cent of the leaf moisture is driven off.

(III) Drying of Midrib of the Leaf

This process consists of nothing but drying of midrib of the leaves. During this process the ventilators are closed for conservation of heat and temperature of the barn is advanced at the rate of 5 degrees per hour until it reaches to 170 to 180°F which is maintained for 18-24 hr for completion of the drying process of the midrib in all parts of the barn. This way the entire process of flue curing usually requires about 84 - 96 hours. After drying, the leaves are generally left in the barn overnight with doors and ventilators opened so that the leaves may absorb moisture from the atmosphere and permit their handling next morning otherwise they will remain very brittle and break into pieces while handling. To achieve stiffness in the leaves, sometimes water is sprinkled over the floor of the barn and doors and ventilators are closed at night. The cured leaves are reducing to about 10-20 % of their green weight which possesses a moisture range of 8-18 percent after drying. A nicely cured leaf should be of bright lemon yellow colour, fine texture, no blemish, mild strength, slow, regular and continuous burning, white ash having pleasant and neutral aroma etc.

Bulking

When the midribs are perfectly dry, curing is completed. The cured leaves are piled up into heaps for fermentation on account of sweating from the moisture contained therein the leaves. Leaves is allowed to cool down, ventilators completely opened and leaves allowed absorbing moisture from the atmosphere in the night so that they become soft enough to be unloaded. After unloading, leaves are removed from sticks, and bulked in small heaps and kept covered for a few days to avoid loss of moisture. During bulking, the colour evens out so that more uniformly coloured leaves are obtained. This way the leaves lose their excess moisture and improve their flavour and aroma. Reconditioning is the process of regulating the moisture of the cured leaves and is supposed to be essential for cigarette tobacco.

Non-FCV Tobaccos

1. Air-curing

In air curing the source of energy is the atmospheric temperature carried through air. The leaves are spread in the shade either in a barn or in a shed for curing. A proper regulation of atmospheric temperature, oxygen and humidity results in a better quality cure. There are two stages of air curing-yellowing of leaves, the first stage and development of brown colour, the second stage. As such this method has very little value for curing of high grade tobacco as the leaves give up their moisture slowly and turn finally into a brown colour.

Air-curing is commonly adopted in West Bengal, Bihar, U.P., M.P., Punjab, Kamataka, Andhra Pradesh and Tamil Nadu. There are three main methods for air-curing viz. ground-curing, rack-curing and pit-curing. The methods adopted for curing of tobacco differ from place to place as explained below:

(i) Ground Curing

In U.P. the popularly grown hookah tobacco is slowly cured in heaps on the ground. At Kampil in Farukhabad district, which is famous for its hookah tobacco the wilted plants are cut and collected into heaps and are left in the field for two days after which the leaves are separated and tied into a bundle of two or three leaves. These bundles are again heaped up and stored on a farmstead and turned occasionally till finally cured.

The chewing tobacco of Kamataka, after whole plants are cut, is allowed to remain on the ground for six days after which they are turned over in the early morning to expose the other side. These fully exposed plants are taken to the curing-shed where they are bulked and rebulked for 15 to 20 days. The leaves are then separated from the plants and tied into bundles according to their length or size.

(ii) Rack-curing

In this method the matured leaves are harvested by priming method and they are tied into bundles containing 4 to 6 leaves according to their size.

These, bundles are then strung on a jute string which is fixed on specially erected bamboo poles in a barn. A required atmospheric temperature and relative humidity of 70 to 80 per cent is maintained by either closing or opening the ventilators or sprinkling water inside the barn. The curing process takes about 5 to 6 weeks time. The rack-curing states in India is mentioned in table 4.2.

Table 4.2 Rack-curing method adapted to various tobacco type

S No	States	Type of tobacco
1	Tamil Nadu	country cigarette, cigar, cheroot, chewing, snuff tobacco
2	Maharashtra	bidi and chewing tobacco
3	Andhra Pradesh	cheroot and bidi tobacco
4	Assam	hookah tobacco

Lanka tobacco leaves in Andhra Pradesh are strung on ropes in shed for 30 to 45 days for curing and developing yellowish-brown colour of the leaves.

(iii) Pit-curing

The pits, prepared in the ground, are used for curing the tobacco leaves however; the method is not very common. Pit-curing is mainly adopted in Punjab, Tamil Nadu, Maharashtra and Andhra Pradesh. The tobacco grown in Jullundhar and Firozepur districts of Punjab is pit cured. In this method the pits are lined with tree leave's or dried straw in order to layer in to small heaps. In Firozepur a layer of Aak (*Calolrop* sp) leaves are placed between the layers of tobacco leaves for increasing the pungency of cured leaves. The top of the heap is covered with a layer of straw and then with soil which should be about 10 to 15 cm above the ground level so that rain water does not percolate into the pit. It takes 7-8 days for curing after which they are twisted into ropes or made into bundles. The pits must be opened at right time as any delay in opening causes overheating and the leaves will get spoiled. Sometimes leaves are placed in the first pit for 24 hour and then transferred to the second pit for 48 hours and then back again to the first for another 24-48 hours. Thus a quick fermentation takes place and the leaves attain a dark brown colour and fruity smell.

2. Sun-curing

Sun curing is common method applied in to chewing and hookah tobacco. In sun-curing the source of energy is solar-heat. Sun curing is commonly adopted in Tamil Nadu, Andhra Pradesh, Bombay, Bihar, U.P., Punjab and West Bengal. In this method of curing either whole plant is strung on string or rack (as in case of chewing and cigar tobacco in Tamil Nadu) or whole plant is spread on the ground (as in case of bidi tobacco in Bombay, hookah and chewing tobacco in U.P. and Punjab) or otherwise primed leaves or separated leaves with pieces of Stalks are tied into bundles and strung on ropes (as in case of Natu tobacco in Andhra Pradesh) or these leaves are spread on the ground (as in case of chewing tobacco in U.P. and hookah tobacco in W. Bengal). Time taken for curing differs from place to place and type to type of tobacco viz. cigar and chewing tobacco in Andhra Pradesh takes about 15 to 20 days and Natu tobacco takes around 45 to 60 days for curing. The leaves are separated from the stalk after the plants are initially wilted in the field and the leaves are then strung on bamboo poles for curing in the sun. The hookah and chewing tobacco in Bihar are harvested and left in the field itself upto 4 to 6 days for drying after which they are heaped for two days. Again they are spread and dried for 6 days then heaped for two days and spread thereafter upto 4 days for drying. Thus after 45 days the leaves attain dark brown colour after which they are stripped and tied into bundles of 25 to 30 leaves for fermenting in bulks.

3. Fire-curing

Fire is used as the source of energy for curing and the important leaf which in cured is Jaffna tobacco (a chewing type) in Ceylon and in Tamil Nadu. In this method the harvested leaves are wilted for few hours in the field, tied into bundles of 3 to 4 leaves and hung on racks in a smoke-hut. They are then smoked for 12 hours by burning coconut husk, leaf, stalks, etc. and stocked for 3 days and again smoked. During the smoking treatment the creosotic substances produced from the smoke are absorbed by the tobacco that imparts a particular taste and antiseptic properties in the leaves. The leaves are fermented into bulks for 3 to 4 weeks after smoke treatment and then treated with salt water from the lagoons or with jaggery for a special or peculiar taste of this type.

Chemical Changes During Curing, Fermentation, Aging and Storage

Curing is a careful and regulated drying of freshly harvested leaves in curing sheds or barns with an object to get bright lemon, yellow or brown colour and good texture while fermentation relates to the changes after curing (death of cells) brought either by enzymatic reactions or by micro-organisms and aging is a mild fermentation that may follow either curing as in cigarette types, or vigorous fermentation as in cigar and in some heavy dark types. The chemical changes differ under curing, fermentation, aging and storage as discussed below:

Chemical changes during curing : Air-curing and flue-curing are the two important methods of curing generally adopted in most of the countries of the world. In very limited cases sun-curing is also practiced. Of these methods air-curing is essentially a slow process of curing while flue-curing is a far quicker process due to use of an artificial heat. Thus the nature and extent of change differ under both the methods.

Changes During Air-curing

It is meaningful to mention here that there are two methods of tobacco harvesting, namely priming and stalk cut. The movement and translocation of the chemical substances take place from leaf to stalk in the latter case and movement of materials is restricted between lamina and midrib in the primed leaves. The most evident event of quantitative change in the leaves, harvested by any method, is the loss of water from the leaves. It is observed that the concentration of the solute in the cell-sap of the leaves under curing was doubled after four days and increased to five times after eight days.

Another significant visible change during curing process is the change in colour of the leaves. Leaves change their colour form green to yellow and finally to brown. Connecticut shade tobacco leaves contained about 0.5 to 4.0 per cent chlorophyll and most of it disappeared in 12 days during air-curing. The yellow colouration in leaves is due to carotene and Xanthophylls and carotene was found to decrease upto 60.0 per cent in stalk-cured leaves.

With the advancement of curing the yellow colour is over powered by brown pigments which are formed partly due to the oxidation of phenols, polyphones and tannin. It is found by that the green leaves contain about 1 per cent rutin (phenol glucosides) that is destroyed during curing process. It is the dark brown coloured product of oxidized rutin probably makes an important fraction of the brown compounds formed during curing. There is also loss of organic solids from the leaves due to respiratory activities during curing; there is about 16 per cent loss of organic solids during air-curing in first eight days when the leaves were yellowing. Almost entire quantity of starch and sugars, amounting to about 4.4 per cent of the initial organic solids, disappeared during the first six days of curing. The other soluble substances (3.4 per cent of initial organic solids) like paraffin hydrocarbons, terpenes, alcohols, fatty acids phenolic acids, waxes, fats and oils, volatile oils, resins, sterols and plant pigments were lost in the first eight days of curing. Chlorophyll consists of a substantial part of these compounds which are lost probably because of the decomposition of chlorophyll. There are evidences for transformation of malic acid into citric acid. Quantitatively it was found that malic acid, equivalent to 9.0 per cent of organic solids, disappeared and citric acid, equal to 6.6 per cent of the solids initially not present, was formed. Later it was found that every two moles of malic acid disappeared formed one mole of citric acid.

Of the inorganic substances nitrogen and its derivatives are the most important substances that undergo various changes. It is observed that more than 50.0 per cent of the total proteins hydrolyses during the first six days of curing but the hydrolysis stopped completely after the leaves became brown Again it was observed that in the yellow leaves, having less initial protein, hydrolysis was small (only 20 to 35 per cent). Hydrolysis of protein results in the formation of soluble forms of nitrogen like amino acids, amides and ammonia.

These reactions result in a loss of about 6.0 per cent of total nitrogen during curing that occurs particularly after the leaves turn brown.

The nitrate nitrogen (50.0 per cent of the total nitrogen) was lost to a negligible extent during curing. It was reported that the inorganic substances like, ash constituents (phosphate, potassium, sulphur and chlorine), oxalic acid and crude fibre undergo a little change except that some of the ash constituents may migrate from leaf to stalk. Calcium and magnesium are immobile.

Chemical Changes During Flue-curing

In case of flue-curing there is no development of brown colour, as in case of air-curing, after yellowing. The leaves used for curing contained 77.4 to 79.5 per cent water most of which was lost during curing. Losses in dry matter ranged from 2.50 to 3.25 per cent on the green leaf basis or 12 to 16 per cent of the original dry weight basis. An extensive hydrolysis and loss to the extent of 76 to 90 per cent of starch was observed. This loss was reduced in fixing of colour and it was negligible during process of curling. Reducing sugars are found to increase during fixing of colour but they are decreased during stem drying. At the end of the process a small quantity of sucrose to the extent of 6.0 per cent is found to increase in case of green leaves, the extent of sucrose formation with the level of phosphate in the leaves reported it to be the effect of phosphorylation. There was more marked increase in ammonia and amides during yellowing in case of leaves having low carbohydrate in them. The respiratory and carbohydrate losses showed that these increases occurred where materials other than carbohydrates were respired. The conversion of malic acid to citric acid was not apparent and evident in case of flue curing, as it is in case of air curing.

Chemical Changes During Sun-curing

In India cigar, chewing and some cheep quality cigarette tobaccos are sun-cured. In sun curing it is observed that there is a considerable loss of starch with production of sugars and formation of soluble nitrogenous constituents at the expense of proteins.

Changes During Fermentation

The freshly cured leaves are not fit for human consumption until it goes under fermentation or aging in which colours are evened out, the rawness and bitter taste of the freshly cured, tobacco leaves are eliminated and a fine taste and aroma are developed. The fermentation, benign exothermic process in nature, generates heat and increases the temperature of the fermenting leaves to the extent, of 110 to 1200°F in 3-4 days. During this process the organic matter gets oxidized and there is fast evolution of carbon dioxide in presence of sufficient oxygen in the air. After 20 to 25 days the leaves become easily separable and they are stripped and after 30-40 days of fermentation the leaves moisture falls around 15 per cent. Now the leaves are fit for safe storage or sale. The experiments indicated a loss of 10 per cent in nicotine content during fermentation process. Only a small portion of nicotine get lost due to volatilization while major portion of loss was due to breakdown of the pyrroline ring and conversion of nicotine into 3-substituted pyridine derivatives. They found formation of compounds like Oxynicotine; 2-3 Dipyridyl; 3 pyridyl methyl ketone; 3 pyridyl propyl ketone; Nicotinamide; and N-methyl nicotinamide during fermentation process.

Quality in tobacco is currently an object of subjective evaluation and best defined as "that for which the buyer pays money". Thus quality determines the Price of the tobacco leaves produced for marketing purposes. Sugars are considered to have a positive influence on quality while nitrogenous compounds a negative one. But it is necessary for a leaf to have too much sugar and some nitrogenous constituents because in absence of either of the two the smoke is insipid and lacks character. Thus the commercial value of a leaf is judged by its suitability for specific purposes of manufacture viz. tobacco used for smoking purposes must have a combustibility and fine aroma. In general the leaf characters, chemical composition of leaves, burning-quality and aroma are the most important factors in determining the quality of a tobacco.

Leaf Characters

Leaf character primarily hereditary in nature but it may slightly be modifies with the growing season, nutrient conditions, planting density, topping and de-suckering etc. Size of leaves is a very important factor in cigar types than in most others because for cigar-wrappers normally the medium and shorter leaves are preferred. The prominence of leaf veins is of a more significance in cigar-wrapper types. Fine vein constitutes an essential characteristic of a high grade cigar wrapper. The angle formed by midribs and principle veins makes a better look of cigars and helps in fixing a higher price. Usually midribs and webs are waste products to the extent of 20-30 per cent in tobacco leaves and therefore, they affect the economy in usage of the products.

Leaf thickness is next important factor for determining the value of the produce as the leaves of different thickness are used for different commercial purposes viz. leaves used for pipe smoking mixtures and cigar wrapper are thinnest of all types, while good cigarette leaves and plug fillers usually are somewhat thicker. Leaf thickness is partly controlled by growing season, cultural practices, etc. Crop grown under drought conditions produces small, thicker and heavier leaves which are relatively less elastic in nature and can't be used as wrapper or filler. Apart from this the, leaves produced under drought conditions have more sticky and gummy secretion which is composed of ethereal oils and resins. Although they don't affect the quality and aroma but the look becomes deceptive. Leaf-colour its depth and uniformity in shade over the leaf surface is equally important in selecting the grade of the leaf within the type. Leaves of light colour being mild tobaccos, may present a high luster while the dark coloured leaves, which are strongest ones, may have a dull appearance. On individual plant the lower leaves are light coloured and younges leaves are of darker shades.

Chemical Composition of Leaves

Tobacco leaves contain a number of chemical compounds (alkaloids) which influence their quality. Of these, polyphenols and carbonyls have an important bearing on flavors and aroma? Cellulose and minerals are deciding base materials for combustion. Glycosides cause bitterness in leaves. Soluble carbohydrates, especially sugars are considered to, have a positive influence on quality and nitrogenous compounds have negative one. The detrimental effects ate further pronounced if the nitrogenous compounds are present in protein forms. If the leaves have low sugar and higher nitrogen content, their smoke becomes insipid, sharper, more bitting and of no taste. But nicotine is the single compound which over rules all other chemicals in governing the quality. Nicotine is produced mainly in plant-roots and is carried through stem to the leaves where it is stored. Smoking enjoyment depends on a definite percentage of nicotine in manufactured products. Individual taste of smokers varies and therefore, manufacturers have to cater for their tastes by blending different types of leaves in their products. The nicotine content varies with different varieties, spacing, fertilization, topping and de-suckering time, number of leaves on plant, other cultural practices and method of curing etc. Heavy soils, drought conditions and higher soil temperature produce leaves with higher nicotine and vice-versa. Nitrogen, being an integral constituent of nicotine, influences the content of nicotine in the leaves while all other nutrients have no or littel effect.

Burning Quality of Leaves

The burning quality includes different elements of combustibility such as fire-holding capacity, the rate, evenness and completeness of the burn. The fire-holding capacity i.e. the property of continuing to glow without appearance of flame depends upon the structure and composition of the leaf. A high grade smoking tobacco must have a good fire-holding capacity with relatively slow bum. Usually thin leaves burn quickly and because of that the cigar-wrapper often burns some what in advance of burning core (filler). In cigarette a free and more rapid combustion is favoured by smaller diameter of the filer. Formation of grains (pimple like projections) on the leaf surface, which are composed of Ca, Mg and potassium salts of citric and malic acids improve the Fire holding capacity and finally the burning quality. Cellulose and related compounds, soluble carbohydrates and nitrates improve the combustion. Potassium, a major constituent of ash, results in a continued burning without flame. Mg and calcium have a controlling influence on the ashing process. The acidic compounds of ash, chlorine, phosphorus, silicon and sulphur, in general, tend to retard the combustion. Chlorine, if present in sufficient quantities, is outstanding in its injurious effects. The concentration of these harmful and beneficial elements may be controlled to greater extent by a balanced nutrition of the crop.

Aroma

The extent and type of aroma is greatly influenced by presence of polyphenols and carbonyles. Fermentation and aging of leaves are also found to affect the aroma. The quality of aroma depends primarily on tobacco species and varieties but soil and climatic factors are major factors in determining the strength of aroma. But the real factors responsible for nature and strength of aroma are yet to be known. The upper leaves on the same plant are found to be more aromatic compared to the lower ones.

Factors Affecting Tobacco Leaf Qualities

The factors affecting tobacco leaf qualities can be divided into three main factors.

(A) Genetical factors
(B) Ecological factors
(C) Cultural factors

(A) Genetical Factors

Curing behavior, flavour, aroma, body, texture colour retention, burning qualities etc. are expected to be genetically controlled to a greater or lesser extent, viz., sugar content is higher in flue-cured tobacco than in any other types. Similarly, carbohydrate content and nicotine content were found to be different in various type of tobacco.

Position of leaves on stalk influences the leaf quality as the bottom leaves are relatively thin light in weight and body, light in colour, mild and have good burning qualities. Passing up the stalk, the leaves become progressively thicker, heavier in weight and body rougher, gummier, stronger, and darker in colour, more aromatic, but in poorer burning qualities. The leaves approaching to the seed head can never reach to full maturity. As regards the chemical composition, in general, passing from the lowest to the top most leaves of fully matured plant there is a progressive increase in content of total nitrogen, protein, nicotine and ammonia, a decrease in total ash, calcium, magnesium and pH-value. While the content of soluble carbohydrates is lowest in the upper leaves and highest in the middle leaves of the plant.

Thus, it is observed from the same plant that on an average bottom 5-6, middle 2-4 and top 1-2 leaves formed the three grades or qualities viz. superior, medium and inferior respectively, which is roughly in the proportion of 6 : 3:1.

(B) Ecological Factors

1. Soil. Tobacco soil must be well drained, properly aerated with low inherent fertility. Soils with light sandy surface and friable sandy loam sub-soil surface, low content of organic matter, acidic reaction (desirable pH-5.0-6.0) and good amount of potash, phosphoric acid and iron are preferred for quality leaf production. Heavy soils produce tobacco with higher nitrogen than the sandy loam or sandy soils. Tobacco grown on saline lands absorbs moisture quickly before and after drying; changes colour rapidly during storage and has poor burning qualities which are attributed to large quantities of Chlorine (more than 0.01 per cent in soil and more than 10 per cent in leaf) which plant takes up from the soil. Tobacco grown on paddy lands and tank beds becomes bleached.

2. Climate

Temperature, humidity, the extent of rainfall and its distribution during the growing period of tobacco have considerable effect on the quality of leaf. High temperature and low relative humidity has harmful effects on leaf burn but the leaves are thick and aromatic, while at lower temperature, due to slower rate of metabolic processes, the leaf does not ripen fully and does not develop the desirable qualities. This desirable range of temperature for quality production of tobacco is 70-90 °F. Rainfall desirably ranging from 50 to 125 cm during growing season of tobacco results in leaves with light weight, thin texture, light colour, less gummy, light body, reduced nicotine content and aroma but of better combustibility, while drought results in the production of leaves having all the reverse qualities. Effect of sunlight, as studied in some of the foreign countries, does not have much effect except that shade or prolonged cloudy weather reduces the thickness of the leaves, increases leaf area without any increase in total dry weight of leaves. Wind causes dryness of plants and may bring a temporary atmospheric and soil drought which may result in a poor quality leaves.

(C) Cultural Factors

1. Crop Rotation

The soils under continuous tobacco cultivation are depleted of humus because tobacco does not add much of organic matter to soil like most of the crops. Such soils produce poor quality tobacco leaves and have possibility of an increased disease and *Orobanche* (flowering parasite) infestation. Therefore, tobacco in rotation with legumes always gives higher and better quality leaves.

2. Planting Material

Hardened and healthy seedlings, obtained from pure seed of the suitable strain, should be planted for better quality tobacco production. The planting material should have the resistance to insect pest, diseases and rough handling while harvesting. It should have uniformity in ripening and stand with fewer, smaller or no suckers.

3. Manuring

Quality in tobacco, particularly in the smoking types, (which are the ones that have been mostly investigated) depends on the balance of the nutrients in the leaf. Nitrogen has the profound effects on larger leaf area, increased moisture content, increased organic acid, enzymic activity and nicotine content but leaves have a poor carbohydrate content in them; light green leaves of good quality are produced from plant partially starved of nitrogen. Phosphorus deficient plants produce narrower and smaller leaves of dark green colour, the rate of growth is slowed down and maturity is delayed, the leaves are dull and devoid of luster on curing.

Potash, in the smoking tobaccos, is responsible for improving burning qualities like completeness of combustion and fire holding capacity that is also influenced by amount of potash in leaf. The physical qualities of the leaf, e.g., elasticity, smoothness and low prominence of veins and the smoking qualities are attributed to the high potash in the leaf.

Chlorine in excess amount injures the growth of the plant and reduces the quality (same as the excess of Calcium) by producing thick and brittle leaves. It has an unfavorable effect upon the burning quality and stability of colour.

4. Time of Planting

Early planting of all the tobacco types other than hookah in late October to early November is better than late planting (late November and early December). Bihar best time of planting of chewing and hookah tobacco is 2nd and third week of October. Planting at appropriate time produce best quality leaves with maximum production.

5. Plant Spacing

Closer spacing gives higher yield of leaves but the closer spacing tends to reduce the size, body thickness and weight per unit area of leaves. While in case of slightly wider spacing the plants produce larger, thicker and heavier leaves which are again not good quality leaves. From yield and quality point of view a spacing of 90 cm x or 120 50 cm x 60 cm is found to be optimum. Spacing again a subjective and solily depends upon tobacco type *i.e.* Bidi, chewing hookah *etc.*

6. Irrigation

This is of general belief that heavy irrigation spoil quality of the leaf. While light irrigation at critical stages of the crop growth gives a sufficient boost without spoiling the quality of leaf. High moisture content of soil results in low content of nitrogen, nicotine, calcium and magnesium in tobacco leaf, although the yield is high. Such leaves are thin with poor burning quality.

7. Removal of Useless Sand Leaves

Bottom sand leaves are very trashy and are of inferior quality, thus it is uneconomical to use them. First 2 or 3 leaves from bottom must be removed once the plants are 30 cm tall and again when plants are 60 cm tall.

8. Topping and Desuckering

Topping results in bigger, thick leaves which are cured as heavy, dark in colour, oily, rough, leathery and stronger. But topping should never be carried out in fields of rank growth, in areas known to produce poor curing as this could further deteriorate production of top grade tobacco. Nicotine content in the leaves consequent to topping depends on the height of topping being inversely related to the number of leaves left on the plants.

a. Height of Topping

Topping height in differ for different type of tobacco. Topping at 14 leaves would maximize the yield of hookah and chewing tobacco without impairing the quality. While in case of 'Motihari' tobacco, topping higher than 16 leaves tended to reduce the yield of first grade leaf.

b. Time of Topping and Desuckering

The best quality leaves are found where the plants are topped at early flowering stage and the quality deteriorated as topping was delayed. This also decreases the sugar and nicotine content in cured leaves. Desuckering is usually done after 6-7 days of topping, otherwise, topping will be useless for better quality leaf production. Generally 2-3 desuckerings are done. Apart from desuckering the topping is also followed by piercing of the main stem. It is done immediately after topping with 20-25 cm long needle. It increases cured leaf yield and yield of first grade leaf.

9 a. Harvesting Method

Correct harvesting helps in production of better quality leaves. Out of two methods of harvesting viz. priming and stalk cut in former method the leaves get sufficient time to attain optimum size till they mature and thus help in increasing more desirable grades leaves. The sand leaves should be separated out from the harvest. Immature leave are gummy and of poor combustibility, on the other hand, over matured leaves, become scrappy and lack the elasticity. Thus the leaves lead to a high percentage of the low grade leaves. Bruised and wilted leaves produce unsatisfactory cures of poor quality. The green leaves should be handled carefully and-these-leaves should not be exposed to sun in the fields for long periods nor allowed to wilt badly by exposure for longer periods in stringing sheds.

9b. Time of Harvesting

Generally Cigar and cheroot tobacco are ready for harvest after 20 to 25 days of topping, as is indicated by yellowing of tip and margins, slight puckering and brittleness of the leaf which breaks on folding. The highest yield of excellent quality leaves are obtained from the crop harvested three weeks after topping than the crop harvested earlier or later.

10 Curing of leaves

Temperature, relative humidity and air supply are the factors affecting curing process, which are responsible for fixation of colour, yellowing of leaves, and better quality of cured leaves. These factors vary from method of curing; thus selection of suitable curing method must be done, to ensumetha to quality of the finished products. A single furnace barn should never be loaded with more than 750 sticks. Rapid drying of tobacco improves quality over the slower drying by old methods because in slower drying the leaf gets discoloured. The cured leaves should be bulked by piling them in heaps for improving flavour and aroma. Also the amount of moisture in the cured leaves must be regulated for better qualities of Cigar tobacco, which is known as reconditioning.

Table 4.3 The nicotine contents of different tobaccos produced in India

Tobacco Type	Nicotine (%)
Virginia tobacco	1.2 - 3.6
Bidi tobacco (Anand)	9.71
HDBRG (Guntur)	3.89
Natu tobacco (Black soils)	2.79
Natu tobacco (Light soils)	3.5
Burley tobacco	1.26
Chewing tobacco (Tamil Nadu)	2.93

Contd...

Cigar tobacco (Tamil Nadu)	1.24
Chewing tobacco (Bihar)	3.7
Cigar filler (West Bengal)	2.05
Cigar wrapper (West Bengal)	1.44
Jati-Chama (West Bengal)	3.69
Jati-Podali (West Bengal)	4.02
Motihari-Hemti (West Bengal)	4.83
Motihari-Bitri (West Bengal)	6.64

Table 4.4 Characteristics of tobaccos grown in India

Tobacco type	Nicotine (%)	Reducing Sugars (%)	Chloride (%)	K (%)	Total N (%)
FCV (Karnataka)	2.53	23.14	0.44		-
Burley	1.70	0.84	0.23	3.13	-
Bidi (Anand)	9.71	2.28	1.80	1.90	3.15
Natu (Guntur)	2.79	2.36	0.42	2.29	2.97
Chewing	2.93	-	5.08	3.20	3.22
Cigar	1.24	0.29	4.52	4.50	2.37
Cheroot	1.01	0.51	3.87	4.51	2.68

Grading

After completion of fermentation tobacco leaves are graded according to their size, thickness colour and aroma. Leaves are graded by sorting leaves into uniform lots according to body, colour and degree of blemish or damage. Most important elements of quality in FCV tobacco are colour, texture, size, blemish, strength, even burning with white ash and agreeable flavour. The colour of leaf ranges from yellow/orange to dark brown with fine silky to coarse texture. A large number of grades are recognized in FCV tobacco for export trade but the individual grower usually sorts

Tobacco grading and packaging

out. The cured leaf into five or six grades: bright, semi brights, brown, light and dark greens and perished leaf. The grades adopted by exporters are called Agmark Grades. Merchants who supply tobacco to cigarette factories may paek according to the requirement of the buyers.

Table 4.5 Tobacco types and grading

Grading	FCV	Non FCV
1st Grade	Bright lemon or golden yellow colour, good texture, without any sponginess and other blemishes and with yellow veins.	Consists of long broad, thick and dark brown leaves with pungent taste and pleasant aroma. It is locally called Moorhan or Jotia.
2nd Grade	Bright lemon or golden yellow colour, good texture, with greenish vein and greenish patch on leaves.	It is locally called male moorhen. It consists of medium size leaves, less pungent than first grade and brownish in colour.
3rd Grade	Yellow colour with greenish and brown patch on leaves.	Chhobuva or Chhootki is the local name of third grade chewing tobacco. Consists of brown and small leaves with low pungency.
4th Grade	Reddish and green colour with greenish brown patch on leaves	It is locally called Kharsan and it consists of sand leaves of very light weight and inferior quality.

Average Leaf Yield

Average leaf yield of FCV tobacco is around 800 kg/ ha and that of natu tobacco 1000 kg/ ha in Andhra Pradesh. Bidi tobacco yields in Gujarat, Maharastra and Kamataka are 1000, 500 and 400 kg/ ha, respectively. Average leaf yield of cigar and cheroot tobaccos in Tamil Nadu is around 1300 kg ha-l and that of chewing tobacco around 1700 kg/ha. In Bihar, West Bengal and Uttar Pradesh, average leaf yields of hookah and chewing tobaccos are 1000,900 and 800 kg kg/ ha respectively.

Packing and Storage

Pack the bundle in grass and gunny cloth, and tie up with jute rope. Make the package of 150- 200 kg each which is commonly known as bales. Before packing some water is sprinkled on the leaf bundles to maintain luster and aroma and to protect the leaves against storage beetle. Bales packing are opened after 10-20 days and repacked. Open the bundle after 10-20 days and repack. Open and pack the bundle at one to two months intervals till the disposal of the leaf. Stored bales preferably in *kachcha* thatched shed and arranged as single layer on wooden planks or bamboo structure. Bales should be checked, opened and repacked at certain intervals. Fumigation may be done if required to protect the leaves against damage of storage pest's viz., storage beetles etc.

Tobacco Quality

Quality in tobacco is determined by a complex phenomenon and it is different for different people. Green, yellow and brown pigments provide qualitative information of leaf and usability. Harvesting the crop at correct (maturity) stage is paramount importance for producing quality tobacco. It has been amply demonstrated that amino-nitrogen is a good index of maturity and mature leaf has a lower content of this compound. Both physical and chemical characters of leaf affect tobacco quality. Physical characteristics include colour, body, texture, graininess and aroma. Graininess is due to grains appearing as minute pimples to unaided eye or papillae raised above the leaf surface.

□□□

5

Tobacco Quality Standardization

Any agricultural produce without desirable quality is just wastage of resources and time, and everything. Tobacco being leafy crop which entered in the world market needs proper quality standerd to fetch good price because it is grown for its aromatic and secondary metabolite i.e. alkaloid known as nicotine along with varying level of nicotine content along with some other physico-chemical properties. Different kind of tobacco has different standard for various compound and physical properties, which is described in this chapter.

Tobacco was produced and marketed with price support since 1939. Price schedules reflected an emphasis on quality, with deep discounts for unripe and immature grades. In the post-buyout environment, profitability will be directly related to the marketing of quality tobacco produced at the lowest cost, being a highly quality concern crop. From beginning much emphasis has been given for the chemical constitutes viz. nicotine, nitrogen, reducing sugar, and chloride content, beside others. Quality ranges or indices were fixed for different types. Thus quality determines the price of the tobacco leaves produced for marketing purposes. Quality parameter in flue-cured tobacco (FCV) has attained a new dimension in the present days because of higher mobility in international market, ever growing sophistication among smokers taste and increasing automation in cigarette industry.

While overall monetary return is a good enough criterion of quality, to a farmer, visible quality like colour, is very important for marketing purposes; but to an industrialist, those manufacturing qualities which maximize profit are most desirable features provided tobacco possesses satisfactory smoking quality which in turn depends upon chemical composition of leaf material. Flue-cured tobacco quality thus is made up of many complex components, viz. physical, chemical, organoleptic and also economic attributes. Smoking quality plus manufacturing capacity equals the suitability of tobacco. So the quality characters of FCV tobacco are divided into visual characters, manufacturing characters and chemical characters. Visual features like colour, body, texture, maturity/ ripeness, graininess, hygroscopicity, shatterability, blemish, elasticity, fluffiness, aroma, leaf size, vein colour etc. which are highly subjective based on which the tobacco leaf is graded and purchased by the trader. Other qualitative criteria have been developed for the manufacturing purpose are filling value, equilibrium moisture content, pore volume, elasticity, shatterability, combustibility, lamina-midrib ratio (Strip yield), No. of leaves/kg, lamina weight/unit area etc. Though tobacco leaf contains hundreds of chemical constituents, only few of them will have dominating influence on quality. Nitrogenous and carbohydrate fractions are the two groups of chemical constituents having outstanding effect on smoking quality. Chloride in leaf is also very important as it influences the combustibility and keeping quality of leaf. Sugars is considered to have a positive influence on quality, while nitrogenous compounds a negative one. But it is necessary for a leaf to have too much sugar and some nitrogenous constituents because in absence of either of the two the smoke is insipid and lacks character. Thus the commercial value of a leaf is judged by its suitability for specific purposes of manufacture viz. tobacco used for smoking purpose must have a combustibility and fine aroma. In general the leaf characters, chemical composition of leaves, burning quality and aroma are the most important factors in determining the quality of a tobacco.

Leaf Characters

Leaf character is primarily hereditary in nature but it may slightly be modified with the growing season, nutrient conditions, planting density, topping and de-suckering etc. Size of leaves is a very important factor in cigar types than in most others because for cigar-wrappers normally the medium and shorter leaves are preferred. The prominence of leaf-veins is of a greater significance in cigar-wrapper types. Fine vein constitutes an essential characteristic of a high grade cigar wrapper. The angle formed by midribs and principle veins makes a better look of cigars and helps in fixing a higher price. Usually midribs and webs are waste products to the extent of 20-30 per cent in tobacco leaves and therefore, they affect the economy in usage of the products. Leaf thickness is next important factor for determining the value of the produce as the leaves of different thickness are used for different commercial purposes. Leaves used for pipe smoking mixtures and cigar wrappers are thinnest of all types, while good cigarette leaves and plug fillers usually are somewhat thicker. Leaf thickness is partly controlled by growing season, cultural-practices etc. Crop grown under drought conditions produces small, thicker and heavier leaves, which are relatively less elastic in nature and can't be used as wrapper or filler purpose. Apart from this the leaves produced under drought conditions have more sticky and gummy secretion which is composed of etheral oils and resins. Although they don't affect the quality and aroma but the look becomes deceptive. Leaf-colour, its depth and uniformity in shade over the leaf surface is equally important in selecting the grade of the leaf within the type. Leaves of light colour, being mild tobaccos, may present a high luster while the dark coloured leaves, which are strongest ones, may have a dull and lifeless appearance. On individual plant the lower leaves are light coloured and top leaves are of darker shades.

Visual Observation

Seeing is beliving are the first criteria and tobacco traders evaluating its quality visually by its characteristics, which is not very objective and such a system of subjective quality evaluation varies with personal fancies and hence cannot be considered as precise.

Physical Criteria

It includes parameters like colour, flavour, texture (graininess), taste etc. Taste depends on proper ratio of nicotine to sugar. Aroma or flavour indicates well matured ripe tobacco of grainy texture, cured under ideal conditions. It includes equilibrium moisture, filling power and combustibility. Equilibrium moisture content indicates the hygroscopicity of leaf and is the moisture content of the tobacco when in equilibrium with 60 per cent relative humidity. Filling power is the volume occupied by a given weight of tobacco under standard pressure. It is negatively correlated with equilibrium moisture content. At 13.5 per cent equilibrium moisture content, the filling power is 3-5 g cm^{-3}.

Filling Value

Tobacco leaves types used in the manufacture of cigarettes and like smoking articles comprise leaf lamina, a longitudinal main stem (rib) and veins extending from the main stem. The main stem and large veins are hereinafter jointly referred to as 'stem'. The stem has substantially different physical properties from the lamina, and it is long-established practice to separate the stem from the lamina at an early stage in the processing of tobacco leaves, the stem and lamina then being processed independently and differently. Filling value is the volume occupied by unit weight of cut tobacco at predetermined moisture level under a specific stress. Filling value may be taken to be a measure of the relative number of cigarettes of a given firmness that could be manufactured from a unit weight of tobacco. Good filling value enables the manufacturer to ensure that cigarettes are made as well filled rigid rods. Poor filling results in poorly made soft cigarettes from the ends of which tobacco shreds fall out easily or else cigarette weight has to be increased in order to restore cigarette firmness. Heavy filling on the other hand can result in too much draw resistance and in sharply altered burn characteristics. Dimensionally, filling value is inversely related to lamina weight. Hence, very thin bodied leaf resulting in 'chaffiness' also shows a high filling value and as such it is of undesirable quality. There is an optimum filling value for high quality leaf.

Shatter Resistance

It refers to resistance of leaf to shattering under such mechanical impact and stress, it receives during drying, cutting and at other post harvest operations. The manner in which stem material is separated from lamina material is generally by means of a complex and large threshing plant comprising a number, eight for example, of serially arranged threshing machines with classification units disposed intermediate next adjacent threshing machines. As well known, the separated stem material, or a proportion of it, after suitable reduction in size, is often added back to the lamina after the lamina has been subjected to further processing. Stem material is often desirable in the tobacco blend to improve fill value. Another important economic factor in tobacco quality is its resistance to breakage during handling. Tobacco is a fragile material that tends to shatter to a great or less degree with each handling. Breakage becomes accentuated under the stress of mechanical processes in the factory. Tobaccos do differ in their relative brittleness due to various factors. Strength in tobacco leaf is dependent on calcium pectate, the cementing material in the cell wall. Although moisture content influences this property to a great extent, other unknown factors may also play the role in controlling brittleness of leaf. A value of 2.0 is considered as high shatter resistance, while below 1.0 is suggestive of poor resistance.

Leaf Combustibility

It is ability of tobacco to burn continuously, evenly and gradually. It is related to glow duration. Porosity of leaf is a physical characteristic. It measures irregular air spaces, which regulates the leaf's capacity to absorb and retain additives during cigarette manufacture.

Strip Yield

Strip yield in flue-cured tobacco is important to manufacturers since it is the lamina portion of leaf that is normally used in cigarette making. Because of the low utility of midrib, a large bulk of exportable flue-cured leaf is dispatched only in the form of strips. This makes strip yield an important criterion in developing varieties as the higher the strip yield, the greater is the economic return. The strip constitutes on an average, about 75% of leaf by weight, usually ranging from 70-80%. Within this range, the higher the strip yield, the better the usability of tobacco.

Elasticity

Elasticity is considered to be a major quality factor in tobacco. Elasticity is the ability of the leaf, when moist, to undergo stretching without breaking. Such tobaccos after being compressed, as occur during cutting in the manufacture of cigarettes, will spring back immediately. Springiness in flue-cured tobacco is thus due to its elasticity. Elasticity is dependent upon water soluble constituents, and stands in direct correlation with the moisture content in tobacco.

Texture or Porosity

Leaf structure and texture is an important physical property of flue-cured tobacco. Texture and grain are synonyms for cigarette tobacco. Graininess in flue cured tobacco is a measure of porosity of leaf which regulates its capacity to absorb and retain additives in the intercellular air chamber. Leaf structure is also defined as degree of cell development of leaf as indicated by its porosity.

Hygroscopicity

Hygroscopic properties of cured leaf is judged by equilibrium moisture content, is an important technological criterion for judging quality. It is the moisture absorbing capacity of leaf which depends on the relative humidity of the surrounding environment. Cured leaf low in hygroscopicity is very difficult to get to 'order' or 'condition' and leaf handling is impeded. High hygroscopicity on the other hand, entails operational difficulty in the cutting and filling machine. Storage of heavily hygroscopic leaf is also a problem. Cigarettes having very low moisture content give harsh taste and smoke; high moisture content on the other hand impairs combustibility. Therefore, it is apparent that cured leaf of acceptable quality should have its equilibrium moisture content within the optimum range i.e. 12 to 14% at 60% relative humidity. Lipophilic colloidal constituents of leaf materials and tissue density (porosity) have direct influence on the water absorption and retention capacities. Reducing sugars are known to increase leaf moisture content. The hygroscopicity of leaf is increased with chloride content of the leaf. The water holding capacity of cured leaf increased with potassium content. The effect of potassium may be seen as to increase in the level of soluble carbohydrate synthesis.

Table 5.1 Quality parameter of Cigar tobacco

Cigar tobacco components	Texture	Qualities
Filler	Thick	Good burning taste and aroma
Binder	Medium to thin	Medium to thin
Wrapper	Very thin	With fine parallel veins and a pleasing luster, finish

Burning Quality

The burning quality includes different elements of combustibility such as fire-holding capacity, the rate, evenness and completeness of the burn. The fire-holding capacity i. e. the property of continuing to glow without appearance of flame depends upon the structure and composition of the leaf. A high grade smoking tobacco must have a good fire-holding capacity with relatively slow burn. Usually thin leaves burn quickly and because of that the cigar-wrapper often burns some what in advance of burning core (filler). In cigarette a free and more rapid combustion is favoured by smaller diameter of the filler. Formation of grains (pimple like projections) on the leaf surface, which are composed of Ca, Mg and potassium salts of citric and malic acids, which improve the fire holding capacity and finally the burning quality. Cellulose and related compounds, soluble carbohydrates and nitrates improve the combustion. Potassium, a major constituent of ash, results in a continued burning without flame. Mg and calcium have a controlling influence on the ashing process. The acidic compounds of ash, chlorine, phosphorus, silicon and sulphur, in general, tend to retard the combustion but chlorine, if present in sufficient quantities, has negative impact on burning quality. The concentration of these harmful and beneficial elements may be controlled to greater extent by a balanced fertilization of the crop.

Aroma

The extent and type of aroma is greatly influenced by presence of polyphenols and carbonyls. Fermentation and aging of leaves are also found to affect the aroma. The quality of aroma apparently depends primarily on tobacco species and variety but soil and climatic factors are major factors in determining the strength of aroma. However the real factors responsible for nature and strength of aroma are yet not known.

Chemical Characteristics of Leaves

Chemical characteristics depend on chemical composition of leaf. Important constituents affecting the quality are carbohydrates, aliphatic and aromatic alcohols, carbonyls, quinines, sterols, carboxylic acids, amino acids, poly phenols, inorganic constituents and alkaloids. Carbonyls are associated with aroma and fatty acids contribute to overall leaf and smoke flavour. Polyphenolic compounds influence leaf colour, quality and strength of smoke. The major alkaloids are nicotine and anabisine, which are involved in carbon-nitrogen balance in the plant. In general, tannins and resins provide strength. Starch, sugars and organic acids give mildness. Different tobaccos type have differing chemical composition. For FCVs, reducing sugars may vary from 12-25 per cent but the limit is 22 per cent. Nicotine content up to 2.0 per cent. Nitrogen content varies from 1.25 to 2.0 per cent and the total nitrogen to nicotine ratio should be less than one (-0.6). Inorganic constituents like K, and Cl are important from the quality point of view. Chlorides may be limited to 2 per cent while K varies from 2.5 to 5.5 per cent. Burley tobacco has high nicotine, nitrogen and volatile acids but is very low in sugars. Chewing and snuff types are high in nitrogen compounds, nicotine and less in sugar content. Snuff tobacco contains 3.5-4.5 per cent nicotine. Bidi tobacco with parrot green colour contains highest nicotine content of 6.5-8. 7 percent. In view of serious health hazards, smoke tests have become mandatory to reduce the hazards to the minimum. Two parameters, total particulate matter (TPM), smoke condensate or tar and nicotine levels in the smoke are used as criteria to judge the mildness of tobacco. The levels of TPM or tar in India are 20 to 30 mg per cigarette and that of nicotine is 1.92 mg per cigarette.

Total Nitrogen Content

It is generally considered that flavour and taste of smoke is correlated with the nitrogenous constituents. Flue-cured tobacco containing 1.6 to 2.3% total nitrogen gives the most satisfying smoke. Higher nitrogen content of tobacco would result in poor quality apart from curing difficulty, deep brown coloured trashy leaf which shatters readily and it has flat-insipid tasting smoke. Generally high level of nitrogen is associated with high level of nicotine. Lower nitrogen content would result in 'washed out', pale coloured leaf, lacking in rich colour characteristic of good tobacco.

Nicotine Content

Nicotine content of tobacco by virtue of its stimulatory effect on the smoker is next important constituent. It is considered that a nicotine level of 1.75 to 2.0% in FCV tobacco is most satisfactory. The nornicotine in acceptable tobacco should not exceed 5% of total alkaloids content. High proportion of nornicotine in cigarette leaf leads to abnormal and objectionable smoke flavour due to pyrolysis of nornicotine into myosmine.

Table 5.2 Limits for the important quality constituents and quality indices in flue-cured tobacco.

Constituent/Quality Index	Acceptable Limits
Total Nitrogen %	1.0 - 3.0
Nicotine %	0.7 - 3.0
Total Sugars %	10.0 - 26.0
Reducing Sugars %	8.0 - 24.0
pH	4.6 - 5.5
Reducing Sugars/ Total N	7 - 13
Reducing Sugars/Nicotine	7 - 13
Total N/ Nicotine	< 1.2
Chloride %	< 1.5
Filling value at 60% R.H. & 20oC	3.3 - 3.8cc/g shreds
Equilibrium moisture content at 60% R.H. & 20oC	11 - 15%
Pore Volume	0.13 - 0.18 ml/g
Combustibility	2.5 - 3.5 mm/min
Leaf burn	3 - 6 sec.
Shatterbility Index	> 3

Nitrogen/Nicotine Ratio

The ratio of nitrogen to nicotine is assumed to give some chemical balance within the leaf. The higher the ratio the less desirable the tobacco because it tends to be light bodied. A ratio in the neighborhood of 1.35 gives way to paleness of colour, slickness of texture and a general lack of desirable physical characters and deficiency in aroma. In fact, a value exceeding 1.0 has been ascribed as unbalanced. Too low a value (below 0.5), on the other hand, frequently may be considered undesirable because the tobacco is heavy bodied and associated with high nicotine content and low level of reducing sugars. A range of 0.6 - 0.7 ratios has been adjudged as most desirable in medium to light bodied matured tobacco.

Reducing Sugars

Higher content of reducing sugars in flue cured tobacco is undesirable as it imparts to the smoke an acidic character. Lower content imparts alkalinity to smoke due to high nitrogenous constituents. During smoking, sugars are burnt out as CO2 and water, thus helping to neutralize free base and increase moisture content in smoke and so act as an emollient, if present in excessive quality.

Reducing Sugars/Nicotine Ratio

The ratio of sugar to nicotine would give a balance of opposing effects and thus serve as a good smoking quality indicator. A high ratio may tend to indicate mildness and smoothness while a very low ratio may be indicative of harsh irritating smoke. If the ratio is too high, it may indicate that the tobacco is too mild to be acceptable to smoker. If cured leaf contains both low level of nicotine and sugars as generally is the case with Indian flue-cured tobaccos, the ratio appear to be comfortable. High sugar content consistent with nicotine level is the most desirable feature for smoking quality in flue-cured tobacco. The desirable ratio is 7-13.

Carbohydrate / Protein Ratio

The ratio of carbohydrate to proteins is known as Shmuk number and the ratio of carbohydrate to total nitrogen is known as Kovalenko coefficient. These ratios will show similar effect of reducing sugar to nicotine.

Table 5.3 Some important quality parameter of tobacco grown under different soil conditions in India.

Character	Traditional black soil	Northern light soil	Southern light soil	Karnataka light soil
Nicotine%	1.0 to 2.7	1.6 to 2.9	0.9 to 1.1	1.5 to 1.8
Total N%	1.5 to 2.5	1.8 to 2.7	2.3 to 2.6	1.9 to 2.7
Reducing sugars%	8.6 to 20.9	13.8 to 23.5	10.1 to 17.5	18.0 to 27.0
Chlorides %	0.5 to 2.0	0.7 to 1.5	0.5 to 0.9	0.1 to 0.2
Moisture equilibrium %	12.2 to 15.4	14.0 to 17.1	11.7 to 13.6	13.0 to 15.0
Pore volume (ml/g)	0.11 to 0.19	0.14 to 0.24	0.13 to 0.15	0.13 to 0.17
Filling value (cc/g)	2.6 to 2.8	2.2 to 2.8	2.6 to 3.5	2.9 to 3.3

Chlorides

High level of chlorine in leaf inhibits leaf burn or combustibility. The chlorine content of leaf preferably less than 1.5% but should never exceed 2%. Chlorine content is positively correlated with deterioration of colour. High chloride content in leaf leads to dull muddy orange colour with sour or linoleum smell. Further, such leaf due to its moisture holding capacity bruises easily and tends to develop 'off-colour'. These characteristics render this type of leaf to be of low value for cigarette manufacture. Chlorine acts as a negative combustion catalyst in tobacco.

Potassium

Potassium content in the cured leaf was found to improve the burning quality of tobacco. An adequate level of potassium in cured leaf tends to off-set the deleterious effects of high chlorine on burning quality. Potassium acts as a mineral catalyst and oxygen carrier in promoting burn of tobacco leaf. Cured leaf lacking in potassium content would result in poor coloured trashy leaf and looses its luster which may not have any commercial value.

The individual chemical constituents alone should not be taken into consideration for quality evaluation. The ratios of the constituents are also very important and should be taken into consideration in toto for quality appraisal of tobacco. In case of tobaccos other than FCV tobacco, all the visual characters mentioned in case of FCV tobacco are also important. The nicotine content is considered to be the important chemical constituent. The quality of different tobacco type one presented in table 5.4.

Table 5.4 Quality parameter of various tobacco type grown in India.

Tobacco type	Total N %	Nicotine %	Reducing sugars %	Chlorides %	K %
Bidi (Anand)	3.15	9.71	2.28	1.80	1.90
HDBRG	2.59	3.89	1.09	0.40	2.02
Natu (Guntur)	2.97	2.79	2.36	0.42	2.29
Irrigated Natu (V.R. Gudem)	3.40	2.50	0.84	1.59	1.53
Burley (Chintapalli)	3.52	1.26	0.79	0.27	4.80
Chewing tobacco (Vedasandur)	3.22	2.93	-	5.08	3.20
Cigar tobacco (Vedasandur)	2.37	1.24	0.29	4.52	4.50
Cheroot tobacco (Vedasandur)	2.68	1.01	0.51	3.87	4.51
Chewing tobacco (Pusa)	3.11	3.70	1.98	1.28	3.25
Cigar filler (Dinhata)	3.51	2.05	1.37	0.38	4.62
Cigar wrapper (Dinhata)	3.45	1.44	1.11	0.46	5.11
Jati-Chama (Dinhata)	2.66	3.69	0.99	0.56	2.79
Jati-Podali (Dinhata)	3.05	4.02	0.82	< 0.44	1.89
Motihari-Hemti (Dinhata)	3.76	4.83	0.71	1.91	1.91
Motihari-Bitri (Dinhata)	4.12	6.64	0.82	0.42	1.66

leaves become scrappy and lack the elasticity. Thus the leaves lead to a high percentage of the low grade leaves. Bruised and wilted leaves produce unsatisfactory cures of poor quality. The green leaves should be handled carefully and those leaves should not be exposed to sun in the fields for long periods nor allowed to wilt badly by exposure for longer periods in stringing sheds.

9b. Time of Harvesting

Generally Cigar and cheroot tobacco are ready for harvest after 20 to 25 days of topping, as is indicated by yellowing of tip and margins, slight puckering and brittleness of the leaf which breaks on folding. The highest yield of excellent quality leaves are obtained from the crop harvested three weeks after topping than the crop harvested earlier or later.

10 Curing of leaves

Temperature, relative humidity and air supply are the factors affecting curing process, which are responsible for fixation of colour, yellowing of leaves, and better quality of cured leaves. These factors vary from method of curing; thus selection of suitable curing method must be done, to ensumetha to quality of the finished products. A single furnace barn should never be loaded with more than 750 sticks. Rapid drying of tobacco improves quality over the slower drying by old methods because in slower drying the leaf gets discoloured. The cured leaves should be bulked by piling them in heaps for improving flavour and aroma. Also the amount of moisture in the cured leaves must be regulated for better qualities of Cigar tobacco, which is known as reconditioning.

Table 4.3 The nicotine contents of different tobaccos produced in India

Tobacco Type	Nicotine (%)
Virginia tobacco	1.2 – 3.6
Bidi tobacco (Anand)	9.71
HDBRG (Guntur)	3.89
Natu tobacco (Black soils)	2.79
Natu tobacco (Light soils)	3.5
Burley tobacco	1.26
Chewing tobacco (Tamil Nadu)	2.93

Contd

6

Tobacco Nutrition

There is no other remunerative and alternative crop to tobacco in many tobaccos growing track in India. Tobacco growers are major beneficiary with agro technology of tobacco recommended from time to time, to brig the gap between potential and existing productivity level of tobacco in India. Although the quality of tobacco is a most important aspect of its usefulness to the manufacturer, it is a difficult property to measure in terms of specific chemical and physical characteristics. The components of quality have not been quantitatively delineated, it is not possible to study absolutely the influence of various growth factors on quality.

Nevertheless, there is a large body of qualitative evidence which provides some indications of these relationships. The accumulated mass of this evidence indicates that the quality characteristics are intimately associated with the nature of growth of the plant from transplanting through final harvest. There is lot of gaps due to lack of effective extension of developed technologies and there is an urgent need to bridge the gap between existing old practices with new research to improve the quantity and quality of tobacco. There is several component which need to be address with due consideration but nutrient management is one of the key cause of poor productivity coupled with inferior quality of tobacco leaves being main economic produce of the said crop. The tobacco plant, from seedling stage until final harvest, is extremely sensitive to relatively small fluctuations in nutritional regimes and environmental conditions. Changes which may appear to be slight by visual standards frequently result in distinguishable modifications in the properties of the leaves and may influence their weight and commercial value. Thus some knowledge and consideration of the total growth and development of the plant are essential to a complete understanding of the nutrition of the plant. The differences in the parent materials, topography, climate, age and vegetation have resulted in the formation of many types of soil, which differ significantly between different regions as well as within the region itself. The availability of plant nutrient is largely dependent on the types of soil, climatic conditions, cropping sequence and several other factor. Indian soils are in general deficient in organic carbon (OC) nitrogen and phosphorus and in patches potashium diffciency has also been seen sulphur & zinc one marjar secondly & micro nutrient depleating very fast from Indian soils the reason could be many more but introduction of high yielding crop varieties, increased use of chemically pure high analysis fertilizers containing negligible amount of micronutrients as contaminant, increased cropping intensity and restricted recycling of farm wastes accrued to widespread deficiency of micronutrients in general.

Nutrient Accumulation

The quantity of each nutrient absorbed by a field grown crop of tobacco varies considerably; it is depend on the class of tobacco, fertilization practices, residual level of nutrient in the soil, plant population per ha, rain fall and other environmental factors. Normally growing tobacco plant contains about 3.15 per cent nitrogen, 3.42 per cent in leaves and 3.12 per cent in stock. However, it varies within wide limits of 2-5 per cent in leaf and 2.5-4.0 per cent in stalk. Nitrogen has profound influence on yield and quality of tobacco.

Higher levels of nitrogen application, generally, increase the yield and quality of non-FCV tobaccos and partially nitrogen starved FCV tobacco crop produce light green leaves of good quality. If the nitrogen content of leaves falls below 2 per cent, deficiency symptoms begin to appear. Normal range of phosphorus (P_2O_5) in tobacco leaf is 0.5-0.9 per cent and in stalk 0.5-0.7 per cent. A broad indication of the functions of the nutrient element will be given, as the exact functions for a number of them are still not definitely known. Each elements is specific in its function in plant metabolism. The essential elements exist as structural components of a cell, maintain cellular organizations, and function in energy transformations and enzyme reactions. Based on their relative requirement by plants these elements may be classified as:

(1) *Macro-nutrients*: Those nutrients which are required by plants in concentrations exceeding 1 ppm are termed as major or macro-nutrients. These are further classified into two categories as:

(a) *Primary nutrients* : C,H,O, N,P,K are the primly elements which are essential right from seed germination. Out of these C,H,O are found abundantly in water and atmosphere and they arc supplied whereas N,P and K are either obtained from soil or supplied through chemical fertilizers.

(b) *Secondary nutrients* : They are secondary in the sense that they are needed only when the plants have started growing they have grown for some period. These are Calcium (Ca) Magnesium (Mg) and sulphur (S).

(2) *Minor or micro-nutrients* : The elements which arc required by plants in concentration less than 1 ppm are put in this category. They arc also called as trace elements, oligo elements, supreme elements, etc. Nicholas (1963) used "Ultra micro-nutrient" for molybdenum and cobalt which arc required as low as one part per billion by most of the plants.

Carbon, hydrogen and oxygen form about 94 per cent of the dry weight of plants. These are the major components of carbohydrates, proteins and fats. Besides their structural role, they provide energy required for the growth and development of plants by oxidative breakdown of carbohydrates, proteins, and fats during cellular respiration. The very fact that the micronutrient elements are required by plants in very low concentration suggests that they all function as catalysts or at least closely linked with some catalytic processes in plants. Evidences are there that at least copper; zinc and manganese are components of certain biological oxidation-reduction systems. Some of the micronutrient elements are essential for and related to the activity of some enzyme systems. The interactions and antagonisms of the elements influence their role in plant metabolism. Such antagonistic effects, between manganese and iron, calcium, copper and also between calcium and boron are amongst many. Manganese toxicity is associated with calcium deficiency and vice versa. A close antagonism between calcium and boron is well known.

Role of Plant Nutrients

Every nutrient plays a specific role in nutrition, growth and development of plants. These roles may be described as under:

Carbon

It is available in abundance from air. CO_2 is used by green plants for photosynthetic activities. It is also required for cell formation in plants. About 45% or more parts of the plant tissues are made up of carbon.

Hydrogen

It is essential for cell and tissue formation in plants. This is obtained from water and is required for energy reactions. It forms about 6% parts of the plant tissues.

Oxygen

Plants take oxygen from air and water. It forms about 43% parts of the plant structure. It is required for photosynthetic and respiratory activities. It helps in formation of tissues and cells.

Mineral Elements

1. Nitrogen

Among the elements essential for the commercial production of tobacco, none has as pronounced effect nor requires the degree of attention in fertility practices as nitrogen because it is a major structural part of the cell. Nitrogen is an essential constituent of metabolically active compounds such as amino acids, proteins, enzymes and some non-portentous compounds. It plays a vital role in various metabolic activities of plants and is a constituent part of amino acids, proteins, nucleic acids, porphyries, flavins, purine and pyrimidine nucleotides, enzymes, co-enzymes and alkaloids. If nitrogen is a limiting factor, the rate and extent of protein synthesis are depressed and as a result plant growth is affected. The plant gets stunted and develops chlorosis. Cytoplasm and the particulate fractions of the cell organelles contain nitrogen in varying amounts which exist in combination with C, H, O, P and S. Primary cells are found to have about 5% of nitrogen. It helps in harvesting solar energy through chlorophyll, as well as in energy transformation through phosphorylated compounds, in transfer of genetic information through nucleic acids. Besides, it is essential in cellular and protein metabolism and, acts as biological catalyst. From the seedling stag through final harvest, the soil nitrogen regime affects the process of plant development more than any other mineral element. With respect to time of absorption, form in which absorbed, concentration in the leaf at various stages of growth, and in numerous other aspects, the role of nitrogen in the development and properties of the tobacco leaf is of major importance. The influence of variations in nitrogen supply on growth of the plant and properties of the cured leaf has been investigated more extensively than have the effects of any other essential element.

Role of Nitrogen

Green Leaves Quality

Total number of leaves produced by a plant is not appreciably influenced by the level of available nitrogen, but results in an increase in the area of the leaf but a decrease in the weight per unit area, the latter effect primarily being due to a decrease in the thickness of the leaf. This effect of nitrogen on the size and thickness of the leaf is of considerable practical importance in relation to the fertilization practices for tobaccos for different commercial uses. For proper maturity it is essential that the rate of nitrogen absorption decrease rather rapidly during the latter portion of the growth period. The growth of a tobacco plant involves the successive development of 16 to 20 leaves. Leaves at the basal portion of the plant are approaching maturity when those at the top are still in a very active stage of growth, thus each leaf on the plant is a different physiological age and need of nutrition.

Physical Quality of Cured Leaves

A deficiency of nitrogen, however, increases the probability of premature yellowing of leaves in the field. Under these conditions the cured leaves are generally pale in color and lack the desired textural properties associated with high quality in tobacco. The delay in maturity which results from excessive nitrogen application prevents normal development of the leaves, and thus when they are cured they lack the desired chemical and physical properties. The characteristics of the leaves may vary somewhat, but generally they are dark brown to black in color and dry and chaffy and decrease the fire-holding capacity of tobacco. The decrease in combustibility is probably due to chemical composition; proteins and related nitrogen compounds tend to adversely affect burning properties of leaves. Since a thin leaf usually has better burning qualities than a thick, close-textured one.

Chemical Quality of Cured Leaves

Nitrogen is considered to be a crucial factor which influences the level of strength in tobacco smoke. Whether the level is too high or too low depends on the use that is made of the tobacco. Nitrogen is an integral constituent of the nicotine molecule, and thus nitrogen is an important factor in nicotine synthesis. The accumulation of nicotinic in the plant is regulated more by the nitrogen supply than any other plant nutrient. Tobacco with high total nitrogen content produces a strong pungent tasting smoke whereas tobacco low in nitrogen has a flat insipid tasting smoke. Where as the concentration of nitrogen in the tissue is positively correlated with nicotine, it is negatively related to the sugar content of the leaf. Leaves at middle and top stalk positions may have a relatively high nicotine level when nitrogen deficiency occurs late in the season whereas an early deficiency results in a relatively low level of nicotine in leaves at these stalk positions. A severe nitrogen deficiency, even though temporary, has been shown to result in a low quality leaf physically and chemically.

Deficiency Symptom

Although apparent nitrogen deficiency symptoms do not generally occur when the concentration in a particular leaf is above 1.5 to 2.0 per cent, a slight yellow color indicated reduction in chlorophyll content and length and width measurements showed a reduction in leaf area. The reduction in area, however, seldom is accompanied by a reduction in the total dry weight of the leaves; instead there is greater weight per unit area with the stress than with the nonstress condition.

Tobacco Response to Ammonium v/s Nitrate

Essentially all the nitrogen absorbed is in the ammonium and / or the nitrate form. Since oxidation of the ammonium to the nitrate form may occur to varying degrees, precise information on the regime on which the plants are grown is generally scanty. Nutrient uptake, nitrogen and carbohydrate metabolism, growth of the plant are markedly different depending on the dominant form absorbed. The relative uptake of the two forms will vary considerably depending on the composition and acidity of the solution and the stage of plant development. Tobacco can also absorb limited amounts of urea N through the leaves under normal cultural practices. From experiments in which plants were supplied ammonium nitrate enriched with ^{15}N, found that young tobacco seedlings (39 days from seeding) absorbed nitrogen primarily as ammonium. Older plants (70 days after seeding), however, absorbed more nitrate than ammonium. Ammonium nutrition that increasing the pH by the addition of bicarbonate to the nutrient media improved the growth of tobacco and caused toxicity symptoms to disappear. With nitrate nutrition, however, bicarbonate always had a depressing effect on growth and high concentrations resulted in chlorosis. For both ammonium and nitrate nutrition, the presence of bicarbonate in the medium increased the content of cations and of organic acids, particularly citric acid, in the plant. The improvement in the growth of tobacco due to the additions of bicarbonate to the ammonium cultures was not only due to changes brought about by the increase in pH, but also to absorption and metabolism of bicarbonate. The relative response of field-grown tobacco to the ammonium and the nitrate form of nitrogen is also closely related to the extent to which nitrification occurs. Where soil conditions were favorable for rapid nitrification, differences in response to ammonium and nitrate applied under field conditions were small and generally nonsignificant. Plant beds fertilized with ammonium and to a lesser extent with urea, produced seedlings with thick, fleshy laves with a pronounced upward cupping of the leaf margins. A tobacco seedbed disorder referred to as "**yellow patch**" is found to be associated with high rates of application of easily decomposable organic fertilizers. The plants were stunted, malformed, and highly chlorotic and many eventually died. The cause of the condition was attributed to the excessive accumulation of ammonium from decomposition of the organic fertilizers.

Organic Nitrogen

It has been a traditional practice in the fertilization of tobacco to obtain a portion of the fertilizer nitrogen from natural organic materials such as compost, FYM, oil seed cakes, tankage, fish meal, or, more recently, urea formaldehyde compounds. These organic sources not only supply nitrogen but also provide several other macro and micro nutrient apart from improving soil physical condition. This practice is favored in the belief that the nitrogen from these materials is released at a rate which will meet the requirements of the plants but is slow enough so that no major quantities are lost from the root zone by leaching and other such type of activities.

2. Phosphorus

Although phosphorus has received major attention in fertilizing tobacco, primarily with studies on general growth effects from rates of applications with the principal attention given to yield and quality of the cured leaf. It was observed that a relatively high application of fertilizer phosphorus resulted in a marked increase in the early growth of plants. Phosphorus plays a vital role as a structural component of cell constituents and metabolically active compounds. Phosphorus is a structural component of cell membranes, chloroplasts, mitochondria, sugar phosphates, viz. ADP, ATP, nucleic acid etc., phospholipids and phosphatides. The most essential constituents of plant cells like esters, phosphatides and phospholipids are synthesized by phosphorus when it combines with different organic acids. It also plays an important role in energy transformations and various metabolic activities of plants. Phosphorus plays an important role in energy transformation and metabolic processes in plants. It stimulates root growth. Being a constituent of adenosine phosphate, phosphoglyceraldehyde and rebulose phosphate, it helps in basic reactions of photosynthesis and activates several enzymes participating in dark reactions in photosynthesis.

Role of Phosphorus in Plant Growth

The responses of tobacco to phosphorus fertilization are more frequent in the early growing season than are in later stages of final yield and quality of the cured leaf because the rapid and vigorous development of young plants is stimulated by a high level of available phosphorus. An early-maturing normal plant develops more desirable properties than one which matures late.

Quality of Cured Leaves

With the increasing level of available phosphorus increased total alkaloids and nornicotine whereas the nicotine content remained unchanged. In contrast, a negative correlation has been also observed between the phosphorus and nicotine contents of the leaf whereas other workers have failed to observe any influence. Studies on the influence of phosphorus on uptake of certain cations and the relation of this to leaf color have given variable results.

Deficiency Symptoms

There are, however, reports that a relatively low rate of phosphorus resulted in higher leaf yields than did heavier applications. These effects may be related to the influence of phosphorus on the absorption of other ions. The major symptoms of phosphorus deficiency in tobacco are retardation in growth, particularly during the first month after transplanting, and a delay in maturity of the plant .The leaves tend to be narrower than normal, high in nitrogen, and abnormally green. Occasionally there may be numerous small white spots on the lower leaves. The leaves from plants that are deficient in phosphorus do not mature normally and are generally of low quality. The cured leaves tend to be dark brown in colour because of their immaturity, may be dark greenish in color and lack the luster of leaves from normal plants. It is generally considered that no adverse effects on the growth and development of tobacco will occur from excessive amounts of phosphorus.

Phosphorus Absorption

Absorption of phosphorus occurs at a fairly constant rate through the growing season although speed is slightly higher during early phase of growth and development. Sources of nitrogen and potassium used in fertilizers are quite soluble in the soil solution and thus readily available for absorption. Phosphorus is only slightly soluble in the soil solution, but as phosphorus is absorbed by the plant a replenishment of the level in solution is likely to occur throughout the growing season. The smaller absorption is a genetically controlled factor, whereas difference in the absorption is primarily due to difference in the relative availability of the elements in the soil. The uptake of phosphorus is inhibited by the hydrogen ion in the very acid region and by the hydroxyl ion in the alkaline region. Tobacco is normally transplanted thereafter, the soil temperature in the root zone frequently does not exceed 59°F; consequently, the young transplanted would benefit from high levels of phosphorus. Phosphorus is the most important and beneficial nutrient for growth of tobacco at low soil temperatures. The absorption of phosphorus by tobacco roots is also influenced by the pH in the rooting media. The early growth response so frequently observed from applied phosphorus may result from the influence of temperature on absorption. It has been shown that the uptake of phosphorus by tobacco is reduced significantly as the temperature decreased from 95°F to 50°F.

Phosphorus Application

A response in earlier growth and higher phosphorus content in the plant has occurred however, by the end of the season there may be no differences in the uptake of fertilizer phosphorus between different placements. An increase in the rate of early growth of tobacco can usually be obtained from preplant applications of phosphorus fertilizer. The major source of phosphorus in the plant in the early stages of growth is from the fertilizer only. One month after transplanting about 40 to 90 percent of the phosphorus in the plant has been shown to come from the applied fertilizer, and remaining portion of the growing season there was a reduction in the percentage of phosphorus in the plant from the fertilizer due to greater absorption from residual soil sources. Repeated applications of large quantities of fertilizer phosphorus coupled with low plant absorption and essentially no losses by leaching have resulted in a substantial increase in the level of this nutrient in the soils. The placement of phosphorus fertilizer is quite influential at the time it is absorbed and the amount absorbed and thus may influence early growth of the plant. It is always advocated that mixing the phosphorus in the row or banding it prior to transplanting over side dressing. Particularly under moisture stress condition, side dressing of phosphorus is inefficient in which root development is restricted. In soil, having low in available phosphorus, band placement too deep in the row has resulted in phosphorus deficiency and an early stunting of plants.

3. Potassium

High potassium content in cured leaves has frequently been used as one criterion of quality, and relatively few, if any, soils on which tobacco is grown are capable of producing the quality or the yield desired without fertilizer potassium. Potassium helps in the maintenance of cellular organization by regulating the permeability of cellular membranes and keeping the protoplasm in a proper degree of hydration by stabilizing the emulsions of colloidal particles. Its salts stabilize various enzyme systems. It plays a catalytic role in activating several enzymes as incorporation of amino-acids in proteins synthesis of peptide bonds etc. Presence of potassium is essential for optimal activation of aldehyde dehydrogenase etc. Potassium increases resistance in plants against drought, heat frost and various diseases caused by fungi nematode and other microorganism. It helps in formation of mechanical tissues, resulting into resistance to lodging.

Physical Properties of Leaves

Potassium is concerned with the physical appearance of the leaf and the burning properties. High level of potassium in the leaf results in a dee orange color in flue-cured tobacco. At high rates of potassium, leaves of cigar and wrapper tobacco appeared to be thinner and more elastic and pliable than potassium is low.

Burning Properties of Cured Leaves

The beneficial influences of potassium were obtained on the rate of burn or the fire-holding capacity. Under low soil fertility conditions, potassium fertilization increased the yield, quality score, and duration of burn of cigar, wrapper tobacco. In case of high fertility soil, only an improvement in leaf burn occurred from potassium fertilization. The ignition temperatures of cigar tobacco were found to decrease significantly as the potassium concentration in the leaf increased.The chloride source of potassium has commonly gives the poorest results in comparison with the carbonate nitrate salts or potassium sulphate, probably due to the unfavorable effects of the chloride ion. Potassium sulphate has resulted in a decrease in the duration of burn, which is attributed to the adverse effect of the sulphate ion.

Chemical Properties of Leaves

With the increase in potassium content of the plant concentration of calcium and magnesium generally decrease. This reduction is sufficient to result in deficiencies depends primarily on the level of the elements in the growth media. An increase in the potassium content of the leaf is generally considered to have no significant influence on the nicotine concentration of tobacco.

Deficiency Symptoms

Owing to the richness in potassium content in general all most in all Indian soils, due to its parent material or heavy applications of potassium fertilizer, deficiency symptoms are rarely observed in commercial production. Deficiency symptom appear when its content in the plant falls bellow 2 per cent. Non application of potassium fertilizer, however, will, on most soils, result in observable deficiency symptoms but some time potassium deficiency may be accentuated by excess nitrogen, particularly in the ammonium form, and by high levels of magnesium or sulfur. It has been observed that potassium deficiency symptoms appear first and are most severe on rapidly growing plants and spread progressively to the upper leaves. During early phase of growth period, potassium deficiency is manifested by a slight mottling and the appearance of brownish yellow specks, especially near the tips of the leaves. In the later stages of plant growth (near topping), the abnormalities may appear first on the upper leaves. At acute deficient condition brownish yellow spots appear alongwith the margins of the leaves. The tips of leaves may curl downward and the margins curve inward until they are almost at right angles to the plane of the leaf.

Potassium Absorption

The high requirements for potassium exist not only after transplanting, but also in the seedling stage. Potassium in combination with either sulfate or nitrate appeared to be the most beneficial elements on root development. Uptake of potassium by the plant generally increases linearly over a wide range of application rates. The potassium content of the leaf of field-grown burley tobacco ranged from a low of 2.0 to a high of 3.2 percent. The rate of potassium absorption is very high during the early stages of growth and diminishes rapidly during the later phase. It is assumed that the reduction in absorption is due primarily to an exhaustion of the supply of available potassium in the soil. Potassium content of leaves varies from 2.5 to 7.0 per cent and in the stalk from 2.5 to 5.0 per cent in the normal plant.

Potassium Sources

The response of tobacco to various sources of potassium viz., the potassium sulphate, potassium chloride, potassium carbonate, and potassium nitrate were tested. In these studies there was a greater difference among the sources on leaf burn than there was on yield or quality score. When alkalinity of the water-soluble ash was used as an indicator of fire-holding capacity, the sulfate source gave lower values than did the nitrate or carbonate source. It was suggested that these differences were due to effects of sources on the sulfur content of the leaf.

4. Calcium

The requirements for calcium are relatively high because it is constitutes one of the principal inorganic constituents of tobacco potassium and calcium are the mineral element absorbed in the largest quantity. Calcium is a constituent of the cell wall, is an activator of different plant enzymes and is essential for the stability of cell membranes. It is a structural part of the chromosomes in which it binds the DNA with protein. It is required by a number of enzymes for their proper functioning viz. lipase, phosphatase D, £ Amylase and pyrase. It makes the stems stiff and thereby reduces lodging in plants. It also neutralizes the organic acids formed within the plant body and eliminates their toxic effects. Values of 1.5 to 2.0 percent calcium in the cured leaves are common. The amount of calcium necessary for the normal growth of tobacco leaves is in excess of 1 percent of the dry weight.

Deficiency Symptoms

Deficiency symptom of calcium are expected of its content in leaf falls below 1.0 per cent. Deficiency symptoms generally occurs in the upper portion of the plant indicates that during stress there is little calcium transportation from the older tissues to the growing points. Leaves of calcium-deficient plants are dark green colour and in advanced stage the terminal bud dies. When this condition occurs early in the development of the plant, the top leaves may become much thickened. Hooking down word at the tips of the bud leaves and ultimately breakdown of the tissue at the tips and margins of the leaves are some prominent symptoms in the early stages of calcium deficiency. Complete breakdown and death does not occur and growth takes place later. Older leaves on calcium-deficient plants generally retain their normal shape.

Leaf Quality

The percentage of calcium is found to be high in the lower leaves and to decrease to a low value in the best leaf grades. The calcium and magnesium contents were lowest in the leaves near the middle of the plant where they obtained the highest quality tobacco. An inverse relationship between the content of calcium and quality of flue-cured tobacco was reported by several workers. In tobacco leaves calcium deficiency is accompanied by a marked increase in the content of free amino acids but little change in their relative proportions.

5. Magnesium

Magnesium is a constituent of chlorophyll and chromosome. Being constituent part of polyrioosemes, it helps in protein synthesis in the plants. It is known to play a catalytic role as an activator of a number of enzymes, most of which are concerned with carbohydrate metabolism, phosphate transfer and de carboxilations. It is involved in photosynthesis and organic acid metabolism. Mg, helps in synthesis of fat and increases oil content in oilseed crops when it combines with sulphur.

Deficiency Symptoms

Magnesium is a constituent of the chlorophyll molecule, loss of green color would be expected to be one of the first symptoms of magnesium deficiency. Discolourations start first on leaves at the base of the plant, and on an individual leaf it begins at the tip and margins and proceeds toward the base and center. The area around the veins of the leaf generally remains green after the rest of the leaf has been decolorized. In extreme cases the leaf may become practically white, suggesting a breakdown of the yellow pigments also. An abnormality in the growth of tobacco, which is known known as **"sand drown".** If Magnesium content in leaves folls below 0.4 percent the dificiency symptoms become visible.

Effects on Leaf Properties

An increase in the amount of magnesium in the leaf decreased the rate of burn of cigar tobacco. By increasing supply of potassium and calcium in the culture solution, the contents or the respective elements increased in the tobacco plant but the magnesium content decreased and magnesium deficiency symptoms appeared. Magnesium deficiency may be expected to occur when the concentration in the leaf is 0.2 percent or less of the dry weight.

6. Sulphur

Sulphur is required for synthesis of sulphur containing amino acids and proteins like cystein, methionine, vitamins (thiamine and biotine), lipoic acid, acetyl coenzyme Aferredoxin and glutathione activity of proteolytic enzymes, and increases oil content in oil bearing plants.

Sulphur Deficiency Symptom

Deficiencies are most likely to occur in the early stages of growth or during dry weather. When they do occur, the first evident symptom is a lighter color of the young leaves compared with the older ones. A distinguishing difference between and nitrogen deficiency is that in the sulfur the plants do not lose their leaves by "firing." Flower emergence is delayed when sulfur deficiency occurs early in plant development. Deficiency symptom of sulphur are expected if its content in leaf is below 0.1 per cent.

Effect of Sulphur on Tobacco Leaves

High sulphur content in tobacco has detrimental effects on leaf burn; effects of sulphur on leaf burn are neither as large nor as consistent as those from chloride or potassium. Increases in the application of sulphur have resulted in decreases in the uptake of chloride by tobacco. The nicotine content of burley tobacco was higher when potassium was applied as potassium sulphate than when added as potassium chloride. Calcium and potassium concentrations in the tissue were inversely related to the sulfate concentration of the substrate. This same relationship is valid for the accumulation of calcium and magnesium in tobacco leaves. Potassium in burley tobacco leaves increased with increasing rates of sulphur in the substrate.

7. Iron

Iron is a constituent of cytochromes, haem, nonhaem enzymes and metalloproteinase like ferredoxin and hemoglobin in plants. These cytochmmes play a vital role in oxidative and photophosphorylations during respiratory electron transport and photosynthesis respectively. The ferredoxin helps in reduction of carbon dioxide, sulphate and of atmospheric nitrogen. It synthesises chlorophyll precursor (protoporphyrin-9) which forms chlorophyll in green pipgments. Its specific requirement has been identified in synthesis of enzymes like oxido-reductase, sulphate oxidase; catalase peroxidase and aconitase etc. Being a constituent part of metabolically active compounds, iron is responsible for all major metabolic processes in plants.

8. Manganese

Being a part of nitrite reductase and hydroxylamine reductase, it helps; in the nitrogen assimilation. Manganese performs some function in photosynthesis, acts as regulator of the intake and state of oxidation of certain elements. It activates several enzymes related to oxidation-reductions (oxidoreductase), hydrolysis (hydrolases), breakdown of phosphates bonds in ATP or ligases. It activates photosynthesis and nitrogen metabolism. It also accelerates enzyme participating in Calvin cycle, helps in chlorophyll and chloroplast synthesis for boosting photosynthetic rates.

Manganese Deficiency Symptoms

Manganese deficiency is most likely to occur where the soil reaction is neutral to alkaline; a condition that seldom exists in soils used for tobacco production. The first apparent symptom is chlorosis of the young leaves. The tissue between the veins becomes light green to almost white, giving the plant a checkered appearance. In the advanced stages, chlorosis may be followed by necrotic spotting and the affected tissue may drop out, producing a ragged appearance.

Manganese Toxicity Symptoms

Tobacco appears to be a luxury feeder of manganese and thus toxicity may be a problem where the level of available manganese is high. The stunted, chlorotic, and necrotic spotting of tobacco are symptoms of manganese toxicity.

9. Zinc

Zinc is concerned with the functioning of sulphydryl compounds such as cysteine in the regulation of oxidation-reduction potential within the cells. It regulates the auxin concentration in plants and helps in synthesis of protean carotene and chllrophyll etc. The average content of zinc in various commercial tobaccos is reported to be between 51 and 84 ppm. The minimum requirement for zinc is less than 10 ppm in the tissue.

Zinc Deficiency Symptoms

Leaf spot disease of tobacco that occurs during wet periods and is attributed to bacteria is reported to correspond in many respects to symptoms produced experimentally by a shortage of zinc.

10. Copper

Copper is an essential component of the oxidizing enzyme catechol oxidase or polyphenol oxidase. The compounds containing copper like plastoquinones and plastocyanins help in, electron transport from chlorophyll to NADP and from water to chlorophyll during photosynthesis. The copper content of tobacco leaves varied from 12 to 60 ppm; the stalks contained about 10 ppm. An abnormality that resulted in a breakdown of the tissue in tobacco leaves as they approached maturity was corrected by the addition of copper sulfate. The affected leaves were high in total and protein nitrogen and low in sugars. Use of copper sulfate reduced the nitrogen and increased the sugar content. The addition of copper sulphate to field-grown tobacco has been reported to give a more even ripening and an increase in the body of the leaf.

11. Molybdenum

Molybdenum is a constituent of nitrate reductase and nitrogenase enzyme and is associated with nitrogen utilization and in nitrogen fixation. It helps in protein and amino-acid synthesis. It accelerates nitrogen fixing efficiency of aerobic (*Azotobacter*), anaerobic (Clostridium), blue green algae, Azolla and symbiotic bacteria. It regulates the carbohydrate metabolism in plants.

Molybdenum Deficiency Symptoms

Molybdenum deficiency symptoms were associated with high nitrate and low protein nitrogen in the leaves. The deficiency of molybdenum would be expected to occur more commonalty on acid soils since its availability is lower under acidic condition. The first indication of molybdenum deficiency is mottling of the lamina in leaves at the mid stalk position. Bending and twisting of the leaf lamina were usually present and were followed by the appearance of small necrotic areas which gradually enlarged until the entire leaf is withered. These symptoms gradually spread to other leaves. Deficiency may cause delay in blooming and in extreme cases a loss of flowers.

12. Chlorine

During photosynthesis it helps in evolution of oxygen. It is a part of anthocyanin and affects protein synthesis. It increases turgor pressure. Chlorine stimulates the activity of some enzymes and influences carbohydrate metabolism. The uptake and accumulation of chloride ion by plants increases linearly over a wide range of concentration in the substrate. Concentrations in the leaf as high as 10 percent of the dry weight have been measured in plants grown in the green house.

Chlorine Toxicity

Under high concentrations of chlorine, the green leaf becomes greatly thickened and exceedingly brittle, the leaf margins curl upward, and the leaf presents a distinctive sleek, glabrous appearance. When ammonium is the principal form, however, at chloride levels as low as 0.95 percent the leaves possessed the color and configuration normally associated with chloride toxicity. Much of the leaf abnormality attributed to excess chloride may have been due to either ammonium toxicity or to an ammonium-chloride interaction. An increase in pH would promote nitrification, thus minimizing the possibility of an ammonium-chloride interaction leading to toxicity.

Effects on Chemical Properties of Leaves

The use of excessive amounts may cause adverse effects on combustibility of the leaf more than any other inorganic constituent. The causal mechanism through which chloride reduces the burning qualities of the leaf is not understood, though the Potassium salts or organic acids were the main factor determining the rate of burn. A decrease in total nitrogen content of the plant has been associated with high chloride nutrition. Nitrogen metabolism also appears to be influenced by chloride. The chloride effect on carbohydrate metabolism is on the amylytic enzyme in the leaf which resulted in the marked accumulation of starch when the chloride content is high.

13. Boron

It regulates development and differentiation of vascular tissues, formation and lignifications of cell-wall. It is associated with reproductive phase in plants and under imbalanced nutrition, it causes sterility and malformation in reproductive organs. It is involved in carbohydrate metabolism, particularly in translocation of photosynthesis. Influences cell development by its influence on polysaccharide formation. It regulates translocation of sugars across membranes and polyphenolase activity. It boosts nodulation in legumes, regulates water absorption and is essential for synthesis of DNA, RNA and pectin. Under normal production practices.

Deficiencies of this element are rare. They are most likely to occur at a reaction in the neutral to alkaline range which renders the soil boron unavailable or on coarse sandy soils during periods of high rainfall. The development of deficiency symptoms is also favored by rapid plant growth.

Boron Deficiency Symptom

Deficiency symptoms have been observed in plants with a concentration of 5.5 ppm of boron. Because minute amounts of boron are needed, and slight excesses have a deleterious effect on the plant. The bud leaves are the most reliable index of boron deficiency in flue cured tobacco, the critical level being 15 to 16 ppm. Boron deficiency affects primarily the youngest leaves and the terminal bud and occurs during periods of rapid growth indicates that little or no transfer occurs from the older plant parts to the younger growing points. The bud leaves become light green in colour and distorted, In advance stages, breakdown of the tissue and death of the terminal bud occurs and the remaining leaves become abnormally thickened. Boron deficiency has been reported to increase the amount of cell wall material in tobacco pith parenchyma and to decrease relative galactan content of these cells.

Boron Toxicity

Application of more than 400g of boron per acre caused boron toxicity on young plants. Prominent symptoms of toxicity are slight stunting of the plant and a slight specking and marginal burning of the leaves, however, did not affect the final yield and value of the leaf.

14. Sodium

It maintains the osmotic pressure. It also regulates water uptake by plants. Deficiency symptoms appear if it falls below 0.3 per cent in the whole plant.

Sodium partially supplemented potassium in the nutrition of the plant. In a greenhouse experiment, the yield was increased and potassium deficiency symptoms were decreased by the addition of sodium to the nutrient media. Burn rate was increased by sodium application in absence of potassium. Sodium ion could partially substitute in some plant process or processes when potassium is low.

Nutrients Management in Tobacco

Tobacco being a heavy feeder of nutrient, its require heavy doses of all nutrients specially NPK and other micronutrients. In general, a normal tobacco crop removes about 225 N, 17 P_2O_5, 365 K_2O, 135 CaO and 17 MgO, kg/ ha. Nutrient management in tobacco starts with the sowing of tiny seeds in nursery bed. Both are separately discussed here.

Nutrient Management in Nursery

For uniform and steady seedling growth in nursery beds it is very much essential to apply organic bulky manures to the soil. Organic manures improve physicochemical properties of soils. Basal application of well decomposed compost or FYM @ 25 kg per 10 m^2 bed and it should be done preferably 10-15 days before seed sowing; fine powdered mustard cake, 500 g SSP 25 g urea and 100 g muriate of potash should be applied.

Nutrients Management to Main Crop (Chewing tobaco of Bihar)

Most of the tobacco growing soils of North Bihar are light textured are low in organic matter available N and P, medium in K content and mostly alkaline in reaction (pH 8.2 to 8.4). The elite and muscovite are the dominant minerals responsible for the enrichment of such soils. Due to intensive cropping system, the potassium status of soils is depleting day by day which is reflected on yield, and the elasticity and colour of the leaf. Available phosphorus content in these soils is low due to high rate phosphorus fixation in the form of di and tri calcium phosphate and low P recovery that hardly exceeds 10 to 20%. The recovery of phosphorus from applied phosphatic fertilizer and its use efficiency can be improved organic manure like oil seed cake. This significantly influences the quality leaf production in general and the spangling and puckering of the leaf in particular. Bulky manures and oil seed cakes are found to improve yield and quality of tobacco leaf. Green manuring in situ with sunhemp in *kharif* season preceding tobacco crop is also found beneficial from yield and quality point of view. Tobacco growing tracks of North Bihar require heavy organic manures including oil cakes. Application of 25 to 30 tones of compost or FYM (well decomposed) per ha 15-20 days before planting is beneficial for tobacco production which provides all most all essential nutrient required for tobacco production. Newly released input responsive chewing tobacco varieties viz., PT-76, Vaishali special and Lichhavi require 250 kg N, 60 to 70 kg/ha each of P_2O_5 and K_2O / ha of which 50 % of nitrogen is applied in the through concentrated organic sources like oil seed cake i.e. mustard and caster cakes and remaining 50 % nitrogen may be applied through inorganic sources viz., ammonium sulphate, urea or calcium ammonium nitrate. It is always advisable to apply half dose of concentrated organic manure through mustard and remaining half through caster cakes to avoid monotony of single source and also to make it available for other uses because mustard cakes is mainly used as concentrate feed for livestock. Placement of entire quantity of manures and fertilizers will be done in furrows before planting is found highly beneficial as compared to broad cast method, the general farmers practice because it enhances nutrient use efficiency and minimizes nutrient loss through volatilization, run off, leaching and soil fixation particularly of P. Suitable sources of nitrogen are urea, calcium ammonium nitrate and ammonium sulphate; where as single super phosphate and dai ammonium phosphate are suitable sources of P. For quality tobacco leaf production single super phosphate and ammonium sulphate are of the good sources of sulphur needed for tobacco production in saline, calcareous soils of north Bihar. Muriate of potash was found economically better source of potassium.

Micro nutrients play crucial role in tobacco production by influencing the quantity as well as quality too. Among micro nutrients Zn, Cu and B showed positive response in chewing tobacco. Increased use of chemically pure high analysis fertilizers containing negligible amount of micronutrients as contaminant, increased cropping intensity and restricted recycling of farm wastes accrued to widespread deficiency of micronutrients in general. The magnitude of micronutrients deficiency viz. zinc, copper, iron, manganese and boron in different broad soil groups of Bihar has been studied. The data reveal wide spread zinc deficiency in north (40- 85%) and south Bihar plain (30 - 60%). The deficiency of copper was recorded in Recent Alluvial and l sub-Himalayan soils with sporadic deficiency in calcareous soils. In other soils, the deficiency of copper was negligible. Iron (10 - 30%) and manganese (5 - 15%) deficiency was mostly confined to calcareous and saline sodic soils of old alluvium belt. The deficiency of boron has been recorded in calcareous and recent alluvial soils (30 - 50%) with higher magnitude than old alluvial and Tal land soils (20 – 30%). Though the micro nutrient deficiency was not realized in tobacco growing field it might due to because of continuous application of heavy dose of organic manures including oilseed cake, as deficiency of these nutrients are more pronounced in areas where high yielding and high fertilizer responsive varieties of cereals and other crops are being grown under high cropping intensity ands are found wide spread and their application has become the part of general recommendation. The micro nutrient required by tobacco for producing quality leaves is mentioned in table 6.1.

Table 6.1 Micronutrients requirements of different crops

Crops	Micronutrients requirements g/ t dry matter)					
	Iron (Fe)	Manganese (Mn)	Boron (B)	Zinc (Zn)	Copper (Cu)	Molybdenum (Mo)
Cotton	106	14	15	16	8	0.77
Cow pea	260	89	53	17	11	1.31
Green gram	170	18	32	13	11	1.05
Sorghum	360	27	27	36	3	0.98
Cabbage	205	28	25	21	8	0.98
Guar	71	15	29	15	9	1.14
Castor	223	41	31	14	2	1.01
Pearl millet	349	44	-	50	12	-
Hybrid Bajra	264	23	27	22	9	0.84
Groundnut	499	39	44	9	5	1.32
Wheat	232	26	18	21	8	0.87
Potato	160	12	50	9	9	0.80
Brinjal	320	36	45	23	12	0.23
Tobacco leaves	692	132	96	21	11	0.60
Lucerne	420	41	36	21	10	2.59
Maize (Fodder)	217	74	-	30	31	-

Source : Fertilizer Statistics, 2004-05, The Fertilizer Association of India, New Delhi.

There is no need to press the panic button in respect to micronutrient application to the chewing and hookah tobacco as sufficient quantities of bulky organic manures viz., FYM/Compost, Mustard cakes are used for its nutrition which cantinas trace amount of these micronutrients. High analysis chemical fertilizer also cantinas some micro nutrients in trace amount measured in ppm, listed in table 6.2 along with some important bulky organic manure like FYM and compost.

Flue Cured Virginia (FCV) Tobacco Fertilization

Flue Cured Virginia (FCV) Tobacco is known for its low nitrogen and high sugar containing in its leaves. The low nitrogen and high sugar content of leaf is conducive for betterment of its quality. High nitrogen produces leaf with high nicotine, low sugars and off colour. Adequate nitrogen, however, is necessary for optimum growth and development. Therefore, nitrogen supply should be controlled within narrow limits. Cigarette tobacco in India is fertilized rather randomly, there being no standard schedule with the farmers. General practice is to apply around 10 t FYM/ ha, some times supplemented with 300-500 kg/ ha groundnut cake. However, fertilizers have almost replaced scarce organic manures in the recent part. Response to applied nitrogen is, generally, overshadowed by season (soil moisture availability) and soil. In dry season where there may not be any rain after the application of fertilizers, there may not be any response to applied fertilizer. In some cases, the effective may be negative. In general, application of nitrogen beyond 30 kg/ha for heavy black soils and beyond 50 kg/ha for light red soils lower both the yield and bright grades. On equal nitrogen basis, all the nitrogen fertilizers are equally effective. Potassium is the next important nutrient and its absorption is higher than any other nutrient. High potassium content in leaf improves colour retention, leaf bum, reduces tar and nicotine in tobacco smoke, thus making it less hazardous. High potassium leads to brighter cured leaf colour and increased seed yield. Quality improvement due to addition of potassium was not significant. Form of potassium is important and in chloride form it gives poor results. Hence, potassium sulphate is preferred for FCV tobacco. Application of phosphorus up to 125 kg P_2O_5 /ha indicated no significant improvement either in yield or quality of FCV tobacco. All the forms of phosphorus are equally effective on equal P_2O_5 basis. Results of experiments on placement of fertilizers are also inconsistent. Among the secondary nutrients, magnesium, calcium and sulphur play an important role in controlling yield and quality of tobacco. Magnesium deficiency (sand down) has been reported in sandy soils. Its deficiency decrease sugars and starch. Calcium deficiency leads to increase in free amino acids. Among the micronutrients, chlorine influence the leaf quality and leaf bum. Tobacco leaf shout not contain more than 1.5 per cent chlorine.

Non- Fueled Cured Virginia Tobacco Fertilization

In the normal culture, cigar and chewing tobaccos are heavily manured with FYM (20 t /ha). These are more liberally fertilized than FCV tobacco. Abundant fertilization improves yield and quality. Limited availability of FYM is forcing many fanners to go for inorganic fertilizers. Hookah tobacco is also heavily manured with the application of FYM (20 t/ ha) supplemented by groundnut cake (300-500 kg/ ha). Results of experiments at Vedasandur (TN) indicated necessity for 10 t FYM/ ha and 100 kg N/ ha for cigar and chewing tobaccos. Response to phosphorus and potassium up to 100 kg/ha has been recorded with positive effects on yield and quality of leaf. Bidi tobacco in Nipani (Karnataka) responds to around 120Kg N/ha over & above 12 t FYM/ha. In Gujarat, response of bidi tobacco to nitrogen is up to 200 kg/ha. Hookah and bidi tobaccos are also not responding to phosphorus and potassium application when FYM is appliedat 20t /ha.

Table 6.2 Micro-nutrient contents of some important fertilisers and manures

Fertiliser/Manure	**Micro-nutrient contents inparts per million**				
	Copper (Cu)	**Zinc (Zn)**	**Manganese (*Mn*)**	**Boron (B)**	**Molybdenum (Mo)**
Nitrogenous fertilisers					
Ammonium Sulphate	Tr-0.5	0.33	70	6.0	0.1
Urea	0-3.6	0.5	0.5	0.5	0.7-6.2
Calcium Ammonium Nitrate	Tr-18.0	8.35	10.50	Tr	-
Phosphatic fertilisers					
Single super phosphate	26.0	50-165	65-270	9.5	3.3
Triple super phosphate	2-12	53-100	175-245	529	9.1
Basic slag	9.2-56.4	4-59	6.89	33.4	10.0
	9.3-85.0	10-29	280	-	-
Rock phosphate	5.6-9.5	24-137	130-320	19	5.6
Bone meal	270	660	500	715	-
Potassic fertilisers					
Muriate of potash	3.0	3.0	8.0	14.0	0.2
Potassium Sulphate	5.6-10.4	2.0	2.2-13.0	4.0	0.2
Complex fertilisers					
Ammonium phosphate	3-4	Above 80	115-220	-	2.2
Manure					
Farmyard Manure	10	43-247	201	17.4	0.13
Compost	300-600	9.40	12.40	5.8	0.10

Note : Tr : Traces.

Source : Fertiliser Statistics, 2004-05, The Fertiliser Association of India, New Delhi.

Fertilizer Schedule

From the results of experiments under different tobacco growing ecosystems, the following generalizes schedules can be recommended for different types of tobaccos (table 6.3). Since the availability of FYM is limited, farmers are applying higher doses of inorganic fertilizers to almost all the non-FCV tobaccos. The above are only general recommendations. These schedules, however, differ depending on location and specific circumstances.

Table 6.3 Recommended generalizes schedules for various type of tobacco grown in India

Types of tobacco	Org. manures (q/ha)	Nutrients in Kg/ha N_2O	P_2O_5	K_2O	MgO	Bo	Times of application
Flue cured Virginia (in heavy soils of Andhra Pradesh)	50-75	22	50	50	-	-	Org. man. 6 weeks & Fert. 1 week before planting.
Flue-cured Virginia (in light soils of Andhra Pradesh)	100-125	30-40	80	80	12-15	-	Org. man. 6 weeks, MgO, $P_2O_5$1/2 N & 1/2 K_2O (one week before planting followed by topdressing of 1/2 N & 1/2 K_2O 3-4 WAT.
Natu tobacco of A.P.	150 to 300 kg G.N. cake	44	20	12-15	-	-	Org. man. 8-12 weeks & Nitrogen I week before planting
Bidi of Kamataka	4-5 qts. G.N. cake	45 as A/S	20	-	-	-	Org. man. 8-12 weeks & Nitrogen I week before planting
Bidi of Gujarat	8-10 qts. of G/ N. cake	90 as A/S	25	-	-	-	Basal 12 weeks before planting
Cigar & cheroot tobacco in Tamil Nadu	2-3 qts. of G/.N. cake	50 as A/S	20	-	-	-	Cake 1/2 N. 2 weeks before planting &1/2 N. as topdressing 6-7 WA T.
Chewing tobacco in Bihar	60 qts. FYM + 6-8 qts. of mustard	112 as A/S	60	60	-	-	Cake 1/2 N. 2 weeks before planting &1/2 N. as topdressing 6-7 WA T.
Hookah & chewing tobacco in West Bengal	300 qts FYM	150	20	150	-	-	-
Wrapper tobacco in West Bengal	170 qts. FYM	125	20	125	-	-	-

WAT-Weeks After Transplanting, Wks-Weeks *Source:* Handbook of Agriculture, 1984 (Published by I.C.A. R.)

N.B.: Bo deficiency causes death of terminal buds and thickening of lower leaves; therefore, the deficient soils must be supplied with required quality of Bo.

Specific Recommendation for Different Situation in Chewing Tobacco in Bihar

There is separate recommendation for different agro climatic and agro ecological conditions prevailed in North Bihar, where tobacco is grown under a wide range of climates, basically tobacco is tropical crop (in origin) but it is now grown in subtropical and temperate region of world. The climatic conditions of north Bihar is quite suitable for growth, development and quality leaf production of chewing and hookah types of tobacco as well as some other types. For reedy reference to the tobacco grower schedule of manure and fertilizer is given in the tabular farm.

Table 6.4 Schedule of manure and fertilizer for light and low fertile soil of North Bihar

S No	Particulars	Dose per hectare
1	Compost or FYM	25ton
	Nitrogen @250 /ha	
2	Mustard cake	1875 Kg
3	Urea (Before transplanting)	200 Kg
	At the time of first irrigation	
4	Calcium Ammonium Nitrate	115 Kg
5	Or Urea	57.5 Kg
	Phosphorus and Potassium 60 Kg each	
6	Single Super Phosphate (SSP)	325 Kg
7	Muriate of potash	68 Kg
	Using DAP in place of Single Super Phosp hate (SSP)	
8	Urea before transplanting	100
9	DAP before transplanting	150
10	At the time of first irrigation	57

Organic manures should be applied 10-15days before the transplanting of tobacco seedlings in the lines. Maintain 8-10 cm distance from tobacco to avoid any direct contact between plant and applied manures and fertilizers. To realize potential production of chewing and hookah tobacco in Bihar proper nutrient management is utmost required apart from other input, in conjunction with good agronomic practices. Leave of tobacco is economic produce which is utilized for various purposes. The chewing and burning quality which is solely depends on certain chemical known as nicotine and aroma producing secondary metabolites, all are heavily depends on tobacco nutrition, hence selection of proper source dose time and method of fertilizer and manures application hold the key for chewing and hookah tobacco production in Bihar efficiently.

Table 6.5 Schedule of manure and fertilizer for medium to high fertility soil of North Bihar

S No	Particulars	Dose per hectare
1	Compost or FYM	25ton
	Nitrogen @300 /ha	
2	Mustard cake	2250 Kg
3	Urea (Before transplanting)	250 Kg
	At the time of first irrigation	
4	Calcium Ammonium Nitrate	138 Kg
5	Or Urea	69.8 Kg
	Phosphorus and Potassium 60 Kg each	
6	Single Super Phosphate (SSP)	325 Kg
7	Muriate of potash	68 Kg
	Using DAP in place of Single Super Phosphate (SSP)	
8	· Urea before transplanting	120
9	DAP before transplanting	150
10	At the time of first irrigation	57

An experiment has been conducted at Pusa Bihar for ascertaining the role of sulphur in tobacco production. Different sources and methods were tested and found that tobacco responses up to 60 kg/ ha. Out of three source tested iron pyrite is prove superior over other two tested sources. The economical dose was worked out to 40 kg/ha with any of the tested sources but once again iron pyrite proves economical.

Table 6.6 Effect of source and dose of Sulphure application to the chewing tobacco (variety Pusa Tobacco -76)

S. No.	Treatments	Tobacco leaves yield (q/ha)	
		First grade	Total leaf yield
	Sources of Sulphure		
1	Iron Pyrite	5.86	1.66
2	Elemental Sulphure	5.44	1.35
3	Gypsum	5.78	1.55
	Average	5.70	1.35
	Doses of Sulphure		
1	0 Kg	5.37	1.37
2	20 Kg	5.60	1.47
3	40 Kg	5.81	1.60
4	60 Kg	6.01	1.57
	Average	5.45	1.50

□□□

7

Water Management

Tobacco economic produce is leaves constitute more than 80 percent of total weight of whole plant. For proper growth and development moisture is one of important input, hence proper moisture regime throughout their growing period. Water requirement is again depends upon the tobacco type, length of growing period along with other agroclimatic and soil factor which has direct and/or indirect influence on growth and development and quality and quantity of produce. Quality of irrigation water is most critical factor which influences the quality of tobacco produced up to great extent. Under proper water regime leaves may hold proper amount of carbohydrate (having positive bearing on its quality) and moisture stress condition (drought) nicotine content may be shoot up beyond the desirable limits. Various asset of water management is deal in this chapter.

Tobacco cultivated area under assured irrigation in India is currently about 47 per cent of total which was below 10 per cent during 1950-51. Tobacco is medium deep rooted crop; it rooting depth is up to 90 cm. The quantity of water applied, to be applied or depth of water is depend upon the depth of root zone, water holding capacity of soil and extent and pattern of dryness of soil. The amount of water should be enough to wet the root zone of tobacco and at the same time moisture beyond the root zone is just wastage of other inputs along with water. Water requirements of any crops are governed by climatic factor, soil factor and the nature of the crop itself. On the basis of above the water management / irrigation scheduling practices should be adopted. Tobacco is crop in which every part is economic produce, for all practical purpose as leaves is a useful product, under such condition efficient water management practices hold key to successful tobacco production.

Table 7.1 Tobacco area under irrigation during 2003-04

State	Percentage area under irrigating
Andhra Pradesh	28.5
Uttar Pradesh	99.8
Gujarat	76.8
Karnataka	6.5
Bihar	79.8
Tamil Nadu	95.4
Maharashtra	9.4
All India	46.2

It is general believe, life originated in an aqueous environment and in the course of evolution it became fully dependent upon water in a number of ways. Water is now said to the liquid of life or elixir of life. In general, water is essential for life. Water is the main constituent of the protoplasm comprising up to about 80 to 90 per cent of its total weight. In the absence of water, protoplasm becomes inactive and is even killed. Different organic constituents of plants such as carbohydrates, proteins, nucleic acid and enzyme etc. lose their physical and chemical properties in absence of water. Water participates directly in many metabolic processes. Inter-conversion of carbohydrates and organic acids depends upon hydrolysis and condensation reactions. Water increases the rate of respiration. Seeds respire fast in presence of water. Water is a source of hydrogen atom for the reduction of carbon dioxide in the reactions of photosynthesis. Water acts as a solvent of and carrier for many substances. It forms the medium in which several reactions take place. Water present in the vacuoles helps in maintaining the turgidity of cells which is a must for proper activities of life. The turgidity of cell helps in the elongation of cells resulting in growth. The difference in the amount of available water during summer and winter seasons are the reason for the formation of annual rings in higher plants. In summer, the turgidity is less and as a result smaller cells are formed. A network of thin layer of water surrounding each cell plays an important role in the entry and movement of dissolved substances. Water helps *(i)* in the translocation of solutes *(ii)* in the mobility of gametes, *(iii)* in the dissemination of spores, fruits and seeds, and *(iv)* provides support to aquatic plants. In tropical plants, water plays a very important role of thermal regulation against high temperatures. Some people think that plants lose about 95 per cent of the absorbed water just to maintain the optimum temperature. Thousands of characters develop for balancing the water content of plants. Even atmospheric moisture affects plant growth. On the basis of water need for plant growth, it can be said that plants needs water because water forms over 90 per cent of plant body weight when they are green; they can absorb nutrients only when the nutrients are dissolved in water; they need water for turgidity of their cell walls; arid lastly they can synthesize food through photosynthesis only in presence of water in their system. Water is absorbed by root hair of the plants from where it reaches xylem via cortex cells and passage cells and through xylem its reaches top of the plant where it is transpired by leaves and used for other metabolic activities. The need for water by plants is further aggravate by the circumstances of its nonuniform distribution, uncertainty of rainfall, intensive cropping, protection of crops from frost, cultivation of high yielding crop varieties and an ever increasing application of fertilizers in agriculture, but, application of excess water at the same time is expansive and equally harmful for plant growth. It is, therefore, essential, for an agriculturist to know how the soil gets water and how it loses the water, soil moisture and' its constant, water requirement of plants and irrigation schedule. The availability of water influences almost all the biochemical and physiological processes in plants which in turn affects the morphology of plants. All the crop plants have an optimal moisture regime and any deviation from the optimum results in adverse effect leading to poor growth, yields and even the quality of the produce. It is, therefore, necessary, that an ideal soil moisture level is maintained for better growth, yield and quality of the crops.

Tobacco is very sensitive crop in respect to water, Extreme situation of water such as water logging and moisture scare is not congenial for tobacco. Tobacco has a well-developed tap root with extensive horizontal lateral roots. Root development is enhanced by withholding water supply during early vegetative period as well as topping and desuckering will also favours root development. It has been observed that water requirement of a crop varies with the stages of its growth. Under limited water supply condition the critical growth stages are taken in to account for irrigation scheduling. The critical growth stages are most sensitive physiological stages of crop growth to shortage of water and if not supplied the yield of crop is drastically reduced. The critical stages are different for different crops, in case of tobacco it is advisable that a moisture content of 50-60 % during whole crop period helps in getting higher tobacco leaf yield. The total water requirements (ETm) for maximum yield vary (with climate and length of growing period) from 400 to 600mm.Normally 75 percent of the water uptake occurs over the first 0.3 m and 100 percent over the first 0.5 to 1.0 m . Grand growth initiation stage *i.e.* 25-30 days after planting and development stage *i.e.* 55-60 days after planting (after topping and desuckening) were found to be most critical stages for irrigation. In general 2-3 irrigations at 30 days interval starting from grand growth initiation stage were found to be essential as it improved leaf yield and quality significantly over that of no irrigation or only one irrigation in case of chewing & Hooka tobacco specific water management of tobacco type are discussed in separatly. Water deficits during the mid-vegetative period (rapid growth) result in reduced growth and smaller leaves. Severe water deficits during the yield formation and ripening periods affect leaf weight and chemical composition which in turn affects the fire-holding capacity. However, a mild water deficit during ripening is desirable to restrict growth of new young leaves. Excess soil moisture in the rhizosphere of tobacco crop is not desirable as tobacco plant is highly sensitive to anaerobic condition which also hinders or stops nitrification. Prevalence of such situation for more than 24 hr. causes yellowing of the leaf due to non uptake of nitrogen and lack of oxygen for respiration of the roots. Such situation at advance growth stages of the crop produces poor quality leaf, devoid of spangles. Interculturing the soil to desirable tilth and application of small quantity of N as top dressing helps mitigate the yellowing problem of the crop. Water must be supplied daily to the young seedlings in the seedbed even when evapotranspiration is moderate. At the end of the seedbed period water is withheld for a few days. Immediately before or during transplanting water is supplied to the plants individually to help them to establish themselves during the first weeks after transplanting. During the rapid growth period water should be supplied frequently.

At pre flowering, little deep irrigation may be sufficient to obtain optimum yields of high quality. When water is limited, water should be applied at transplanting, during the period of rapid growth and during the early yield formation period. To obtain maximum total production under limited water supply, management should be directed towards increasing the area and partially meeting the crop water requirements rather than meeting full crop water requirements over a limited area. Surface and sprinkler irrigation are mostly practiced. The quality of the water is important in selecting the most suitable irrigation method, e.g. sprinkler irrigation should be avoided when only low quality water is available. Fortunately the irrigation water available in most tobacco growing districts of north Bihar is of good quality with very low chloride content i.e. below 30 ppm. Even in northern districts of Bihar where tobacco is grown after harvest of paddy crop. It has been observed that tobacco cannot tolerate water logging at any stage of growth but at the same time the plants may not withstand drought also. Therefore, a compromising situation would be to give light and frequent irrigations to the crop. However, the interval and number of irrigation depends upon soil type, weather of cultivation and type of tobacco to be grown. The field should be irrigated when the soil moisture falls about 20 per cent field capacity because higher soil moisture reduces the leaf quality. Usually hard water (saline) from wells is found to give better results but it must not have more than 50 ppm of chloride, otherwise the leaves get burnt and they have poor burning quality. The flue-cured Virginia (FCV) tobacco grown on black cotton soils is very rarely irrigated whereas the one grown on light soils is given 5 to 7 irrigations. The cigar, cheroot and chewing tobacco in Tamil Nadu and chewing tobacco in Bihar are normally irrigated and about 15 to 18 irrigations are given. The Bidi tobacco in Karnataka, Maharashtra and the hookah and chewing tobaccos of West Bengal are grown rain fed but under drought one or two irrigations are desirable for production of better crop. The hookah and chewing tobaccos grown in U.P. are irrigated and generally 12 to 15 irrigations are given.

Tobacco Type and Water Management

Tobacco is, generally; cultivated as rainfed crop during rainy season and during post rainy season on stored soil moisture. If the non-FCV tobaccos are cultivated on light soils during *rabi,* supplemental irrigation provided, depending on necessity. As such, systematic work on scheduling irrigation and crop water requirements are scanty, especially under Indian conditions. Water needs of tobacco can be considered in phases. In the first 2 to 5 weeks after planting, water needs: small and the plant is sensitive to excessive soil moisture. After the plant has been watered at planting, water should be withheld for 4-5 weeks. The second phase covers rest of the development phase (emergence and expansion of remaining leaves). During this phase, crop water needs are highest and needs frequent rainfall, irrigation or adequate stored soil moisture. The third phase extends over the remainder of maturation phase when the water requirement declines due to removal of leaves and the crop ages. Such a soil moisture regime during the crop growth leads to high yield with acceptable leaf quality. A crop with frequent irrigations or adequate available soil moisture gives high yield, however, at the cost of leaf quality.

1. FCV Tobacco

In Andhra Pradesh, the FCV tobacco on black soils is normally not irrigated, except pot watering at planting. However, the crop on light soils are given up to six irrigations depending of water retentively of soil. In black soils also, under adverse conditions, one irrigation around 40 days after planting is recommended. Critical stage for irrigation is knee-height to bloom. On an average, the yield increase due to irrigating crop on light soils ranges between 10 and 20 per cent. Results of experiments indicate that irrigation at 50 per cent DASM is ideal for improving the yield. The daily water use varies from 0.15 to 0.6 cm depending on the growth stage and the crop growing environment. The CTRI, Rajahmundry estimated the crop water requirement at 500 mm, adequate care should be taken to avoid excessive soil moisture and water logging. Quality of irrigation in respect of chlorides is of paramount importance as it determines the strength and mineral constituents of leaf. The safe limit in irrigation water is 30 ppm chloride and this should not exceed 50 ppm.

2. Non-FCV Tobaccos

2.1 Cigar, Cheroot and Chewing Tobacco

The cigar, cheroot and chewing tobaccos in Tamil Nadu and chewing tobacco in Bihar are normally irrigated. Following the transplantation, light irrigations are given for the first 3 or 4 days and subsequent ones at intervals of 4 to 5 days. In all, these types receive 20-25 irrigations.

2.2 Hookah Tobacco

In Punjab, hookah tobacco is grown as an irrigated crop. After transplanting, first irrigation is given within 2-3 days. Subsequent irrigations are given once in a week or 10 days depending on weather conditions. In Bihar, stored soil moisture at planting is adequate to support crop growth up to 6 weeks after planting. As such, the crop needs one or two irrigations depending on the level of stored soil moisture. In the absence of irrigation facilities, pot watering is done 3 or 4 times for both *N. rustica* and *N. tabacum.*

2.3 Bidi Tobacco

Bidi tobacco in Karnataka, Maharashtra and Gujarat is usually not irrigated. However, the crop is irrigated during years of drought. Only one or two light irrigations are given if the crop is subjected to acute water deficit.

Length of Irrigation Channels for Tobacco Irrigation

The water use efficiency is greatly influenced by depth and slope of irrigation channels because both factors govern the rate of water flow, seepage rate and spilling loss of water from the channel. Thus an ideal channel of proper depth, slope and length will result into many-fold increase in the efficiency of irrigation water which will ultimately reduce the cost of cultivation of heavy water demandity crops.

Table 7.2 Recommended furrow length of different soil types

Furrow slope (%)	Furrow length (m)						
	Clay water application (cm)		Loam water application (cm)		Sandy water application (cm)		
	7.5	15	5	10	5	7.5	10
0.05	300	400	120	270	60	90	150
0.10	350	440	180	330	90	120	190
0.20	370	470	220	370	120	190	250
0.30	390	500	280	400	150	225	280
0.50	380	500	280	370	120	190	250
1.00	270	400	250	300	90	150	220

Systems and Methods of Tobacco Irrigation

The selection of systems of irrigation depends upon several factors *viz.* type of crop to be grown, topography and texture of the soil climatic conditions, water resource and economic conditions of the cultivator. The irrigation system could be divided into following groups:

(a) Surface irrigation system,
(b) Subsoil irrigation system,
(c) Sprinkler irrigation system and
(d) Drip irrigation system.

Surface irrigation system : In this method the water applied on the soil surface but method by which the water is applied on the soil depends upon topography, texture, type of crop grown, quality and quantity of available water, source of available water and energy and labor required for irrigation. Accordingly surface irrigation could be applied in following methods :

(1) *Wild or uncontrolled flooding*: This is most primitive and wasteful method of irrigation. This method is followed in areas where unlimited water is available and a vast area like pasture is to be irrigated. The water is let into the field from a higher gradient and spreads slowly allover the cropped area depending upon the slope. Thus the depth of irrigation varies from place to place due to varied topography. It has many disadvantages like greater loss of water and nutrients, more soil erosion, uniform distribution of water and welting of soil, greater percolation in certain pockets and salinity at higher spots. The method is rarely adopted.

(2) *Controlled flooding*: The field to be irrigated is divided into several plots of convenient sizes of more or less even surface. The water is let into the field through main and sub-channels for irrigating plots one after another. The method needs lesser volume of total water, there is uniform wetting, no soil erosion and it also leads to lesser loss of water and nutrients. This could be done in following ways:

(a) *Check basin method*: This is also called as check flooding method or basin method. The method is practiced in areas where small stream of water is available, ground is nicely leveled and the crop needs a careful distribution of water. Rectangular or square beds of 10 to 100 sq. m area are prepared. The water is conveyed through main channel and branch channels which used to connect two consecutive parallel rows of beds or basins. Though there is water saving in this method yet it is very expensive as preparation of ridges for making channels takes much labour and a fairly large area is occupied by ridges and channels which remains uncropped.

(b) *Border strip method* : The method is suitable in areas having a slope of 0·5 to 1 %, close growing crops are to be grown and the soil has shallow to medium depth. In this method the fields are divided into strips of 3 to 15 meters width and 50 to. 300 meters length depending upon type of soil and extent of slope exceeds 1 % slope the contour borders are prepared. The area remains prone to erosion in cases of semi-graded, graded and contour borders; therefore, it requires higher degree of skill in applying irrigation.

(3) Furrow Method

The method is used for root and tuber crops which are sensitive to saturated soil conditions at the root zone. The crop is sown on the ridges and the irrigation water is applied in the furrows so that most of the roots remain above the saturation zone and there is no soil compaction. In this method the field is divided into ridges and furrows along or across the slope and the furrows are connected with main channels. The depth of furrows is decided based on the length of the roots of the crop to be grown and the infiltration rate of the soil. Optimum length of the furrows depends upon several factors like slope, infiltration rate, rainfall or water application intensity and furrow spacing etc. which may be calculated by using following equation:

Optimum length of furrow: L= 232.5/ (I-A) W.S.

Where,

L =the optimum length of furrow in m.

I =r ainfall intensity or water application in cm/hr W = the furrow spacing in m.

A =the infiltration rate in cm/hr.

S =Slope of furrow in per cent.

Sometimes V shaped furrows are prepared along the slope of the land which is called as corrugated or corrugate ion method of irrigation. Here water is allowed to spread though these corrugations and seep into the adjoining areas.

(4) Drip Irrigation System

Water is delivered to each plant at its root zone, through a net work of tubing. Under medium to low pressure a required quantity of water is given daily so that plants may not suffer from water stress. The lateral pipes are laid at uniform intervals according to the spacing of the crops. This is very expensive and sometimes clogging of equipment makes it ineffective but it has several advantages like water saving to the extent of 30-50%, reduced weed growth, reduced labour cost, constant water supply, early maturity and superior quality of crop produce. In India the drip-trickle irrigation system is still at advance experimental stages whereas in countries like Australia, Mexico and USA it has become a common system of irrigation.

(5) Sprinkler Irrigation System

It is also called as overhead system of irrigation as water is made to spray through nozzles fitted in the stands of water conveying pipe over the foliage of the crop plants as artificial rain. The system is most suited to areas of scarce water supply, for ground having uneven - topography, for light textured open soil types where percolation loss of water is more but the system is expensive arid sometimes the evaporational loss becomes very high. Fertilizers, insecticides, etc. can also be sprayed easily through this system. The entire system being detachable and portable can be carried at any location and may be shifted to other location after the irrigation is over.

The effect of water management on tobacco is marked as mentioned in table given below All most all yield attributes and yield were improved with irrigated farming over rainfed cultivation. Maturity score and nicotine content was higher in case of rainfed culture as compare to assure irrigated condition. However spangling score and puckering scores has not affected due to moisture regime.

Table 7.3 Effect of irrigation on chewing tobacco growth yield and quality

S. No.	Particulars	Irrigated	Rainfed
1	First grade leaf yield (q/ha)	3.55	1.45
2	Total leaf yield (q/ha)	20.55	8.20
3	Leaf length (cm)	53.1	41.2
4	Leaf Breath (cm)	89.3	65.9
5	Number of leaves / plant	34.0	30.0
6	Leaf area / plan	3.62	1.89
7	Leaf area Index	4.66	3.49
8	Spangling score (X5)	3.0	3.0
9	Puckering score (X5)	3.0	3.0
10	Maturity score (X5)1	2.5	14.5
11	Nicotine content	5.25	5.63

A long term experiment (1962-63 to 1982-83) was conducted at Pusa Bihar on tobacco under irrigated and rainfed situation and results revealed that total cured lead as well as first grade leaf yield is always significantly higher under irrigated conditions over rainfed tobacco.

Table 7.4 Chewing tobacco production as influenced by irrigation

Year	Total cured leaf (Kg/ha)		First grade leaf yield (Kg/ha)	
	I_0	I_1	I_0	I_1
1962-63	1751	1993	678	856
SEm $\pm$	9		45	
CD .05	25		150	
1967-68	2524	2511	1399	1657
SEm $\pm$	71		30	
CD .05	NS		160	
1973-74	2056	2075	1086	1095
SEm $\pm$	118		108	
CD .05	NS		NS	
1977-78	1969	2028	1033	1059
SEm $\pm$	98		98	
CD .05	NS		NS	
1982-83	2514	2697	1625	1844
SEm $\pm$	64		54	
CD .05	NS		NS	

Effects of long term (1963-64 to 1973-74) tobacco cultivation on same piece of land with rainfed culture and assured irrigated condition were conducted at Pusa with chewing tobacco and find that There is no adverse effect on soil properties except organic carbon content during 1973-74 table 7.5.

Table 7.5 Effect of long term irrigation and rainfed cultivation of tobacco on soils properties.

Treatments	pH		EC m mhos/cm		Organic carbon (%)		Available P_2O_5 (Kg/ha)		Available K_2O (Kg/ha)
	1963-64	1973-74	1963-64	1973-74	1963-64	1973-74	1963-64	1973-74	1973-74
Rainfed	8.4	8.6	0.22	0.24	0.55	0.45	25.5	64.6	455
Irrigated	8.5	8.6	0.20	0.23	0.55	0.47	33.4	63.4	437
CD 5%	NS	NS	NS	NS	NS	0.009	NS	NS	NS

□□□

8

Tobacco Based Cropping Systems

There is no such remunerative and alternative crops as tobacco in many tobacco growing tracks in India , there is urgent need to develop suitable production and protection technology of various tobacco with improve cultivar to make Indian tobacco more competitive in world market. The major objectives of adopting agricultural system of intensive cropping an approach of raising more than single crops in sequence or simultaneously on same portion of land as a mixed crop to harvest some thing and to reduce the risk of complete failure of sole crops due to natural calamity and outburst of insects pest and diseases.

In the era of intensive agricultural production system, diversification of cropping system have become necessity, obviously due to limited and non expansible arable land, the goal of growing more than one crops in agricultural year has entirely changed from yield stability of mixtures to increase total factor productivity in sustainable manner in the light of availability of improved new techniques of crop management. The availability of short duration high yielding varieties, with photo & thermo insensitive nature in range of crops, also provides ample scope for experimenting with several crops combinations to develop efficient tobacco based cropping systems for different agro ecological condition of tobacco producing areas to scale up the economic status of tobacco farmers.

Cropping System Approach

Cropping system is the cropping pattern used on a farm and their interactions with farm resources, other farm enterprises and available technology that determine their makeup. Cropping system management is basically the design of spatial and temporal combinations of crops in an area. With the introduction of high yielding crop varieties and latest technical know how the farmers in India are now considering agriculture as a business. Since the farmers get major share of income through crop yields, it is, therefore of paramount importance to choose a right crop, in an appropriate area, in suitable cropping system and with a maximum profit. There are certain rules evolved based on agronomic consideration and soil health management necessity. General principle of crop rotation (The repetitive cultivation of an ordered succession of crops and fallow on the same land. one cycle may/complete in several years) *viz.*, the crops with tap roots should be followed by those which have a fibrous root system. This helps in proper arid uniform use of nutrients from the soil and the roots do not compete with each other for the uptake of nutrients.

The leguminous crops should be grown after non leguminous crops because legumes fix atmospheric nitrogen into the soil and also add more organic matter to the soil; while on the other hand, non-legumes are fertility depliting crops. Apart from this, the nutritional requirements of these crops are different, e.g. legumes need more phosphate and less nitrogen while non-legumes need more of nitrogen and relatively low phosphorus. Thus, a combination of these crops helps the farmers in reducing their cost of cultivation. More exhaustive crops should be followed by less exhaustive crops because crops like potato, sugarcane, etc need more inputs such as better tillage more fertilizers, greater number of irrigations, more insecticides, pesticides and better care than crops like oil seeds, pulses etc. which need relatively less or little of the above mentioned inputs. Selection o the crops should be demand based i.e., the crops which are needed by the people of the area and by the family members should be chosen so that the produce can be easily sold at a higher price. The crops of the same family should not be grown in succession because they act like alternate hosts for insects, pests and disease pathogens. Thus, the control of insects, pests and diseases becomes a perennial problem. Apart from this, types of weeds found, are associated with various crops and selection of the same type of crops in rotation encourages weed problems in the field viz. Johnson grass grows with gramminaceous crop throughout year.

An ideal crop rotation has have advantages like all agricultural operations can be done timely for all the crops because of less competition. The supervisory work also becomes easier. Soil fertility is restored by fixing atmospheric nitrogen, encouraging microbial activity, avoiding accumulation of toxins (HCN etc.) and maintaining Physico -chemical properties of the soil. Its also helps in controlling insects pests and diseases. It also controls the weeds in the fields. An alternate cropping of berseem and wheat helps in controlling kasani as well as wild oat. Proper utilization of all resources and inputs could be made by following crop rotation. The family and farm labour, power, equipment and machines are well employed throughout year. The farmer gets a better price for his produce because of its higher demands in the locality or market.

Systems of Cropping

System of cropping is the way in which different crops are grown. Sometimes a number of crops are grown together or they are grown separately at short intervals in the same field. For easy understanding the cropping system may be divided into two main groups:

(a) Mixed Cropping

Mixed cropping is the process of growing two or more crops together in the same piece of land. This system of cropping is generally practiced in areas where climatic hazards such as flood, drought, frost etc., are frequent and common. The farmers always fear that their crops will fail. Under mixed cropping, the time of sowing of all the crops is almost the same, however they may mature either together, *e.g.*, wheat + gram or wheat+barley or wheat+ mustard or they may mature at different times, *e.g.*, arhar + jowar + mung and til or groundnut+ bajra etc. Mixed cropping may be classified into the mixed crops, companion crops, guard crops and augmenting crops, based on their method of sowing.

(b) Intensive Cropping

Intensive cropping is the process of growing a number of crops on the same piece of land during the given period of time. In other words, when the area is limited and the number of crops to be grown is increased within a definite period of time, this cropping method is termed as intensive cropping. The main objective is to increase the income/unit area within a specified period of time. Multiple cropping and relay cropping methods have been developed to make intensive cropping a success.

For successful multiple cropping programme depends upon the availability of most suited high yielding and short duration, crop varieties. These varieties must be responsive to input doses and they should also be thermo and photo non-sensitive so that at least three crops could be grown every year. Availability of genetically superior quality seeds or planting materials so that a required genotype could be grown and they may express their full yields potentials. An excellent physico-chemical properties of soil like well leveled land surface, better structure, aeration, water holding capacity, permeability should be maintain. The soil should be rich in soil fertility and organic matter so that it may sustain high intensity cropping with greater, productivity.

(c) Intercropping

This is a process of growing subsidiary crops between two widely spaced rows of main crop. The main object of this type of cropping is to utilize the space left between two rows of main crop and to produce more grain per unit area. Parallel cropping, companion cropping, multistoried cropping and synergistic cropping are major kind of intercropping system.

The main principles of intercropping are crops grown in association should have complementary effects rather competitive effects. The subsidiary crop should be of shorter duration and of faster growing habits to utilize the early slow growing period of main crop and they must be harvested when main crops starts growing viz., faster sesame, sawan, urd or mung grown with red gram complete their life cycle by September, the period since when red gram picks up branching and attains rapid growth. Autumn planted sugarcane remains dormant after germination until February- March during which potato, berseem, Lucerne, mustard. etc. could be taken successfully as companion intercrops. The component crops should have similar agronomic practices. Erect growing crops should be intercropped with cover crops like pulses so that soil erosion and weed Pupation could be reduced or check. The component crops should have different root depths so that they do not compete for nutrients, water and root respiration among them. A standard plant population of main crop should be maintained whereas that of subsidiary crops the plant population could be increase or decreased as per demand of the situation. Generally component crops of similar pest and disease pathogens and parasite infestations are not included in cropping system.

Agronomic Management

Agronomic considerations for intercropping systems include all practices controlled by the farmer that contribute to crop growth and productivity. Initial considerations for the farmer include which species to plant, in what proportions, on what dates, and with what final objectives. The farmer must decide the type and timing of land preparation, application of manures and fertilizers before or during crop growth, and whether to irrigate or not to irrigate the crop if that option is available. Pest management methods including that of weeds, sowing/planting dates of two or more crops, plant densities, spatial arrangement etc are to be chosen. If the planting pattern will include a relay planting of another crop, this should be decided prior to initial planting and finally, the plant for harvest or multiple harvest must be made, including the storage and/or sale of the products.

In mono cropping where optimum density of single crop species are the rule, the strict management of the physical factors of the crop ecosystems usually sufficient in order to obtain maximum yield. This includes optimum levels of fertilizer, water, solar radiation interception and other factors that can be manipulated to the benefit of crop output. With additional weed, insect and disease management, the monoculture cropping system can lead to substantially high crop yields. As the advantages of multiple cropping become increasingly recognized, the need for a deeper understanding of biotic components of plant interactions becomes very important. Such information is essential for designing and managing complex multiple cropping systems capable of giving yield advantage and stable income.

Interaction in Intercropping

a. Plant Population and Crop Geometry

Plant population defines the number of plants per unit area or the size of the area available to the individual plants. Geometry defines the pattern of distribution of plants over the ground or the shape of the area available to the individual plant. In intercropping, both total population (all components) and component population (each component) have to be distinguished. Taking the optimum population of sole crop as 100, component populations can be expressed in a simple relative basis. A simple intercropping treatment having half of the sole crop optimum of each of the two components is described as 50: 50 component populations. In a replacement series, proportional populations or proportions are related to sole crops of the series whatever their population and the two proportions must always add up to 100. All the intercropping situations can be described using optimum sole crop population for one crop and 50 per cent for the other can be described as 100: 50 (additive series or addition series). With regard to geometry, the proportion of area allotted to each crop should be considered. Thus in a 50: 50 series, having the component crops in alternate rows, will mean allotment of areas in 50: 50, proportion. This can also be obtained by altering the components in the same row or by having double rows of each component alternated.

b. Sowing Time

The relative time of sowing of component crops is an important management variable manipulated in cereal-legume intercropping systems but has not been extensively studied. Intercrops need not always be sown simultaneously. Sowing may be staggered to increase the temporal differences, which might result in higher yield advantages .But it should be seen that the earlier sown crop has as little adverse competitive effect on the late sown crop as possible. In situations where two continuous crops cannot be taken in sequence due to the short season, staggered sowing may be advantageous.

c. Genotype Suitable for Intercropping

Genotypes suitable for sole cropping may not be suitable for intercropping. Hence attempts should be made to select genotypes in actual intercropping situations. For a dominant crop, the criterion of selection for sole cropping will be adequate for intercropping also. In general, genotypes of determinate growth habit (with quite distinct vegetative and reproductive phases) should show a plastic response to population changes and genotypes of indeterminate growth habit (with overlapping vegetative and reproductive phase) should show a rapid recovery from competition in the early stages by compensation in the later stages of growth.

Interaction in Sequential Cropping

Competition for light, water and nutrients as seen in mixed crop communities does not occur when sole crops are grown in sequence. In relay cropping where there is a short span of overlapping between two crops in sequence, the relay sown crop may have competition for light from the standing crop resulting in lanky growth of seedlings. This competition can be minimized by proper choice of crops and varieties and adjustment of time and method of planting.

a. Solar Radiation

Solar radiation differs from the other resources in being a flux of energy; water and mineral nutrients exist in the environment as 'pools' of substances. In intensive multiple cropping involving two or more crops in sequence, the main objective is to harvest as much solar energy per unit area time as possible. Competition for PAR does not occur when sole crops are grown in sequence as in the case of intercropping though intraspecific competition for PAR is not considered here. On an average, about 500 cal cm-^{2} day-1 of solar energy is received on earth's surface in the tropics. Assuming 50 per cent cloudiness and 50 per cent reflectivity, only 375 cal cm-^{2} day-^{1} would actually reach earth's surface. A photosynthetically efficientC_4 plant like maize would convert about 6 per cent solar radiation into biomass. This works out to storage of 700 cal cm-^{2} month-^{1} or 180 q ha-^{1} month-^{1} dry matters. It is obvious that what is actually achieved in the field is far below this calculated value.

b. Effect of Preceding Crops on the Following Crops

The basic precept of sequential cropping is that the farmer manages only one crop at a time. From the soil management point of view, experience from single crop stands is not entirely applicable to sequential cropping systems because of the influence of a previous crop on soil physical properties, water and nutrient availability to the succeeding crop. Soil management practices should be geared up to the crop sequence or rotation rather than individual crops. This includes complementary effects such as release of *N* from the residues of the previous crop, particularly legumes, to the following crops and carryover effects of fertilizers applied to the preceding crops and effects such as allelopathy, temporary immobilization of *N* due to wide *C/N* ratio of the residues and carryover of pests and diseases.

c. Crop and Varieties for Sequential Cropping

During recent years the objectives have changed to intensification of cropping with simultaneous achievement of yield stability. Reduced risk from intercropping, for example, is not universal and much depends on crop combination, cultivars and agronomic management and any generalization on intercropping without due recognition of these variables can be misleading For the success of the multiple cropping programme from biological point of view, proper choice of crops and varieties is a pre-requisite. Recently introduced short duration photoperiod insensitive genotypes are the best suited for intensive cropping systems. It was emphasized earlier that selection of genotypes should be done in actual intercropping situations, particularly for the dominated components. All the recently introduced high yielding varieties have a plastic response to population levels and geometry of planting. Yield does not vary much in a range of populations and at optimum population; row spacing has little influence on yield. This is because these varieties have been evolved for population performance and the traditional varieties for individual performance. Inclusion of high yielding varieties will give scope for adjustment of population and geometry of planting so that an additional crop could be raised as an intercrop.

8. Nutrient Management in Cropping System

8.1 Role of Soil in Nutrient Management

Once the nutrient requirement of a cropping system is workout before deciding on the quantity of fertilizer application we have know out how much the soil can contribute to meet the requirement by routine soil analysis. In many intercropping, it has been found that only the base crops need fertilization and the intercrops do not require extra fertilization. This is probably because of the temporal differences in their peak requirements of nutrients and spatial differences in foraging for the nutrients in the soil. In such situations, the soil test data for the base crop as sole crop can be applied to it as intercrop also. Some crops are surface feeders and soil samples can be collected in the surface layer (0-15 cm). If Intercrop feeds in the deeper layers, then soil samples have to be collected from these layers also for analysis.

8.2 Requirement and Uptake Pattern of Nutrient by Component Crops in Cropping System

The amount of nutrients which a given component crop requires to complete a productive life cycle is valuable information in designing fertilization practice for a cropping system. Each crop included in the system should be studied with respect to its nutritional needs. Mono crop fertility requirements are a good basis for initial studies on cropping system management needs. As the intensity of crop interactions in multiple cropping systems increases, the over all uptakes may change and optimum management techniques must be developed with care, based on component species characteristics and the objectives of the multiple cropping system. Nutrient competition can be minimized in intercropping systems by selecting species with different rooting patterns, different nutrient requirements, different times of peak demand for nutrients and plant spacing One advantage of multiple cropping systems is greater total uptake of nutrients from the soil, although this may be a reflection of greater dry matter production due to better use of light or water rather than nutrient use.

The time period of maximum nutrient demand of one component crop may not coincide with that of another. Information available on nutrient uptake pattern for different species can be made use of, though the pattern may be modified by the presence of another species. The legume component is capable of fixing atmospheric N under favorable conditions and this is believed to reduce competition for nitrogen. In the absence of an effective N-fixing system compete for available soil N. Phosphorus is a major nutrient that determines the production potential of most grain legumes usually intercropped with cereals. Legumes are poorer competitors for P when intercropped with grasses or cereals, this being attributed 10 differences in root morphology.

8.3 Response to Applied Nutrient in Cropping System

All crops included in a cropping system are not equally responsive to application of all the major nutrients. Based on the response data of individual crops, they can be classified as non-responsive, moderately responsive or highly responsive to specific plant nutrients. It would, then, seem appropriate to apply fertilizer to those crops which respond to it and little or none to the non-responding crops. Sequential cropping requires adequate moisture, sunlight and temperature regimes during the growing season to support back-to-back activities. Crops grown in different cropping sequences remove varying quantities of nutrients from soil which affects its fertility status. In multiple cropping systems, use of heavy doses of fertilizers and manures might lead to the accumulation of large amounts of available nutrients in the soil. On the other hand, depletion of nutrients might occur if intensive cropping is practiced without the external supply of nutrients.

8.4 Role of Legumes and Organic Manures in Cropping System

Legumes are widely used for food, fodder, shade, fuel, timber and green manure and as cover crops. Nutrient self-sufficiency is a desirable characteristic of agronomically sustainable cropping systems. Legumes have the potential for self-sufficiency for N, the nutrient most limiting to productivity, by symbiotically fixing atmospheric N. In fixing atmospheric N, legumes contribute to the N content of soil either as sole crops in rotation, or as intercrops. In such systems, legumes may either increase the soil N status through fixation, excretion, or in the absence of an effective N fixing system, compete for N. The quantity of N fixed by the legume component in cereal legume intercropping depends on the species, morphology and density of legume in the mixture, the type of management and the competitive abilities of the component crops. Legumes of indeterminate growth are more efficient, in terms of N fixation than the determinate types. Organic manures have the potential to correct several soil problems, at the same time supply organic matter, create favourable air and water regimes around plant roots and act as carriers of some micronutrients besides its influence on the microbial and faunal activities in the soils. Due to occurrence of multiple nutrient deficiencies and deterioration in soil physical properties, organic manuring has assumed greater importance. Farmyard manure (FYM), compost and green manures are the organic materials commonly used. Because the availability of farmyard manure and compost is limited, green manures offer greater potential as a feasible and cheaper substitute for fertilizer N.

8.5 Integrated Approach

The complementary use of chemical fertilizers and organic manures will augment the efficiency of chemical fertilizers as well that of organic manures and microbial inoculants in order to maintain a higher level of productivity over a period of time. The integrated use of organic manures and mineral fertilizers has been found to be promising in maintaining stability in crop production on certain soils through correction of marginal deficiencies of secondary and micronutrient elements in the course of mineralization of organic manures on the one hand and providing favourable physical, biological and soil ecological conditions on the other. Integrated use of organic and inorganic fertilizers has been found to be promising in obtaining sustainable crop productivity on a long term basis under modern intensive cropping, besides meeting the nutrients turn over in soil-plant systems. Integrated nutrient management (INM) is the improvement and maintenance of soil fertility for sustaining increased productivity through optimizing all possible organic, inorganic and biotic sources of plant nutrients required for crop growth and quality in an integrated manner appropriate to each cropping system and farming situation in its ecological, social and economic possibilities. The basic principle behind this concept is to use both the chemical fertilizers and organic manures in most efficient manner because of their limited availability and higher prices.

9. Water Management in Cropping System

General principles applicable to sole crops will also apply for mixed population. With the proper water management, it might become possible to grow two crops where only one was grown previously under rainfed conditions. Scheduling of irrigation and water application may have to be carefully done under intercropping situations, if one of the components is sensitive to excess water while the other may need frequent irrigation. With the continuous crop cover and presence of stubbles and residues, the rainfall impact on soil surface and run-off are likely to be reduced. If a quick growing intercrop is planted in the wide space between the rows of the slow growing species, the evaporation loss will be reduced but the intercrop will lose some water through transpiration. In general in case of widely spaced, slow growing crop, a large area of soil surface is exposed HI direct incidence of solar radiation resulting in high evaporation loss. The differences in root systems, depth of rooting, lateral root spread, and root density are factors that affect competition for water between component crops. The use of different parts of the soil profile by root systems of different crop species minimizes the degree of competition for water. When component crops compete for available water, the cereal, with its higher growth rate and more extensive root system, is generally favoured Intercropping has the twin objectives of reducing evaporation loss and the wasteful use of water by weeds particularly where weed control is not effective, weeds are likely to grow in the inter space losing water through transpiration if inter cropping is not done. In sequential cropping, the choice of a cropping system may depend on the total quantity of water required for raising the crops, especially with the receding ground water level and increasing cost of pumping out the water.

9.1 Irrigation Scheduling in Cropping System

Scheduling irrigation for different crops depends on the total quantity of water required by the crops, the critical physiological stage at which water is required most, the soil type and the climatic factors. When two or more crops are grown together, this becomes a complex problem since the total quantity of water required by the component crops may vary widely and the critical periods of water requirement for the crops may not coincide. If irrigation is given to one of the crops based on its requirement, the other crop may suffer due to excess water or water stress, sometimes resulting in total failure of the crop. To overcome under irrigation and excess irrigation, skip furrow method of irrigation is advocated. In this method, the ridge of the irrigated furrows is extended to the location of skip furrows and the crop, which would have been at the extended ridge, is planted at the other side of the irrigated furrows (similar to pairing of base crop rows). Alternate furrows are irrigated at one irrigation. In the next irrigation, the earlier unirrigated furrows are irrigated, skipping the earlier irrigated ones. The intervals between irrigations are shortened a little so that the accelerated depletion of moisture in the alternate furrows is replenished early. In this method, there is a saving of about 50 per cent water without affecting yield simultaneously producing additional intercrop yield. Intercropping does not greatly increase the water requirement and as a matter of fact, improves the water use efficiency both under rainfed and irrigated conditions. Adopting paired row planting and irrigating only the base crop rows result in considerable saving of water. Scheduling of irrigation should be based on the moisture depletion in the base crop rows only. Presence of crop cover throughout the year results in reduced erosion losses and increased soil storage of moisture.

10. Weed Management in Cropping System

There is a general belief that intensive cropping reduces weed problems. In some situations, weed infestation is reduced while in others, it is not so. This depends on the crops, planting density and cultural operations done. Besides, herbicide selectivity and residual toxicity are critical in intensive cropping. Weed management for any cropping system involves the use of many kinds of biological, physical, chemical and mechanical techniques to promote crop dominance over weeds. Crop density is one of the most easily manipulated factors affecting crop production and it is well known that increasing seeding rates can promote crop dominance over weeds in monoculture cropping systems. Highest yields for many intercrop combinations grown under weed-free conditions are obtained with increased crop population densities. Similarly, maximum intercrop yields and weed suppression are obtained with total crop densities significantly higher than those used for monocultures (additive crop mixtures). Smother intercrops and 'live mulch' intercrop are high density, additive crop mixtures that appear to offer great promise as a means to weed control.

10.1 Weed Control in Intercropping

Weed control in intercropping system is relatively difficult than in sole cropping of the component crops. The predominant method of weed control in intercropping is manual or mechanical. Mechanical weeding may not be possible in random sowing or in closely planted rows. Most herbicides are crop specific. So, it is difficult to find chemicals that will control a broad spectrum of weeds without causing damage to the component crops. A herbicide that does not harm component crops usually does not control a large number of weeds. A limited number of experiments conducted have helped to identify some herbicides.

10.2 Weed Control in Multiple Cropping

Continuous presence of crop cover, residual toxicity of herbicides applied to the previous crop on the succeeding crop and changing weed flora with the seasons all need a different approach in weed management in multiple cropping, which includes sequential cropping and intercropping. With intensive cultivation of land round the year, the weed flora shows a shift towards more persistent perennials. With intensive cropping, minimum tillage is generally adopted. This helps the weeds to emerge easily, if soil moisture is not limiting. Frequent cultivation of the field is not possible because of the time factor and so manual or mechanical removal of weeds and use of herbicides have to be combined for effective weed control. The weeds were effectively checked by pre-emergence application of selective herbicides at low doses followed by a late weeding. Hence, weed competition was minimized and crop yields stabilized with minimal tillage and pre-emergence application. With the use of this technique, the turn-around time, cost of land preparation and labour requirement were reduced. Ploughing the field well ahead of the planting season, either at the end of rainy season or after summer rains helps to control most of the weeds and reduces the weed problem in the ensuing season. When an upland crop is rotated with a lowland crop, the total number of weeds in either is lower than under continuous lowland or upland culture. Now days selective herbicides are available for sole crops and efficient weed control can be obtained by using them, How ever every care should be taken before using them for multiple cropping particularly introspect to residual effects of these agrochemical.

11. Insect Pest Management in Intercropping System

There is in general believe that the outbreaks of pest are to be less common in mixed stands due to crop diversity than in sole cropping system. The reduced incidence of pests may be the result of increased parasitoid and predator populations, availability of alternate food for natural enemies, decreased colonization and reproduction of pests, chemical repellency, masking and/or feeding inhibition from non-host plants, prevention of pest management and/or emigration and optimum synchrony between pests and natural enemies. In intercrop systems, the choice of a tall or short, early or late maturing, flowering or non-flowering companion crop may magnify or decrease the effects on particular pests). The inclusion of a crop that bears flowers during most of the growing season can condition the build up of parasitoids, thus improving biological control. Similarly, the inclusion of legumes or other plants supporting populations of aphids and other soft bodied insects that serve as alternate prey/hosts can improve survival and reproduction of beneficial insects in agroecosystems. The presence of a tall associated crop such as maize or sorghum may serve as a physical barrier or trap to pests invading from outside the field. The inclusion of strongly aromatic plants such as onion *(Allium cepa),* garlic *(Allium sativum)* or tomato *(Lycopersicon escuientum)* can disturb mechanisms of orientation to host plants by several pests. Multiple cropping is a useful strategy in the control of several pests in various agroecosystems. Trap crops are plant stands that are grown to attract insects or other organisms like nematodes to protect target crops from pest attack. It is advisable that when pest incidence is high, resistant or tolerant genotypes may be raised. In many intercropping systems involving pulses, pod borer damage is often severe. To minimize the damage, farmers sometimes harvest at green pod stage (especially with cowpea) and sell it as vegetable. The limitation here in India is the availability of the market.

12. Insect Pest Management in Multiple Cropping Systems

The cultivation of several species in multi-crop systems is an agronomic practice with many purposes that is carried .out by small farmers all over the world. Not only does this practice diversify the diet, protect the soil, reduce the risks of low prices, make better use of varying soil fertility and distribute labour more efficiently, it also creates a greater diversity of genotypes and phenotypes, lowering the risk of the rapid spread of pests and diseases. Some crops have few pests and others have large pest complexes (e.g., Solanaceous crops). When new crops are introduced, exotic pests may also get introduced. Planting crops in off-seasons may result in the hitherto minor pests becoming major pests. In long duration crops, the natural enemies of pests sustain themselves well and regulate the pests. On the other hand, early maturing crops escape damage by pests. Continuous cropping with little turn-around period between two successive crops may cause increased pest outbreak in areas where this is newly introduced. After a few years, there is stabilization because the natural enemies of the pests also increase in number and keep the pests in check. When a particular cropping system is followed in a particular region, it would be ideal to have synchronous planting of the crops. Management of insect pests in multiple cropping involves a diversity of control methods. Though pesticides have to be used, their use should be need based. Residual toxicity on the succeeding crop, especially with persistent chemicals, should be borne in mind. Integrated pest management is key to succeed in the multiple cropping systems.

13. Disease Management in Cropping System

Crop rotation is another multipurpose cultural practice which lowers the level of inoculums of many pathogens. Plant density plays a vital role and affects the disease incidence. With air and soil borne disease spores, the wider the separation between individual hosts plants, the less likely it becomes that organisms infect new plants. Intercropping now days used as a means of increasing the income and productivity and raising crops in sequence or as intercrops is effective in lowering populations of soil borne diseases. Similarly, raising cover crops reduced the movement or spread *of pathogens in soil* in through surface water and antagonistic microflora. Cover crop might reduce the soil splash during rainy days and the movement of disease propagules. It is also possible that the build up of microbial population suppressive to *Phytophthora* might be responsible for the lower disease incidence under the cover crop conditions. The disease causing organisms, however have resistant stages such as spores that can persist and allow the population to survive for several years under host-free conditions. To control of seed borne disease, pathogens can be checked by seed treatment with fungicides. Removing the affected individual plants will reduce the inoculums source. Planting resistant genotypes, particularly those with multiple gene resistance, can be done to minimize disease incidence.

In India, systemic rotation of crops on tobacco soils has never come into general use. In fact, in most places, tobacco is a monoculture crop. In most types of tobaccos, the crop has been grown continuously on the same field for many years without any obvious deleterious effect on yield or quality. Tobacco can be grown alone as sole crop but it can also successfully cultivate in close association with other crops. Tobaccos being wide spaced crop one can go for additional crop in between as this crop occupy the field for long duration. Tobacco is best fitted in different cropping system viz., sequential/ multiple or intercropping. Some of the experiment were carried out at our research center are discussed here.

14.1 FCV Tobacco Based Systems

The FCV tobacco is largely grown in Andhra Pradesh and to a limited extent in Karnataka as rainfed crop. In Andhra Pradesh, the FCV tobacco is grown without any rotation. However, rotating tobacco every third year with sorghum is beneficial in blank soils of Guntur and Prakasam districts. The cropping intensity can be double by taking sesame crop during *kharif* followed by FCV tobacco during *rabi* in Krishna, Guntur and Prakasam districts. For East and West Godavari districts, *kharif* rice followed by *rabi* FCV tobacco appears to be ideal.

In Karnataka state, finger millet or sorghum is grown once in three years to control root parasite *orobanche*. In red soil region, finger millet or sesame is taken during *kharif* followed by FCV tobacco during *rabi*. Pearl millet- tobacco and cotton-tobacco systems are also common in Karnataka.

14.2 Non-FCV Tobacco Based Systems

In Tamil Nadu, cigar, cheroot and chewing tobaccos are, generally, grown on the same land year after year, with one or two crops like pearl millet, finger millet or sorghum in between. In West Bengal, summer rice motihari tobacco is the common system. However, *kharif* rice-rabi tobacco-summer rice, a three crop system are highly remunerative in the recent past (ICAR 1997). In bidi tobacco tract of Gujarat, tobacco-pearl millet and cotton-tobacco systems are most common where sin Nipani (Karnataka) area sorghum-tobacco is the usual system. In Bihar, a two-year rotation of sunnhemp (green manure)-potato in the first year and sunhemp (green manuririg}-tobacco in the second year is ideal for hookah tobacco. In Uttar Pradesh, winter tobacco-maize and summer tobacco-wheat are the usual copping systems. In West Bengal, where the low lying areas are submerged during monsoon, *N. rustica* is grown on year after year.

14.3 Sequential Cropping with Tobacco

Growing at least two crops in sequence with tobacco on the same field in a year is commonly known as Sequential cropping. The succeeding crop is planted after the preceding crop has been harvested. Sequential cropping systems utilize higher inputs and income maximization is a more important objective because there is no intercrop competition in sequential cropping systems and are customarily encountered where resource endowments than in the case of intercropping. Some of the tested and found suitable and recommended are given below.

14.4 Maize-tobacco+garlic-maize

This system was developed in early seventies for assured irrigation areas. A maize variety suitable for summer cultivation is after harvesting sown in the second fortnight of February or first week of March of tobacco. Maize crop is fertilized with its full recommended dose in two splits and is irrigated 2-3 times or as per requirement followed by interculturing and earthing up operation. Summer maize gives grain yield of about 50-60 q/ha which is almost double of the kharif maize yield (25-30 q/ha).

Tobacco + garlic in sequential cropping

This system is recommended for limited irrigation water areas or rainfed areas. Manurial and cultural requirements are followed as per recommendation of individual crop in the system. This is one of the best systems evolved for highly intensive cropping for small and marginal farmers of Bihar state. In this system Maize crop is taken as kharif crop and normal tobacco is planted as per varietal requirement viz 105X90 cm for PT-76 and for Pusa Vaishali 90X75cm. Garlic and black gram or green gram should be taken as intercrop. In this case after harvesting of garlic crop black gram or green gram crop find place in this cropping sequence.

Maize-tobacco + garlic-black gram / green gram

14.5 Maize in Sequential Cropping with Tobacco

Other important sequential cropping systems recommended for north Bihar conditions are as follows.

- Tobacco – summer maize – dhaincha (GM)
- Sesame-Tobacco-Sunflower
- Sesame-Tobacco- Onion
- Tobacco – Winter Maize – Sesame

15. Intercropping Systems

Raising two or more crops simultaneously on the same field in definite proportion and row arrangement is known as intercropping. In intercropping system crop intensification is in both temporal and spatial dimensions Intercrop components like garlic and potato act as insurance against hail storm damage to tobacco crop in the year of its occurrence. Following intercropping systems are recommended in north Bihar.

15.1 Tobacco+Sugarcane Intercropping System

In this system tobacco crop is planted as per its normal manorial and cultural schedule and sugarcane is planted in between tobacco rows in furrows 8 to 10 days after tobacco planting. Sugarcane crop does not affect tobacco growth or development adversely as sugarcane growth remains very slow during winter season i.e. from October to February. After harvest of tobacco crop by the end of February or 1st week of March, all the tobacco stubbles are removed and sugarcane is top dressed with 1/4th of its recommended nitrogen i.e. 30 kg nitrogen /ha is applied as 2nd top dressing after 3 months from the 1st top dressing in June. This system is recommended for upland sugarcane growing area of north Bihar.

15.2 Tobacco+Papaya Intercropping System

In this system tobacco is planted as per its recommended normal manurial and cultural schedule in the 1st week of October and papaya saplings are planted in alternate row of tobacco maintaining 6'X6' or 2.0m X 2.0 m spacing both ways. After harvest of tobacco crop in the last week of February or 1st week of March, its stubbles are removed completely and papaya crop is given a light irrigation followed by interculturing. After two to three months *i.e.* in May-June papaya crop is top dressed with 40 kg nitrogen /ha. Papaya crops comes in fruiting 9-10 months after its planting and fruits get ready for harvest in another two months time. In 2nd and 3rd year papaya crop is grown as pure crop without tobacco. This system is suitable for upland tobacco growing area of north Bihar having sandy or sandy loam soil with good internal drainage. The farmers are following tobacco+ papaya intercropping system in certain areas with good margin of profitability.

- Tobacco + Garlic
- Tobacco + Rajmash
- Tobacco + Potato
- Tobacco + Winter Maize
- Tobacco + Green onion
- Tobacco + Radish
- Tobacco + Coriander
- Tobacco + Fenugreek
- Tobacco + Black cumin (spices crops)
- Jowar+ Moth (fodder) – Tobacco-groundnut
- Jowar+ Moth (fodder) – Tobacco- black gram
- Jowar+ Moth (fodder) – Tobacco- cowpea

Tobacco + Rajmash inter cropping

15.3 Tobacco+Rajmash in Intercropping

Some of the above sequential and intercropping systems have already become very popular among tobacco growers of north Bihar obviously due to more remunerative and better protection measure against natural calamity and outburst of insect pest and diseases mentioned under suitable heads.

□□□

CTRI (Central Tobacco Research Institute). 1999. Status Report on Research Programme on Alternative Crops to Tobacco. CTRI, Indian Council of Agricultural Research, Rajahmundry, Andhra Pradesh.

D'Alessandro A., N.L. Benowitz, G. Muzi, M.D. Eisner, S. Filiberto, P. Fantozzi, L. Montanari and Abbritti G. 2001. Systemic nicotine exposure in tobacco harvesters. *Arch Environ Health*. 56: 257-263.

Dahl Dale C. and Jerome W. Hammond. 1977. Market and Price Analysis: The Agricultural Industries. New York: McGraw-Hill, 323 pp.

Das S.K., P. Das and S.K. Roy. 2006. *Handbook of Agriculture*. ICAR, New Delhi.

Decaux F. 1961. L'utilisation du tabac dans la médecine d'autrefois. *Rev. Phytothérapeutique*. 8: 66-71.

Del Piano L., M. Abet, C. Sorrentino, F. Acanfora, E. Cozzolino and F. Acanfora. 2000. Genetic variability in *Nicotiana tabacum* and *Nicotiana species* as revealed by RAPD procedure. *Beit. Tabak. Forsch. Int.* 19: 1-15.

Denev, B. Denva, K. Balchvarova. 2007. The bio synthetic origin of germination stimulants for Orobanche ramosa (L. in tobacco and Arabidopsis. *Biotechnol. & Biotechnol.* EQ 21: 2007/1.

DES (Directorate of Economics and Statistics, Ministry of Agriculture). 2009. Cost of Cultivation of Principal Crops in India.

Dickson S.A. Panacea or Precious Bane. 1954. Tobacco in 16th Century Literature. New York Public Library, New York.

Doctor P.B., V.N. Gokani, P.K. Kulkarni, J.R. Parikh and H.N. Saiyed. 2004. Determination of nicotine and cotinine in tobacco harvesters urine by solid phase extraction and liquid chromatography. *Journal of Chromatography* 802: 323-328.

Duft Kenneth D. 1979. Principles of Management in Agribusiness. Reston, Va: Reston. 470 pp.

Dwivedi S.S.L., R. Singh and A.K. Pandey. 1984. Response of chewing tobacco to nitrogenous fertilizers and organic manures. *Tob. Res.* 10: 118-22.

Einhellig, F.A. 1995. Mechanism of action of allelochemicals in allelopathy. pp 96-116. In K. Inderjit, M.M. Dakshini and F.A. Einhellig (ed.), *Allelopathy: organisms, Processes, and Applications*. American Chemical Society, Washington, D.C.

Epp Donald and John W. Malone Jr. 1981. Introduction to Agricultural Economics. New York: Macmillan, 352 pp.

ERC Statistics International Plc. 1998. India. pp. 138. In Wesl K. [illegible] 1998. Evolution trotz Artbeständigkeit. Zeitschr. indukt. Abstamm. u. Vererb. 8: 325-332.

PART-II

Crop Protection

(Stress Management)

9

Weed Pests Management

Weeds are posing serious problems in the tobacco production. Tobacco is attacked by numerous weed pest, some of them are parasitic and bound to be complete their lifecycle with tobacco viz., Broom rope or orobanche. Weed infestation starts with sowing of the crop in the nursery bed. Use of execessive chemical might be harmful and deteriorate its quality. Weed management start well before the sowing of seeds or transplanting seedling in the main field. Various aspects of weed management including good agronomic management is deal in this chapter.

Weeds are no strangers to man and its management is as old as agriculture itself. Weeds are plant out of place; they can also be referred as unwanted and undesirable plants which interfere with land, water added resources. Weeds are important factor in the management but their effect is greatest on tobacco production, In a estimate, of the total annual loss of agricultural produce from various pest in India, weeds accounts for 45%, insect 30%, disease 20% and other pest accounts for less than 5%. Reduction in crop yield has direct correlation with weed competition, because they compete with standing crop plants for nutrition, moisture, sunlight and other added inputs. Competition for resources is depends on the type of weed species, their severity and duration of infestation. Weeds also serve as alternate host to several crop insect pests, nematodes and pathogens. Weeds also reduce human efficiency through physical discomfort caused by allergy and poison. All the crop grown in the field are subject to weed competition, weed competition is maximum during early phase of crop growth, Weeds adversely affect the yield and quality of tobacco, weeds problems in tobacco growing track are acute, both in the seedbed and transplanted main crop. In transplanted main tobacco field weed, competition is maximum during early part of growth period (up to two month).Weed management in tobacco cultivation involves about 10 % of its total cultivation cost. It is estimated the losses due weed infestation is in the order of 20-25 percent of total losses due to different crop pests (Insects, Diseases, rodents and other pest) which is estimated to the tune of 17-21 percent. *Cyprus rotundus and Cynodon dactylan* is most serious competitor during early stages of tobacco growth and weed management is most needed. *Aunia pinnatifida, Cnicus arvensis* and *Chenopodium album* are other weed found to in tobacco crops in more or less severity during later stages.

Good Agronomic Practices (GAPs) for Better Crop Management

Any cultivation practices without considering good agronomic practices (GAP) may be wastage of time and resources resulting futile and some time may pose serious concerned owing to leather effect. Some of the practice which is adopted by tobacco farmers which not only enhance the productivity and quality of produce but also help in reducing the cast of cultivation, which again a eco friendly and sustainable in nature. These GAPS are very much helpful in managing weeds, insect pest and disease causing agent below the economic injury level with out danger of loosing biodiversity and ecological threats.

Summer Ploughing

Deep ploughing in summer could able to control egg masses of *Spodoptera* insect pests, root knot nematode, fungus and weeds by exposing to the scorching sun of summer. It can also effectively controls *Orabanche*, a parasitic weed of tobacco crop.

Interculture for clean cultivation

Sanitation

Many of the problemetic weeds in tobacco are the result of seed produced in the field during the preceding crops or seed blown into the field or on to seed from adjacent areas such as fence rows or ditch banks. Preventing weeds from producing seed in these areas may aid in reducing weed problems in succeeding tobacco crops. Destroying weeds around the nursery bed area as well as utilizing natural or artificial windbreaks will reduce weed infestations.

Crop Rotation

Mono cropping of tobacco leads to increase in the incidence of pests and diseases especially orabanche and root knot nematodes. The benefits of crop rotation on reduction of diseases, insects, and nematodes have been well documented. In addition, crop rotation may also be important for reducing weed problems. Many weeds, especially the large seeded broadleaf weeds can be more effectively controlled. Growing black gram and green gram in Kharif is beneficial in black soils and maize, sorghum, sesame and sesbania in alluvial soils.

Trap Cropping

Raising castor in one row, surrounding the nursery and main field and daily collection and destruction of egg masses and larvae from the trap crop will control the pests.

Selection of Correct Varieties

Select correct varieties suitable under particular agroclimatic conditions and according to their needs and choice, which also resist the pest problems specific to the situation.

Healthy Seedlings

For transplanting always use healthy seedlings of resistant varieties of timely sown crop. Before sowing seed must be treated with suitable agrochemical to boost their resistance power against insect pest and diseases.

Intercultural Operations

About 20 days after transplanting, when the plants get established their young rootlets need very fine and pulverized sub-surface soil for their rapid growth and further expansion. Inter cultivation and occasional hand-hoeing are to be, an important part of a tobacco to maintain sanitation in the main field. It s also provides good aeration to the tobacco. Intercultural operations always help to minimize the weed, insect pest and disease. Clean field always ensure good acreage because there is no completion for nutrition, space and moisture. Inter cultivation helps in mulching, removal of weeds, proper aeration, etc. In case of flue-cured, tobacco mulching with 3,600 kg of paddy straw after first inter culture helps in preserving moisture and keeping down weeds but care must be taken to protect the crop from termites. The intercultural operations may be done effectively by khupri or hand hoe in smaller areas and by harrows or cultivators in the large areas. The crop should be intercultured at an interval of 15 to 20 days or after every irrigation till the leaves are not damaged. This way the crop needs 3 to 4 intercultural operations for proper moisture preservation, better crop growth and yield.

Planting Time

Timely sowing of seeds in nursery and planting of tobacco in main field can always protect the crop form insect pests, diseases, weed, and nematodes and up to some extend from parasitic pest because tobacco gets better time and environments to fasten their growth and developments.

Plant crop at correct time

Planting Distance

Always use recommended planting distance between row to row and plant to plants. Over crowding provides favorable environment to buildup of pests & diseases and nematodes. Planting at optimum distance can control pests and diseases by providing favourable sunlight, air and natural enemies.

Fertilizer Management

Apply only recommended doses of important inorganic fertilizers like N, P and K. Balance between N & P should be maintained to protect the crop from the increase incidence of pests and disease. In case of deficiency of micro nutrients suitable ingredients should be supplemented as per prescription. Luxurious consumption of nitrogen in plants leads to increases the attack of different pests, hence recommended dosage of nitrogen has to be applied.

Water Management

Timely irrigation in nursery can control damping off disease.

Weed problems in tobacco are acute both in the seedbed and in the transplanted crop. *Orobanche* is also a predominant parasitic weed very common in tobacco.

Weed Management in Seed Bed / Nursery

Tobacco crop is very sensitive to weed management which starts right from nursery beds, because the tobacco cultivated through tiny and minute seeds which is very delicate during its early establishment phase. Weed management in nursery can be done by sterilization of nursery beds sterilization take care of weeds, seeds of other undesired crops, seed borne diseases and pests, thus it provide protection to young tiny and tender tobacco seedlings. This can be done by rabbing and chemical treatments. Any non selective per emergence herbicide should be applied for mass killing the existing flora. Deep summer ploughing twice in April-May expose soil pathogens, weed roots, eggs and larvae of nematodes etc. The waiting period before tobacco seeds can be sown in the treated nurseries varies from 2 to 30 days, depending on the compound used. Methyl bromide, a volatile liquid is applied beneath a plastic gas proof covering. Exposure below the cover should be for 1-2 days. The beds should be aerated 2-4 days before sowing. Calcium cyanamide is applied 2-3 months before sowing. Metham is applied 3 weeks before sowing followed by watering. Diphenamed (2-3 kg) and Trifluralin (0.5-1.0 kg *a.i.*) as PPI incorporation into soil control most annual weeds in the nursery. Most common method of weeding in tobacco nurseries in tropical and subtropical countries in hand pulling. In developed countries, however, chemicals are used for effective weed control. Some of the temporary soil sterilants used in tobacco nurseries is Methylbromide (8-10 kg), Metham (2-5 kg). The quantity mentioned in parenthesis is sufficient to 100 M^2.

Weed Management in Main Crop

Critical period of weed competition in the planted crop is the first 60 days after transplantation. In a transplanted crop, inter-row cultivation is feasible due to wider row spacing. Hence, one or two cultivations followed by manual weeding give adequate control of weeds. In standing main crop or in a cropping sequence where tobacco has to planted an integrated approach accompany with good agronomic practices is need of time along with methods of weed management which include control and eradication of weeds from tobacco field are grouped in to mechanical and chemical, these methods are applicable for both pre infested or preventive control and post infested control and eradication of weeds.

Mechanical Weed Control

Mechanical or physical methods of weed control are old age practices, which include tillage, hoeing, hand weeding, digging, cheeling, sickling, mowing, burning, flooding, mulching *etc.*

Cyprus rotundus in tobacco

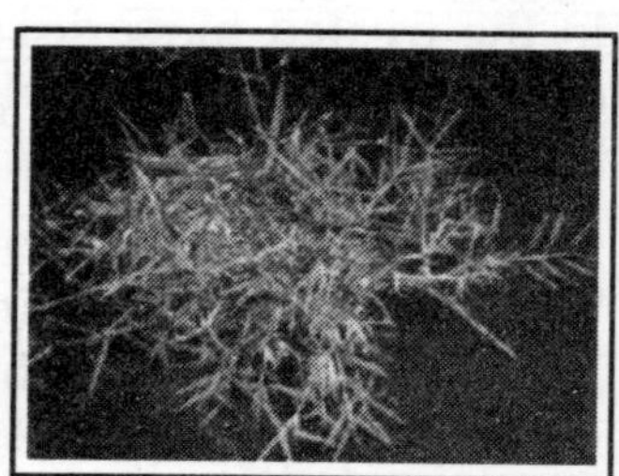

Cynodon dactylan- a common weeds

Intercultural operation about 20-25 days after transplanting, when plants get established their young roots need very fine pulverized sub surface soil for their growth and development. It also helps in mulching removal of weeds and proper aerations. Weed control is an important production practice in tobacco since weed competition for moisture and nutrients may result in yield losses. Weed control in tobacco is best achieved by utilizing all available methods of weed control in an integrated program. Under North Bihar situation interculturing with grubber and three tined bullock drawn cultivator a fortnight after planting followed by one weeding with khurpi (hand hoe) after first irrigation cut down the labour cost by 25 mandays per ha and enhanced the efficiency of cultivation by 26 per cent over the traditional practice of two hand weeding with khurpi alone.

Chemical Control

The use of herbicides in tobacco has become widely accepted by growers and offers several important benefits which supplement other components of the weed management program. Good early-season weed control with herbicides can be extremely important in reducing competition and allowing the rapid establishment of tobacco, resulting in more effective cultivation. Labor requirements for hand-hoeing and cultivation can be reduced by proper use of herbicides.

Chenopodium Album Annual Weeds

Herbicides may supply some insurance against fields becoming weedy during wet periods following transplanting. All weeds do not respond the same to all herbicides; therefore, the weeds expected in the field should be known when planning a weed control program. As tobacco is sensitive to most POST herbicides, PPI herbicides provide effective weed control. Herbicides commonly recommended before planting the seedlings in the main field are Pronamide (1.5-2.0), Fluchloralin (2-3) Isopropalin (1.5-2.0) and Isoxaben (1.5-2.5).

- To control weeds in nursery bed treat the soil when is moist. Application of methyl bromide (4-8 kg /100m^2) and cover with plastic film. Leave for 48 hours, then remove cover and aerate for 48-72 hours. (Controls most weeds including nematodes and soil borne insects and diseases except some hard seeded legumes.
- Under main field conditions weeds may be controlled by application of metachlor @ 0.75 kg / ha + one hand weeding will be needed to destroy previously germinated weeds. Pendimethalin @ 1.0 kg/ha can be applied to main field only for the control of emerged annual grasses.

Broom Rape Management

Tobacco crop is very often infested by broom rape also known as orobanche (*Orobanche cernua*) a phanerogamic, total root parasite. It is an endemic every year, graining epidemic proportion in certain year. It is very common in all tobacco growing tracts in India. It debilitates the plant to the maximum, resulting in stunted crop growth and very poor yields. It is reddish brown or yellowish brown unbranched herb growing to a height 60 cm with leaves scale like ovate to lanceolate, 6-20 mm long, acute, sessile and devoid of chlorophyll. The stem is pubescent. In presence of tobacco plant the seeds of orobanche germinate in the second week after tobacco transplanting.

Orobanche flower

Orobanche – in tobacco field

Seeds of orobanche can be germinated any time between Novembers to February. Orobanche can damage yield up to 60%

in severe cases. It is advisable that as soon as they appear in field collect them and burnout as this is quicker method of control of orobanche.

Fumigating with DMTT granules @ 300-500 kg/ha about 30-40 days before transplanting tobacco seedlings is effective. Pre planting application of EPTC @ 6.0 kg/ha (one month before planting) fallowed by post planting application of 2-4,5T; 2-4, D and 1% allyl alcohol has also been recommended for effective control . However, the treatment may not be feasible on large scale POST herbicides Glyphosate (300-400 g/ ha), MHT (1.5%o spray) have been promising. The most effective method in controlling *orobanche* is growing trap crops in rotation with tobacco.

Allelopathy and Weed Management in Tobacco

The word allelopathy derives from two Greek words. They are *allelon* which means "of each other" and *pathos* which means "to suffer". Allelopathy refers to the chemical inhibition of one species by another. The "inhibitory" chemical is released into the environment where it affects the development and growth of neighboring plants. "Allelopathy refers to any process involving secondary metabolites produced by plants, microorganisms, viruses and fungi that influence the growth and development of agricultural and biological systems". Allelochemicals refer mostly to the secondary metabolites produced. These substance can be present in any part of the plant viz., leaves, flowers, roots, fruits, or stems. They can also be found in the surrounding soil. Target species are affected by these toxins in many different ways. The toxic chemicals may inhibit shoot/root growth, they may inhibit nutrient uptake, or they may attack a naturally occurring symbiotic relationship thereby destroying the plant's usable source of a nutrient. Allelopathy provides basis to sustainable agriculture and has been increasingly recognized as an important ecological mechanism which influences plant dominance, succession, formation of plant communities and climax vegetation and crop productivity. Tobacco is known to have root allelotoxins. In general, they contain allelochemicals, are principal sources of toxicity, but sometimes they contain fewer and less potent or smaller amounts of allelochemicals than stem and leaves. Tobacco root exudates inhibited seed germination and seedling growth of maize, mustard and foxtail millet. Barley is excellent smother crop due to its extensive root growth and root exudates, which inhibited the germination and growth of tobacco, chick weed and shepherd's purse. However, its root exudates had no inhibitory effect on wheat plants. The root exudates from living plants contained the alkaloid *'gramine'* and were inhibitorier than aqueous leachates of dead roots, this proved active metabolic secretion of allelochemicals. Crop rotations remove the ill effects of soil sickness arising from continuous cropping of tobacco. The seed leachates of tobacco particularly during germination liberate inhibitors which may decrease or increase the seed germination and growth of amaranth.

Seed of *Orobanche cernua*

Allelopathic effects of tobacco is not always negative but its have some stimulant effects also particularly in case of *Orobanche.* Orobanche (*Orobanche cernua*) a phanerogamic, total host specific root parasite of tobacco. Tobacco has some stimulant effects on Orobanche, germinate after the tobacco root system is fully developed and secrets germinator stimulant for orobanche.

Alectrol Orobanchol

Chemical structure of germinator stimulant for *Orobanche*

Alectrol and ***orobanchol*** are two major germinator stimulant in the tobacco crop has been identified. Generally in Bihar tobacco is transplanted in October – November and Orobanche come in to picture during December January. This plant completes their life cycle before the harvest of main crop and dormant seeds remain in the fields which germinate again next year after tobacco.

Biotechnology Consideration in Weed Management

The potential applications of biotechnology in addressing selected issues related to weed management and herbicide technology are quite promising. Bioherbicides, naturally produced herbicides, and constructed microbes with novel derivative ability have been already developed and. used commercially. In addition, the concept of using genetic engineering techniques for the generation of herbicide resistance or tolerance in sensitive: crops have created great excitement.

The commercialization and marketing of the mycoherbicides DeYine® and Collego® illustrates the great potential of the use of phytopathogenic agents as bioherbicides to control selected weeds. At present, bioherbicides are viewed as complementary adjuncts to current weed management practices rather than as alternatives to chemical herbicides. In particular, the future potential of mycoherbicides is seen in areas that are currently served inadequately by chemical herbicides. According to Templeton et al. (1986) such areas include (1) control of parasitic weeds; (2) control of weeds closely related to crops (crop mimics), in which case a high degree of selectivity is necessary; (3) control of weeds resistant to chemical herbicides; and (4) control of weeds infecting small, specialized areas where development of chemical herbicides would be too costly. Advances in our knowledge of the interactions of plants and phytopathogenic organisms at the molecular level will increase our understanding of the role of toxins in pathogenesis, toxin production, and its genetic control by the pathogen. Such knowledge could lead to the commercialization of new bioherbicides or the synthesis of new chemical herbicides developed by means of a biorational design based on the chemistry of natural toxins. It is expected that in the near future, chemical herbicides would be developed by new approaches, based on the principles of biotechnology and chemistry, rather than by the currently dominant process of empirical synthesis coupled with biological screening. The commercial exploitation of the microbial toxin, L-phosphinothricin, which is the active ingredient of the microbial herbicide bialophos and of the chemically synthesized herbicide glufosinate, illustrates this point.

Plant cell cultures and fermentation by the microorganisms are expected to have a major impact in agriculture through the novel production of agrochemicals such as herbicides used in weed management. Additional advances are also expected in the area of biocatalysis benefiting from the use of plant or microbial cell enzymology, which is presently superior to the skills of organic chemists in performing complex biotransformation. Learning more about the plasmid-determined resistance of soil microbes to relevant herbicides will allow us to employ effectively such resistance in several microbial related processes such as the decontamination of soil residues of persistent herbicides or the development of microbial herbicide safeners. Advances in genetic technology will also facilitate our efforts to understand, manage, circumvent, and exploit the resistance of plants to selected herbicides. Isolation and characterization of plant or microbial genes coding for mutations altering the sensitivity of specific target proteins to herbicides or for herbicide-detoxifying enzymes will improve our understanding of the biochemical and genetic basis of plant resistance or tolerance to herbicides. Such information coupled with advances in recombinant DNA technology will enable us to engineer herbicide-resistant determinants and develop herbicide-resistant crop plants. At a theoretical level, progress in this area has been enormous. However, at the practical level, success has been only partial. Presently, only one herbicide-resistant crop, atrazine-resistant canola, developed from classic breeding techniques rather than sophisticated genetic engineering technology, is marketed in Canada. The development of tobacco and petunia plants resistant to the herbicides glyphosate and chlorsulfuron by means of gene transfer and transformation techniques indicates that genetic engineering of herbicide resistance is feasible. The full utilization of biotechnological procedures as tools for incorporating herbicide resistance to major agronomic crops (e.g., cereals and legumes) is presently limited by a number of unresolved problems such as a limited pool of genes of interest, lack of appropriate vector systems for gene transfer, and inefficient methods for the regeneration of selected crops from cell or tissue cultures. Future research, hopefully, will address and solve these problems.

For a more fruitful application of biotechnology in engineering herbicide resistance several considerations should be examined or reevaluated in future investigations. A given herbicide should be considered only when it more cost effectively controls specific problem weeds in a given crop better than other presently available herbicides. All biochemical mechanisms conferring plant tolerance or resistance to herbicides should be considered and exploited for practical application. If we are looking for resistance at the mode of herbicide action level, we need more information on the biochemical mechanisms involved and their genetic basis. There is a need for the development of better culture systems from which economic crops (e.g., corn and soybean) can be regenerated. The in vitro selection systems already available should be tested to determine whether plant improvement can be accomplished. In isolating plant mutants, learn from the plant breeders: isolate many mutants and evaluate their cost and effectiveness. There is a need for advances in technology for transferring resistant genes into plants: in addition to Ti plasmid vectors, new vectors for gene transfer to monocotyledonous plants or cell organelles (e.g., chloroplast) are currently needed. Herbicide resistance incorporated into a sensitive crop should not cause potential yield reduction or alteration of the quality of the marketed product. In achieving the.

Overall goal of engineering herbicide-resistant crops, a close cooperation between genetic engineers and plant breeders is imperative. Once the required techniques have been developed, it may be easier and less costly to transfer herbicide resistance into crop plants than to develop new herbicides. Biochemical and genetic advances may facilitate also the development of chemical additives that could regulate the activity of currently marketed herbicides. Such bioregulators could either potentiate the activity of a herbicide on target weeds (synergism) or reduce the activity of a herbicide on nontarget crops (antagonism or safening). The commercialization of tridiphane, a potent inhibitor of plant glutathione sulfotransferase enzymes, as a synergist of triazine herbicides on grass weeds and of safeners such as flurazole, which acts as a chemical regulator of the gene coding for the GST I enzyme and confers tolerance on corn against chloroacetanilide herbicides illustrates the practical ramifications of this approach.

Apart from main areas, weed management is emerging as promising fields for applying biotechnology, other areas of weed management or herbicide technology may be considered for such applications. Among them, the utilization of monoclonal antibodies and immunoassays in trace analysis of herbicide soil residues is becoming widely appreciated. Immunological assays offer considerable potential for rapid, inexpensive, and sensitive methods of analysis of agrochemicals or their metabolites and they could be used either as supplements or as alternatives to instrumental methods of analysis.

□□□

10

Insect Pests Management

The attentions of agricultural scientist to find out suitable measure for efficient tobacco production without disturbing the ecological balance and biological diversity as well as to keep the pest population below the economic injury level (EEL). Contribution to the damage due to insect pest is 20 percent of total damaged (18 per cent) caused by various pests viz. diseases, memals rodent etc. Attack of insect pests badly affect quality of produce (i.e. tobacco leaves) due to use of harmful pesticide to control/manage them. Integrated approach is best way to produce quality tobacco which can able to attract global market and fetch good price.

Tobacco crop is attacked by number of pest during its life cycle right from in nursery, main crop to storage pest, which pose main hurdles of for successful cultivation of tobacco crop.

A. Nursery Pests

As a preventive measure before sowing seed beds, chloropyrifos 1.5% dust is mixed @ 40 g per 10 sq. m. bed in the top soil to prevent damage to seeds and tiny seedlings from soil pests like ants, mole crickets, earthworms *etc*. The pests that infest in the nursery are

(1) Caterpillars and Cut-worms (2) Stem borer (3) Grass-hoppers and (4) Whitefly.

Tobacco Caterpillar (*Spodoptera litura* Fab)

It is the most important pest of tobacco in the nursery when the seedlings are small. Four species of Caterpillars *viz., Spodoptera litura* F., *Laphygma exigua* HB. *Agrotis ypsilon* Rott. and Plusia signata F. are known to infest tobacco nurseries. Of these 4 species *Spodoptera litura* F. commonly known as tobacco caterpillar is the most important pest. The caterpillars act as cutworms by cutting the tender stems. They voraciously feed along the veins of leaves when the seedlings are grown up. If left unchecked, they can defoliate the whole nursery in a few days.

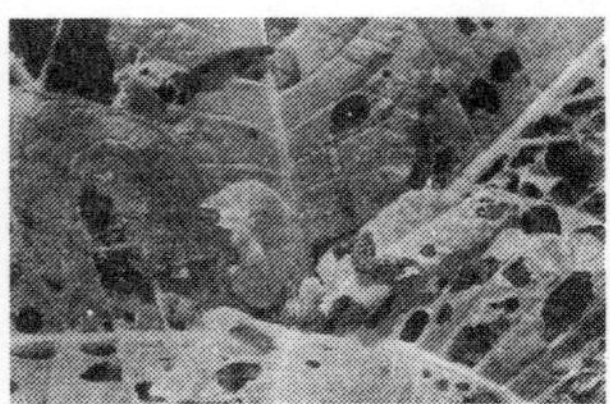

Tobacco caterpillar in nursery, Courtesy by CTRI

Control Measures

This pest can be controlled by spraying Endosulfan 35 E.C. at the rate of 30 ml or Carbaryl 50% WP at the rate of 50 g or Monocrotophos 40 WSC at the rate of 22 ml in 22 liters of water for 100 square meter area nursery, 30 to 45 liters of spray fluid is required. First spray should be given when the seedlings are three to four weeks old. Give another spray after 8-10 days. Spraying has to be repeated 8-10 days afterwards, if necessary.

Semi Looper (*Plusia signata* Fab)

The green caterpillars with looping movement are seen feeding on the underside of the leaves in the nursery beds.

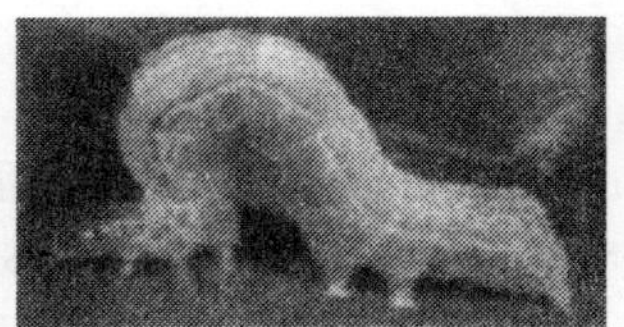

Semi Looper on tobacco leaf

Control Measures: Same as for tobacco caterpillar.

Stem Borer (*Gnorimoschema heliopa* Lower)

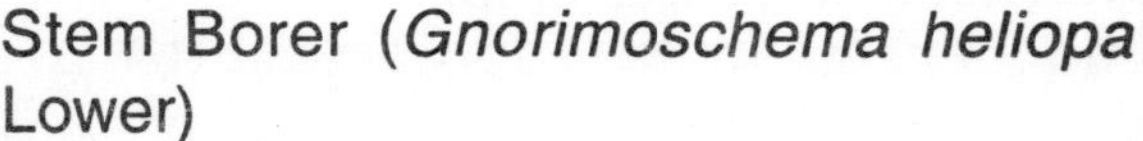

Stem borer are whit shed brown in colour and the moth are small and brown in colour. The caterpillar feeds by mining the through leaves and midribs and ultimately on the tissues inside the stem.

Damage done by Stem borer Courtesy by AINRP Tobacco

As a result, swelling appears where the borer stays. Swelling indicates the presence of this pest inside. It causes stunting and unusual branching of the seedlings. It is seen in the nursery mostly and spreads to the planted crop through the affected seedlings.

Control Measures

Same as for tobacco caterpillar, Final spray should be given just before pulling out of seedlings. Stem borer affected seedlings should be removed and destroyed.

Grasshoppers

These are green or brown jumping insects. There are 3 species commonly noticed in tobacco nurseries viz., (1) *Acrida exultata* Walker, (2) *Cyrtacanthacris tartarica* Linn. and (3) *Atractomorpha crenulata* F. They feed on leaves making circular holes.

Control Measures

Cultural practices like keeping the nursery area clean free of weeds and grass helps to keep away the grass hoppers. Spraying the surrounding vegetation with chlorpyrifos 20% E.C. in 10 liters of water or dusting the area with chlorpyrifos 1.5% dust @ 40 grams per 10 sq. m. area controls the pest.

B. Field Pests

The major pests that affect tobacco crop in the field are ground beetles, whitefly, leaf eating caterpillars, tobacco aphid/green peach aphid and capsule borer. Control measures for these major pests are detailed below

Ground Beetles (*Mesomorphus villager* Blancnch)

Ground beetles are the pest of main field when the fields are transplanted, ground beetles migrate from the bunds and attack freshly planted seedlings and cut the tender stems of seedlings at the base. When the cut is deep, the seedlings die causing gaps in the field.

Tobacco infested by Ground Beetles

Control Measures

Application of *Pongamia* cake powder or neem cake powder at the rate of 5 g mixed in a at handful of sand the base of each plant immediately after transplanting has been recommended for effective control of these beetles. Preplant soil application of aldrin 5% or chloredane dust @ 15-20 kg/ha. Spraying of Endosulfan 35 EC at the rate of 1.5 liters or Chforopyrifos 20% E,C, at the rate of 1.8 liters in 1100 liters of water per hectare at the time of transplanting may be done.

Green Peach Aphid (*Myzus persicae* Sult.)

They are polyphaugs pest These aphids are louse-like insects which are pinkish to brown or green in colour with three dark lines on the back of the abdomen. These small insects suck the plant sap with their tiny beaks. These are seen in hundreds in planted crop in the underside of the leaves. They develop very fast in to big colonies. By constant feeding of aphids on the sap. Aphid's secrets the honey dew on which sooty mould developed. The leaves look sickly and become unfit for curing. The quality and yield of tobacco are badly affected as they transmit Tobacco Mosaic virus by acting as carrier.

Tobacco leaves attracted by Green Peach Aphid
Courtesy by AINRP-Tobacco

Control Measures

Application of Endosulfan 35 EC @ of 500 ml per hectare in 500 liters of water or Phosphamidon 85 SL @ 125 ml per hectare or Oxydemeton methyl 25 EC @ 400ml per hectare in 500 liters of water.

White flies are seen in the lower side of tobacco leaves
Courtesy by CTRI

White Flies (*Bemisia tabaci*)

These sucking insect pest harm the crop by nymph and adult, suck the sap from the leaves and causes deformation of plants. Secondly adult transmit tobacco mosaic virus and leaf curl virus.

Control Measure

Application of systemic insecticides viz. endosulfan 35 EC @ of 500 ml or Oxydemeton methyl 25 EC @ 400 ml per hectare in 500 liters of water.

Tobacco Capsule Borer (*Helicoverpa armigera*)

Being a polyphagous pest the caterpillars feed on the tender shoots and inflorescence of the plant. It bores into the developing capsule and feeds on the immature seeds. They also feed on the top leaves, if the infestation is heavy.

Feeding view of Capsule borer

Control Measures

(1) Caterpillars can be controlled by topping the plants and destroying them.

(2) Spraying of any of the pesticides recommended for tobacco caterpillar on the flower heads helps in controlling this pest.

Storage Pest

Cigarette Beetle (*Lasioderma sericomi*)

The cigarette beetle, commonly known as the cigarette weevil, is a serious pest of stored tobacco. Cigarette beetle infests the cured leaf in storage. It attacks principal types of cigarette, cigar, chewing and snuff tobaccos as well as most forms of manufactured tobaccos. The larvae of the beetles feed on tobacco leaves making large galleries and when infestation is heavy, turn the tobacco into powder.

Life cycle of Cigarette Beetle

Control Measures

Since the beetle can breed even in small quantity of tobacco dust or waste, the godowns and surroundings should be kept free from tobacco bits and dust. Store tobacco leaves at 16-18°C. Tobacco leaves should be re dried. Fumigate tobacco with aluminum phosphide tablets at the rate of 20-30 tablets for 28 cubic meters for 96 hours. Keep the godown open for 72 hours after fumigation.

❑❑❑

11

Disese Pests Management

Economic produce of tobacco is leaves and most of disease infested on leaf portion which impaired the quality of produced, hence crop protection play vital role. Damage done by disease is to the extent of 25-30 percent of total loss on account of various pests. More than 20 diseases are reported on tobacco crops in India. 5-6 are very common on tobacco.

Tobacco due to its wider adaptability it can be grown in range of agroclimatic conditions in India and abroad as well, it is grown as rainfed and assured irrigate condition. Due to huge variation on the resistant quality of the varieties grown, there may be different pathogen found on tobacco in different areas. Tobacco is crop produced by tiny seeds; before seedling are shafted to main field, it is grown in the nursery with proper care for the period of 7-9 weeks. A number of disease attacking on the tobacco crops right from seedling stage in nursery to main field.

Damping Off (*Phytophthora palmivora*)

Damping off is a common disease occurring in tobacco nurseries. It causes extensive damage in the tobacco tracts. The damage in some areas can be hundred per cent, the absence of protective chemical sprays, the nursery seedlings are killed out entirely. The damage by this disease in some areas can be upto hundred per cent, the absence of protective chemical sprays, the nursery seedlings are killed out entirely. With only a few seedlings distributed sporadically in the nursery bed, the infection may spread rapidly and damp off the entire bed of seedlings within three or four days.

Damping off in nursery
Courtesy by AINRP-Tobacco

Control Measures

The disease is controlled by providing adequate drainage in the soil. In nurseries beds are raised about 15 to 45 cm. above the ground level, so that rain or irrigation water does not stagnate. The seed is sown thinly to avoid overcrowding of the seedlings, which favours the pathogen. Partial sterilization of the seed bed by burning trash on the surface destroys the pathogen on or near the surface. Spraying the nursery beds with 1 per cent Bordeaux mixture, Wetcop or Blitox at fortnightly intervals from about three weeks after sowing, is most-effective in controlling the disease. Ridomil 25 NP at 0.5 per cent gave good control of damping off in seed beds.

Collar Rot (*Scleroderma Rolfs*)

Just like damping off, sudden death of seedlings in patches is noticed in seed beds. Blackening of the collar region, wilting and rotting of leaves are the symptoms.

Control Measures

Spraying 0.4% Bordeaux mixture or 0.2% Ziram or 0.2% Blitox, 3 weeks after seed germination. This is to be sprayed twice or thrice at weekly intervals depending upon weather conditions.

Black Shank (*Phytophthora parasitica var nivcotianae*)

This is also a common disease of tobacco in many parts of India, particularly in the areas with heavy rainfall. The diseased plants show characteristic black lesions on the stem. Starting as minute black

Symptoms of Black shank on leaf

spots it spreads along the stem causing irregular patches, and often girdling it. The lesions also spread downward to cover the roots. The infected tissues shrink, leaving a depression, and as the disease advances, the stem shrivels and the plant wilts. In some cases the pathogen may attack the leaves, producing large water-soaked spots, which enlarge to blight the leaves. The disease may appear sporadically, affecting a few plants in the field but during rainy and humid weather it spreads rapidly to a majority of the plants in the field.

Control Measures

The plant residues and debris must be collected and burnt. Drenching the soil with Cheshunt compound, Perenox or other fungicides before planting, also helps to reduce infection. The plants may be sprayed with one per cent Bordeaux mixture three or four times, starting from about a month after planting and repeating at fortnightly intervals, so as to cover the basal stem thoroughly, to reduce the disease intensity. Adequate drainage to prevent water from stagnating in the field should be provided. BSF-1, 2, 3 and 4 are varieties resistant to the disease.

Frog Eye Spot *(Cercospora nicotianae)*

This is a common disease of tobacco reported from most tobacco growing tracts of India and from other countries. The damage caused, however, varies with the variety and climatic conditions prevailing in the tract. The lower and more mature leaves are often affected by the disease. The spots are brown with ash grey centres. Often the centres may turn white and dry up. The typical spot has a white centre, surrounded in succession by grey and brown portions, surrounded by a dark brown to black margin, resembling the eyes of a frog, hence the name. The spot size may vary from about 5 to 10 mm in diameter. One to over hundred spots may appear on each leaf blade and several spots may coalesce towards the leaf tip and margin causing the leaf to dry up from the margin and wither prematurely. Both the yield and quality of the tobacco are reduced by this disease.

Frog eye spot in nursery bed

Frog eye spot on leaf
Courtesy by CTRI

Control Measures

The disease is controlled by spraying the nursery. Seedlings and the main transplanted crop with one per cent Bordeaux mixture Perenox, or any other effective fungicide at 2-3 week intervals. During a rainy season the spray should be given more frequently.

Powdery Mildew (*Erysiphe cichoracearum* Var. nicotianae Comes)

This is a severe disease of tobacco in many countries. In India it occurs in almost every tobacco growing area, causing severe damage to the crop. The disease is characterized by the appearance of powdery, ash coloured growth on the leaf surface. First it appears in small patches about 5 mm in diameter, but these rapidly enlarge into irregular patches covering the major portion of the leaf blade. The infection appears on the lower leaf surface first, but soon spreads to the upper surface, and as the disease advances the leaves turn brown and the affected portions show a scorched appearance. The disease causes-severe reductions in quantity and quality of the yield.

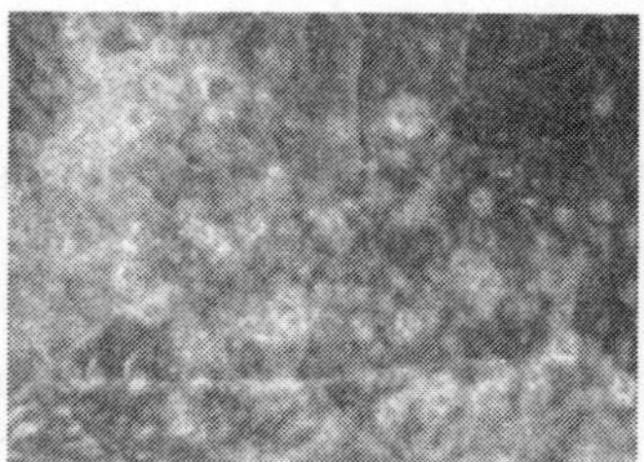

View of Powdery mildew disease
Courtesy by CTRI

Control Measures

Dusting finely powdered sulphur controls the disease @ 40 kg per hectare, which combined with the removal of older leaves at the base of the plant, helps to further check the disease. Usually sulphur is mixed with equal quantities of ash and applied to the soil between rows 4 to 6 weeks after planting, at the rate of 100 to 120 kg per hectare. Overcrowding the plants in the field should be avoided and balanced fertilizers, low in nitrogen, should be applied to reduce primary infection.

Tobacco Mosaic Virus (*Marmor tabaci*)

Tobacco mosaic, the first recognized virus disease of plants, is worldwide in distribution. It is most common on tobacco crops in India, appearing in every tobacco tract of the country. The same virus occurs on many other host plants and over 20 different virus diseases occur on tobacco. The mosaic disease is the most important. The first-

Tobacco mosaic virus-Field view

symptoms are light discolouration along the veins of the youngest leaves. Soon the leaves develop characteristic light and dark green pattern, the dark green areas usually associated with the veins. This typical mosaic symptom is sometimes accompanied by a blistered appearance of the leaf, due to more rapid growth of the dark green tissues. The virus is easily transmitted in the field by mechanical means and by wind and water. Various cultural operations cause wounds on the plant, which become the portals of entry of the virus. Such operations as topping or clipping the shoots, spreads the virus from diseased to healthy plants. Even the workers in the field may transmit the virus from chewing tobacco and snuff, to the standing crop. Thus, their is unlimited scope for the virus to set in and cause damage to the crop.

Control Measures

Because of the various physical and biological properties, which impart highly resistant qualities to the virus, and also because of its varying modes of spread, it becomes almost impossible to control the disease. The only method seems to be to grow resistant or at least tolerant tobacco varieties and practice field sanitation.

Tobacco Leaf Curl (*Ruga tabaci*)

Field view of tobacco mosaic virus

This virus disease is widespread in India and occurs in severe forms; affecting up to 15 per cent of plants .The disease usually appears in the field about 4-6 weeks after transplanting. It is characterized by the downward curling of young leaves, leaf margins turn downwards and come together at the bottom, exposing the middle upper surface of the leaf blade. The usually thickened leaf blade exhibits vein clearing symptoms. As the disease advances the plant becomes dwarfed and most of the leaves are curled. The inflorescence is greatly condensed and the veins thickened and turned green.

Control Measures

The plants must be sprayed with systemic insecticide regularly to control the white fly vector, starting in the nursery and continuing in the field, at 10-15-day intervals.

Root Knot (*Meloidogyne javanica)*

This is a common disease of tobacco in many areas and the disease seems to be on the increase. It is particularly severe in irrigated crops and in soils where tobacco and other related plants are grown repeatedly in rotation on an intensive scale. The disease is characterized by knotting of roots, stunting and chlorosis of the plant and reduction in crop yield, as is common for similar root-knot diseases on other crops. It is caused by the root-knot nematode *Meloidogyne incognita* and by *M. Chitwood* in some places.

Tobacco roots infested by Root knot

Control Measures

The disease is effectively controlled by crop rotation and by fumigating the soil with a nematicide such as D.D. and methyl bromide at least a fortnight before planting. Furadon 3G, Mocap lOG and Hexnema 5G at 4-kg/ha.give effective controls. The variety NC 95 is resistant to M. *incognita.*

Leaf Blight *(Ascochyta phaserum* Sacc)

The disease can occur on all stages of plant growth. The spots are circular with grey to brown centres. Centre of spots may fall out giving a ragged leaf spot appearance. Scattered black pycnidia may be found in the centre of older lesions. Spots may coalesce to form large dead areas.

Leaf blight in nursery stage

Control Measures

Sow only disease free seeds from healthy fields. Collect and destroy the crop debris after harvest. Transplant only healthy seedlings or remoye the spotted basal leaves from seedlings before transplanting and destroy them. Spray nursery beds with Dithane Z- 78 @ 2.5 g liter of water at 10- 12 days intervals. Spraying the crop with Bavistin or Benlate @. 2.5 g /10 liter of water or Cercobin (0.1%) using 500-600 liters of water per hectare gives good control of the disease.

Antharacnose (*Colletotrichum tabacum*)

All the above ground parts of the seedlings are attacked. The young leaves have small, irregular, whitish spots which coalesce to form bigger patches. The larger patches are oval in appearance and have a dark brown centre. Affected leaves become wrinkled, distorted an5l finally dry. On stem, elongated brown lesions are seen girdling the stem. Affected plants turn pale and collapse. The fungus survives in the infested crop residues in the soil.

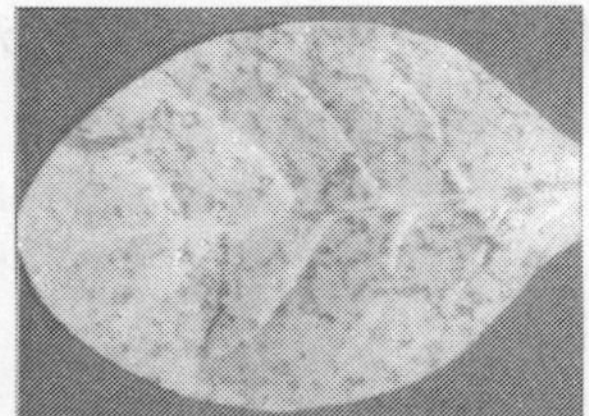

Tobacco leaf infested by Antharacnose Courtesy by CTRI

Control Measures

Rabbing (burning the soil surface with paddy husk or dry twigs) seed bed before sowing minimizes the pathogen population in the soil. Drenching the seedbeds with Daconil/ Fytolan/ Blitox-50 @ 2.g/ liter of water two days before sowing and spraying of Dithane Z- 78/ Difolatan/Cuman @ 2g/litre of water will reduce the disease infestation. During wet weather spray the corps twice a week immediately after rainfall.

Brown Spot (*Alternaria alternata* (Fr.) Keissl)

This disease occurs in the field conditions. The spots appear on lower leaves as small light brown lesion with dead tissue in the centre. The spots are marked by the concentric rings. Under favorable wet conditions the spots appear on upper leaves, petioles and even on stalks with typical concentric brown rings. The fungus survives as mycelium in woody plant residues like tobacco stems.

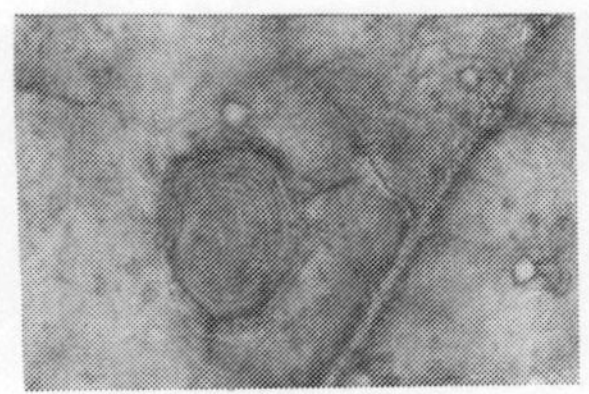

Brown spot on tobacco leaf Courtesy by AINRP-Tobacco

Control Measures

Sow only disease free seeds from healthy fields. Collect and destroy the crop debris after harvest. Transplant only healthy seedlings or remove the spotted basal leaves from seedlings before transplanting and destroy them. Spray nursery beds with Dithane Z- 78 @ 2.5 g/ liter water at 10- 12 days intervals. Spraying the crop with Bavistin or Benlate @. 2.5 g/10 liters of water per hectare gives good control of the disease.

Bacterial Wilt (*Pseudomonas solanacearum* E.F Smith)

The lower leaves droop and wilt. Often only a half leaf wilts. If the disease progresses slowly, affected leaves become lighter green and may gradually turn yellow. In case of severe infestation the whole plant wilts and dies within a few days. Dirty white slimy bacterial ooze exudes when the stem is cut in the form of glistening beads. Roots become dark brown to black and rotted. The bacterium survives in the soil.

Devastating view of Bacterial wilt
Courtesy by AINRP-Tobacco

Control Measures

Growing of resistant varieties is the best way to control the disease preferably transplanting should be done only with disease free seedlings. Follow suitable crop rotation for 4-5 yeas. Practice good nematode control. The crop debris should be collected and burnt.

□□□

12

Drought Management

The growth and development of tobacco plants are adversely affected by a range of abiotic stresses which include drought, desiccation, high salt, extreme temperature and heavy metal toxicity etc. leading to substantial reduction in crop productivity. Improvement of tobacco plants for the tolerance of abiotic stresses has been the priority area; however introgression of complex traits often limited due to their complex and polygenic nature. Identification of numerous genes and putative functional transcripts involved with threshold tolerance abiotic stresses has given new diversion to the rational engineering of crop plants against the abiotic stresses.

Tobacco is grown under a wide range of climates but it has been observed that tobacco cannot tolerate water logging at any stage of growth but at the same time the plants may not withstand drought also. Therefore, a compromising situation would be to give light and frequent irrigations to the crop. Drought is a common adverse environmental condition that seriously affects crop productivity. Dehydration stress is one of the most serious yield-reducing stresses in agriculture. Drought stress causes an increase in solute concentration in the environment, leading to an osmotic flow of water out of plant cells. This in turn causes the solute concentration inside plant cells to increase, thus lowering water potential and disrupting membranes along with essential processes like photosynthesis. These drought-stressed plants consequently exhibit poor growth and yield. In worst scenarios, the plants completely die. Certain plants have mechanisms to survive under low water conditions. Most of these adaptations to drought have disadvantages. A genotype of short duration usually yields less compared to that of normal duration. The mechanisms that confer drought resistance by reducing water loss (such as stomatal closure and reduced leaf area) usually result in reduced assimilation of carbon dioxide. Osmotic adjustment increases drought resistance by maintaining plant turgor, but the increased solute concentration responsible for osmotic adjustment may have detrimental effect in addition to energy requirement for osmotic adjustment. Consequently, crop adaptation must reflect a balance among escape, avoidance and tolerance while maintaining adequate productivity. These drought resistance mechanisms have been grouped into three categories, viz. drought escape, drought avoidance and drought tolerance. Generally crop plants use more than one mechanism at a time to resist drought.

Drought Escape

Drought escape is defined as the ability of a plant to complete its life cycle before serious soil and plant water deficits develop. This mechanism involves rapid phenological development (early flowering and early maturity), developmental plasticity (variation in duration of growth period depending on the extent of water-deficit) and remobilization of preanthesis assimilates to grain.

Drought Avoidance

Drought avoidance is the ability of plants to maintain relatively high tissue water potential despite a shortage of soil-moisture, whereas drought tolerance is the ability to withstand water-deficit with low tissue water potential. Mechanisms for improving water uptake, storing in plant cell and reducing water loss confer drought avoidance. The responses of plants to tissue water-deficit determine their level of drought tolerance. Drought avoidance is performed by maintenance of turgor through increased rooting depth, efficient root system and increased hydraulic conductance and by reduction of water loss through reduced epidermal (stomatal and lenticular) conductance, reduced absorption of radiation by leaf rolling or folding and reduced evaporation surface (leaf area). Plants under drought condition survive by doing a balancing act between maintenance of turgor and reduction of water loss.

Drought Tolerance

The mechanisms of drought tolerance are maintenance of turgor through osmotic adjustment (a process which induces solute accumulation in cell), increase in elasticity in cell and decrease in cell size and desiccation tolerance by protoplasmic resistance.

Physiological Adaptations to Avoid Water Stress

Water stress avoidance of a plant is the ability to maintain a favourable water balance and turgidity even when they are exposed to sever drought conditions. Thus the drought forces the plants to bring about certain changes in the physiological process which result into changes of morphological and anatomical characteristics of plants A favourable water balance in drought affected plants can be achieved by following two mechanisms:

(i) Mechanism for conserving water in plants
(ii) Mechanism for improving water uptake from soils.

(i) Mechanism for Conserving Water in Plants

The drought affected plants develop certain changes in the anatomical characters of plants for an effective management of the cell sap against an accelerated rate of transpiration due to rise in plant body temperature. These changes can be descrived as under.

(a) Stomatal Mechanism

In certain plant types, it has been found that they open their *stomata* only for few hours early during morning when moisture stress are minimum and they close the stomata for rest of the day hours which result in an efficient photosynthesis with least loss of water. These types of plants survive for longer period under drought than those which open their stomata *late* in the morning and keep them opened for rest of the day hours that favours more water *loss* due to transpiration. The stomatal mechanism of opening and closing differs greatly from crop to crop or even within the varieties of the same crop. However, the mechanism of closing of stomata or stomatal regulation of water loss is not operative beyond a certain thresh-hold level of water content in plants and beyond this level of water the closure of stomata would lead to reduced photosynthesis which may result into drought induced starvation injury and subsequently in death of the plants.

(b) Increased Photosynthetic Efficiency

Closure of stomata to reduce the water losses through transpiration (low water availability results in a lower rate of photosynthesis and best possibility to overcome this problem is an increased CO_2 assimilation by plants for given stomatal opening. The plant types CO_2 assimilation capacity (C_4 group of plants) like maize, soybean, finger millet, grasses etc. *Cyeedon dectylon.* sudan grass have capacity to fix most of the C_4into the form of organic acid like malic acid, aspartic acid etc. and convert it into carbohydrate durng the day -time, these called C_4 pathway. The carboxylating enzyme: Which has a high potential activity and a high affinity for CO_2, helps in a high rate of CO_2 fixation and maintenance at relatively restricted stomatal opening? These physiological changes use to inhibit the photorespiration and thereby the net photosynthesis is enhanced.

(c) Position of Stomata and Its Functioning

The stomata present on the leaf surface or on slightly elevated position come in direct contact with hot air current and transpire more water for the protection of plant's life from desiccation but stomata situated in cavities or in depressions use to retard the transpirational loss of water from the leaf surface. The duration of the stomatal opening also governs the extent of water transpired from the leaf surface. The plants, affected by drought, use to bear less nummer of stomata, which are located either in depressions (in cavities) and they

(ii) Mechanism to Accelerate Water Uptake by Roots

If the plants bear well developed and extensive root system then they can withdraw water from relatively deeper layer of the soil but contrary to this the shallow rooted plants are often victim of *drought.* The mechanism of water uptake totally depends upon the root system of the plants which may be described as under:

Effective Root System

Roots are one of the efficient adaptations of plants to dry condition. Roots when well develop with full laterals and extensive root hairs are more beneficial to fight the drought. Roots are good indicator for the drought resistance however the water uptake from the soil partially controlled by the Presence of salts concentrations in the soil solution.

High Root to Top Ratio

As described above that deeper the roots better is the drought resistance in plants. The water movement to deeper layers in soil results in more root elongation and hence the root growth rate exceeds considerably from that of the stem. This type of growth results in better uptake of water because of deeper roots and also in reduction of water loss through transpiration as volume of the top or shoot remains lower with less number of stomata and cuticles per plant.

Differences in Osmotic Potential of the Plants

The water movement in plants depends upon the osmotic pressure. Draught affected plants withdraw water from soil even after wilting point. therefore, plants grown under dry conditions use to have high osmotic pressure and they, survive for relatively longer period.

Development of Water Economy in Plants

Plants use to transpire water with normal speed under regular water supply but a restricted water supply or reduction in availability of water to plants results in a reduced rate of transpiration by plants. Thus the water spenders become as reserves to fight better with drought. This change comes automatically in the plants and drought resistance is developed in plants.

Physiological Adaptations in Plans to Induce Drought Tolerant

When plants are exposed to moisture stress conditions they develop tolerance to drought for mitigating the actual stress:

Mitigating Moisture Stress

Some plants, when exposed to drought conditions, develop adaptation by marinating high turgor or the internal water potential even under very high drought. Thus by maintaining the turgor meseplants avoids secondary drought induced stress caused by dehydration. These drought affected plants behave in the following two ways for mitigating drought:

Resistance to Dehydration

The morphological adaptations of drought affected plant help in accumulation of sufficient solutes which help in maintaining highly cell turgor pressure. Thus the plants keep on growing in normal way even under severe drought conditions.

Prevention of Leaf Collapse

The drought induced morphological adaptation like thick cell wall, sunken stomata, lipid or waxy coating over leave by reducing the transpirational loss of water. With further fall in soil moisture status the period of stomatal opening is reduced and other morphological adaptations to reduce water loss become operative. Thus the leaf collapse is avoided and plants are saved.

Resistance to Metabolic Setback

The drought tolerance of a plant depends upon the ability of the stomatal guard cells to remain turgid even when the leaf cells are wilted, which is possible due to decrease in the solution in the guard cells. The dehydration compensation point (it is water potential at which the volume of CO_2 evolved during respiration and CO_2 absorbed during photosynthesis becomes equal) is reduced with increase in drought tolerant plant. At dehydration compensation point the photosynthesis is completely stopped and subsequently the water loss is also reduced. Under this stage opening of stomata would be harmful. The plants which reduce their metabolic activity below dehydration compensation point would need lower carbohydrate and hence they would be more droughts tolerant at lower net assimilation rates. These plants remain alive and do not starve suddenly. The protein loss is avoided by drought resistant plants by two important processes like decreased rate of breakdown which attained due to prevention in the loss of RNA and the plants continue to synthesis of protein, more prolene protein breakdown and these plants become more droughts tolerant and keep plant growing under moisture stress conditions. Under this stage opening of stomata would be harmful. The plants which reduce their metabolic activity below dehydration compensation point would need lower carbohydrate and hence they would be more droughts tolerant at lower net assimilation rates. These plants remain alive and do not starve suddenly.

Escaping the Drought

This is directly related to the life span of a plant. The plants with shorter span of life or the plant of shorter duration can escape drought but with the longer duration the plants would be caught during some or the other growth phase.

Classification of Plants According to Their Capability to Resist Drought

Plants differ greatly in their ability to face the drought and those which can tolerate or resist drought can keep on growing while others die. Based on their resistance power to drought conditions the plants may be classified as under.

Ephemerals

In real sense these pants are not true drought resistant types. Under semi-arid regions the seed of such plants germinate and the plants complete their life cycle within a few weeks time. The new crop of seeds survives the inter dry period until rainy season comes. Such plants are termed drought escaping or ephemerals.

Succulents

Succulents contribute a considerable proportion of the vegetation of the semi-arid regions and are frequently found in locally dry habitats. The most conspicuous succulents belong to the cactus family (*Cactaceae*). The other important family is *Euphorbiaceae, Liliaceae*, etc. which include a number of such plants. These succulents are of a distinct group of plants not only in structure but in metabolism and water economy. Such plants are able to survive dry periods because of the relatively more reserves of water accumulated in the inner tissues of the fleshy stems or in the fleshy leaves.

Drought Enduring

Neither the ephemerals nor the succulents can be regarded truly drought resistant in the sense that their cells can ensure a severe reduction in water content for extended period of time without injury but the plants of drought enduring group can only resist the injuries caused by drought. The example of such among higher plants is creosote bush (*Lerrea tridentata*) which is dominant plant throughout large areas in the semi-arid regions.

Understanding Drought Mechanisms

To develop novel phenotypes that are conferred by drought stress root response, termed drought rhizogenesis, as a developmental adaptation to drought stress. Genetic screens using Arabidopsis were devised and the DIG (for Drought-Induced rhizo Genesis) loci that control drought rhizogenesis were defined. Characterization of the dig mutants revealed that these mutants also exhibit altered drought stress tolerance, indicating that drought rhizogenesis is intimately linked to drought adaptation of the entire plants and can be used as a trait for researchers to access the elusive drought tolerance machinery. Molecular cloning of several DIG loci revealed that certain chloroplast-targeted proteins are of critical importance to drought tolerance. Furthermore, cell metabolites are found to function as signals in regulating drought stress response and drought tolerance. There are multiple mechanisms for drought adaptation, which is consistent with the notion that plant drought tolerance, is a complex and multigene trait. Discovery of these novel drought tolerance determinants will be very useful to understand drought tolerance mechanisms and to breed or bioengineer drought resistant crop plants.

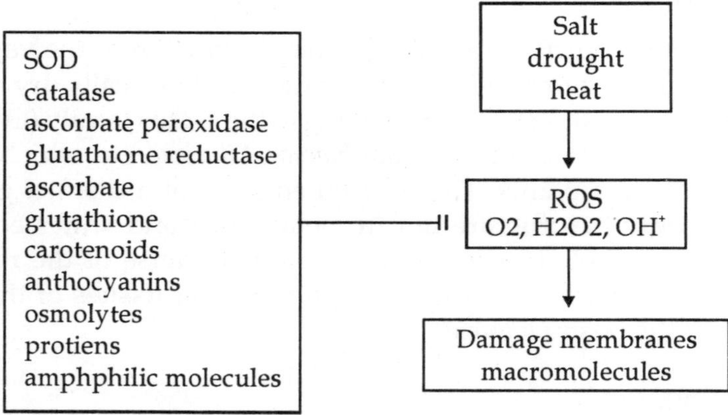

Fig. 12.1 Reactive oxygen species pathway. (Source: http://dragon.zoo.utoronto.ca-B03T0301D/: Drought Tolerance in Agriculture)

Pathways related to drought and other stresses are reactive oxygen species (ROS), which are molecules formed by the incomplete one-electron reduction of oxygen. Under stress, ROS formation is usually exacerbated. Under water deficit conditions, drought stress leads to the disruption of electron transport systems; the main sites of ROS production in the plant cell are organelles with highly oxidizing metabolic activities or with sustained electron flows: chloroplasts, mitochondria, and microbodies. ROS are generally damaging to essential cellular components, and plants have evolved various ROS scavenging mechanisms. These include the enzymes superoxide dismutase (SOD), catalase, and peroxidases, as well as oxidized and reduced glutathione.

Tobacco Adaptation to Drought Stress

Abscisic Acid (ABA)

Although our understanding of plant tolerance mechanisms to these stresses is incomplete, the plant hormone abscisic acid (ABA) is believed to be an essential mediator between environmental stimulus and plant responses. The plant hormone ABA levels increase when plants are subjected to drought, salt, or temperature stress, and these increased levels have been shown to be essential for stress tolerance. ABA is a sesquiterpenoid hormone that plays an important role in many physiological processes, including seed development and dormancy, vegetative growth, and responses to environmental stress. Modulation of ABA levels prior to the onset of stress also provides plants with better stress tolerance. ABA is synthesized by the cleavage of a C40 carotenoid precursor followed by a two-step conversion of xanthoxin via an ABA-aldehyde Several Arabidopsis genes involved in ABA biosynthesis have been identified. One of these is NCED3, a gene encoding a 9-cis-epoxycarotenoid dioxygenase (NCED) that catalyzes production of xanthoxin, a key intermediate in ABA biosynthesis. The recent isolation of an Arabidopsis thaliana mutant, era1, is hypersensitive to ABA and has been shown to also be tolerant to conditions of water deprivation. ERA1 has been identified as a. beta. subunit of farnesyl transferase. Farnesyl transferase is a heterodimeric enzyme that provides the specific addition of a farnesyl pyrophosphate moiety onto the substrate target sequence. The target sequence is defined as a sequence of four amino acids which are present at the carboxy terminus of the protein and is referred to as a CaaX motif in which the "C" is cysteine, "a" is any aliphatic amino acid and "X" is any amino acid. The. alpha. subunit is common with a second prenylation enzyme, geranylgeranyl transferase, that has a different beta subunit and adds a geranylgeranyl isoprenyl pyrophosphate moiety to the target sequence.

Cytokinine (CK)

The impact of water deficit progression on cytokinin (CK) levels was followed in upper, middle and lower leaves and roots of *Nicotiana tabacum*. As drought progressed, the bioactive CK content in leaves gradually decreased, being maintained longer in the upper leaves of all tested genotypes. Under severe stress (11 d dehydration), a large stimulation of cytokinin oxidase/ dehydrogenase (CKX) activity was monitored in lower leaves, which correlated well with the decrease in bioactive CK levels. This suggests that a gradient of bioactive CKs in favour of upper leaves is established during drought stress, which might be beneficial for the preferential protection of these leaves. During drought, significant accumulation of CKs occurred in roots, partially because of decreased CKX activity. Simultaneously, auxin increased in roots and lower leaves. This indicates that both CKs and auxin play a role in root response to severe drought, which involves the stimulation of primary root growth and branching inhibition.

Vitamin C

Vitamin C provides link to a plant's ability to withstand drought conditions Researchers at the University of California–Riverside have found yet another way to improve a crop's ability to survive drought, raising hope that more food can be grown in arid regions of the developing world. This discovery will assist farmers who depend on rainwater for their crops during those years when rainfall is low and should help farmers who grow crops in arid areas such as exists in many third-world countries. Reducing a tobacco plant's ability to recycle vitamin C causes it to scale back the amount of water that escapes from its leaves. That, in turn, allows the plant to better survive drought conditions. Tobacco was used in the study because it's very sensitive to drought, but "our discovery should be applicable to most if not all crop species, as the role of vitamin C is highly conserved among plants Improving plants' ability to survive drought has long been a Holy Grail for plant researchers. And the UC–Riverside discovery is just one of several promising biotech breakthroughs.

Nitrate Reductase

Over expression of Nitrate Reductase in tobacco delays drought-induced decreases in nitrate reductase activity and mrna1 were recorded in transformed tobacco (*Nicotiana plumbaginifolia* L.) plants. Tobacco plant constitutively expressed nitrate reductase (NR) in comparison to untransformed controls when subjected to drought for 5 days.

Genetic Engineering

In addition to morphological and physiological changes, biochemical change involving induction of compatible solute biosynthesis is one way to impart drought. Under drought, plants try to maintain water content by accumulating various solutes that are nontoxic and do not interfere with plant processes and are, therefore, called compatible solutes. Some of them are fructan, trehalose, polyols, glycine betaine, proline and polyamines.

Genetic engineering has been successfully applied to identify and transfer different genes responsible for biosynthesis of compatible solute (metabolites) from different organisms to crop plants through a targeted approach. Several gene transfer approaches have been employed to improve the various stress tolerance in tobacco plant, which involve either product of an effected gene or a regulatory gene. There is a need of a comprehensive approach to pyramid a number of genes involved in the stress tolerance or simultaneous regulation of many stress responsive genes by using single gene encoding stress inducible transcription factor, so as to enhance tolerance towards more complex stresses including drought, salinity, temperature, and freezing for increasing the crop productivity. Further investigation is the need of the hour to have better understanding of genetic control of morphological and physiological traits contributing to drought resistance. The gene TPS1 found in yeast encodes for trehalose-6-phosphate synthetase and is involved in biosynthesis of trehalose. Tobacco has been transformed with the yeast TPS1 gene. By determining the water loss of detached leaves or by determining the effect of withholding irrigation on the death and damage of leaf, it has been shown that the transgenic plants have increased drought resistance. Another gene, P5CS, encodes for pyrroline-5- carboxylate synthetase which is involved in praline synthesis, and the over-production of praline confers drought resistance. Transgenic tobacco over-expressing P5CS gene transferred from mothbean exhibited a high level of enzyme and produced 10–18-fold more *proline* than control plant. The over-production of proline enhanced root biomass and flower development under drought condition. The bacterial gene SacB found in *Bacillus subtilis* encodes for levan sucrase, which takes part in fructan synthesis. When this gene was transferred to tobacco, the transgenic plant produced fructan and showed better performance in comparison to control under PEGmediated drought condition59. The genes betA encoding for choline dehydrogenase and betB encoding for betaine aldehyde dehydrogenase are involved in the biosynthesis of glycine betaine and accumulation of glycine betaine confers drought resistance. Scientist transferred betB gene from *Escherichia coli* to tobacco. With the availability of genes responsible for polyamine biosynthesis such as ADC (encodes for arginine decarboxylase), ODC (encodes for ornithine decarboxylase) and SAMDC (encodes for S-adenosylmethionine decarboxylase), it is now possible to manipulate polyamine content using sense and antisense constructs of these genes in transgenic plants. Transgenic tobacco plants with ODC gene from yeast61 and mouse62, ADC gene from oat63 and SAMDC gene from human64 have been reported, but sufficient studies have not so far been carried out to observe whether the transgenics show any tolerance to drought. Only expression level of polyamines in plants has been studied. However, SOD (superoxide dismutase) gene from pea65 has been transferred to tobacco and transgenics were found to be drought resistant. Some of the Successful introduction of genes for developing abiotic stress tolerance is listed in Table 12.1.

Table 12.1 Successful introduction of genes for developing abiotic stress tolerance in tobacco

Sl No.	Gene/source/ gene product	Product synthesized	Stress tolerance
Compatible solutes			
1	*mtLD/E.coli/ Mannitol-l-phosphate dehydrogenase*	Mannitol	Salt stress
2	*bet A/E.coli /Choline dehydrogenase*	Glycinebetaine	Drought
3	*pScs/Vigna Aconitifolia/ Pyrroline S-carboxylate synthase*	Proline	Drought
4	*tps /A. thaliana/ Trehalose 6-phosphate synthase*	Trehalose	Drought and salinity
5	*osm-I-osm4*	Osmotin	Drought, salinity
6	*Apple, s6pdh, (Sobitol] -6 hosphate dehydrogenase)*	Sorbitol	Oxidative stress salinity
Scavenging enzymes			
1	*SOD/(Cu/Zn)/Pea/Superoxidase dismutase*	Superoxide dismutase	Chilling and drought

Transgenic plants can exhibit a greater rate of growth under drought conditions, or may exhibit enhanced drought-recovery capabilities in conditions that involve a drought period followed by re-hydration. Transgenic plants can have an altered phenotype as compared to a corresponding control plant that either lacks the transgene or does not express the transgene. A polypeptide can affect the phenotype of a plant (e.g. a transgenic plant) when expressed in the plant, e.g., at the appropriate time(s), in the appropriate tissue(s), or at the appropriate expression levels. A phenotypic effect can be an altered growth rate under a stressful environmental condition relative to a control plant. For example, a transgenic plant can exhibit a greater rate of growth under drought conditions. A transgenic plant can exhibit a greater rate of growth following re-hydration immediately preceded by drought. The transgenic plant can exhibit a greater rate of growth under drought conditions relative to a corresponding plant of the same genetic background that lacks the nucleotide sequence encoding the polypeptide. The transgenic plant can exhibit enhanced drought-recovery relative to a corresponding plant of the same genetic background that lacks the nucleotide sequence encoding the polypeptide. Transgenic plants also may exhibit a lower transpiration rate compared to control plants of the same genetic background. Transpiration rate is another physiological parameter that is indicative of how well a plant can tolerate drought conditions. Plants with a low transpiration rate are expected to lose water more slowly than plants with higher transpiration rates and therefore would be expected to better withstand drought conditions (i.e., have better drought tolerance). When a polypeptide described herein is expressed in a transgenic plant, the transgenic plant can exhibit a transpiration rate that is reduced by about 1 to 100% as compared to the transpiration rate in a corresponding control plant.

Tobacco Varieties cv. Bel B & cv. M51 - with Increased Drought Tolerance

Transgenic tobacco cultivar viz., cv. Bel B and cv. Has been identified and rrlease in USA for cultivation with increased tolerance to drought condition. The M51P5CSF129A cDNA and the nptII marker gene were used for tobacco (*Nicotiana tabacum* L. cv. Bel B and cv. M51) transformation via *Agrobacterium tumefaciens* strain LBA4404. Twenty transformed tobacco plants were obtained after transformation of leaf discs. Presence of the transgene was confirmed by polymerase chain reaction (PCR) analysis. Physiological responses to water stress were compared in transgenic and wild-type tobacco plants. Transgenic plants of both cultivars accumulated high levels of free proline. They did not exhibit dry mass relocation or chlorophyll content reduction. Neither precocious senescence, nor leaf necrosis or morphological changes were observed in control and stress conditions (RWC decrease by 7–8%). Transgenic plants with elevated accumulation of osmoprotectants seem to be better adapted to water stress, providing a perspective for future research of stress effects that have a principle role in the functional activity of plants. This study confirmed P5CSF129A to be a candidate gene in crop engineering for enhanced water stress tolerance.

□□□

PART-III
Crop Improvement

13

Botany and Genetics of Tobacco

For any successful crop improvement programme knowledge of systematic botany and its genetics is utmost needed, fortunately tobacco is one of the crop which botany and genetics has been studied extensively by several worker since the deispelling started working. The breeding programme to evolve noble genotype has been started well in time during eighteen century. In this chapter focus has been concentrated on genetic and botany of this important crop.

Origin and Distribution

Tobacco belongs to Genus *Nicotiana,* A genus of large herbaceous annuals or perennials, most of them natives of North and South America and a few of Australasia. Two economically important species viz., Nicotiana *tabacum* is believed to be originated from South America and that of *N. rustica* is believed to be *Peru.* The red Indians use to inhale its smoke from burning leaves. Presently it is grown in subtropical and temperate regions of the world. It is grown from 60°N lalitude in Sweden to 40°S latitude in New Zealand, China, India and USA. An species of plants in the genus *Nicotiana,* or the cured leaves of several of the species, used after processing in various ways for smoking, snuffing, chewing, and extracting of nicotine. Only two species *Nicotiana tabacum* and *Nicotiana rustica* are cultivated extensively in India. *Nicotiana tabacum,* which is known as desi type covers the largest area whereas rustica (known as Vilayati and Calcuttia) is limited to Punjab, Uttar Pradesh, Bihar, West Bengal and Assam. The *tabacum* types are used for cigarette, cigar, cheroot, bidi, hookah, chewing and snuff but rustica varieties mostly used for hookah, chewing and snuff purpose. Indian tobaccos are classified into two species, namely *N. tabacum* and *N. rustica.* Howards described 20 types in *rustica* and 51 in *tabacum* of which 5 are described as petiolate and 64 as sessile. Later on 19 types added to those of Howards. Morphological characters were used as a basis for the classification of this species.

Classification

The Genus *Nicotiana*

The genus *Nicotiana* is considered to have evolved from an ancestral reservoir of pre-*Nicotiana* elements with the basic chromosome number n=6, which evolved along two lines to give rise to a Cestroid complex and a Petunioid complex. The high degree of compatibility shown by even widely separated species of Nicotiana suggests that natural hybridization may have been an important factor in speciation. The present preponderance of species with n= 12 (28 out of 56 species investigated) and a hence of species with n = 6 is said to be consistent with the hypothesis that allopolyploid species have a higher survival potential in changing habitats than their diploid parents. The genus includes about 60 species, of which 30 are restricted to S. America and 9 to N. America.; 6 species are common to both. About 15 species occur in Australia and the S. Pacific. The tobacco genus, *Nicotiana* has been divided by earlier workers into four sections: *Tabacum, Rustica, Petunioides, and Polidiclia.* Latter two sections have been combined to from *Petunioides* by East in 1912. East conclusions were reached by crossing *N. bigelovii,* of the Petunioides section, with *N. quadrivflwis,* which was formerly placed in *Polidiclia* section. *N. quadrivalvis* produces four celled capsules and is a smaller plant than *N. bigelovii.* As the F_1 hybrid was entirely fertile, there seems no good reason for placing these forms in different sections. The four-celled capsule proved to be a partially dominant character. Thus the genus has been divided into 3 sub-genera.

The *Tabacum* section

Sub-genus *tabacum,* to which *N. tabacum* belongs. Includes 6 species in 2 sections. The *Tabacum* section is represented by numerous varieties of the species *Nicotiana tabacum.* These are natives of the New World. All commercial tobacco grown in the everywhere in the world including India belongs to this species.

The *Rustica* Section

Sub-genus Rustica, to which N. rustica belongs, includes 9 species in 2 sections and contains the most primitive elements of the genus. The *Rustica* section includes all the yellow-flowering species and varieties. These are of commercial importance in some countries. In India for example, they are successfully grown commercially and for some purposes prove more desirable than the *tabacum* varieties. Among these rustica forms are three groups; *(a)* one in which the pistil is longer than the stamen and, therefore, one which must be artificially pollinated by hand or crossed by the aid of insects; *(b)* an intermediate type; and *(c)* forms in which the stamens and pistil are so arranged that self-fertilization is the usual rule.

The *Petunioides* Section

The third sub-genus, Petunioides, is a large one containing the remaining 45 species grouped into 9 sections. The *Petunioides* section contains numerous varieties and species. Many of these are grown as ornamental flowering types.

Classification of Genus *Nicotiana*

Division Magnoliophyta (Angiosperms)
Class Magnoliopsida (Dicotyledons)
Order Polemoniales
Family Solanaceae

Division Magnoliophyta (Angiosperms)

A division of seed plants consisting of about 250,000 species, which form the bulk and most conspicuous element of the land plants. Often called flowering plants or angiosperms, they have several unique characteristics, the most prominent of which are their reproductive structure, flowers, and covered seeds. The other obvious woody land plants are the gymnosperms, which have cones instead of flowers and have naked seeds. Another trait distinguishing the angiosperms is the presence of double fertilization, which results in the production of stored food (starch or oils) within their seeds. Angiosperms range from some of the smallest plants known to large forest trees, and they occur in all habitats, including the oceans, where they are only a minor element in most marine ecosystems. Some are capable of growing directly on rock surfaces as well as on the limbs of trees. The angiosperms are usually considered to be the most highly evolved division of the subkingdom Embryobionta. Their highly specialized and relatively efficient conducting tissues, combined with the protection of their ovules in an ovary, give them a competitive advantage over most other groups of land plants in most regions. The angiosperms may be characterized as vascular plants with roots, stems, and leaves, usually with well-developed vessels in the xylem and with companion cells in the phloem. The central cylinder has leaf gaps or scattered vascular bundles; the ovules are enclosed in an ovary; and the female gametophyte is reduced to a few-nucleate embryo sac without an archegonium. The male gametophyte is reduced to a tiny pollen grain that gives rise to a pollen tube containing a tube nucleus and two sperms; one sperm fuses with the egg in the embryo sac to form a zygote, and the other fuses with two nuclei of the embryo sac to form a triple fusion nucleus that is typically the forerunner of the endosperm of the seed. Among plants with alternation of saprophyte and gametophyte generations, the angiosperms represent the most extreme stage in reduction of the gametophyte, which in effect is reduced to a mere stage in the reproduction of the saprophyte. The pollen grain, with its associated pollen tube, and the embryo sac represent the male and female gametophyte generations; the endosperm is a new structure not referable to either generation; and the remainder of the plant throughout its life cycle is the saprophyte. Many angiosperms can also propagate asexually by means of creeping stems or roots or by other specialized vegetative structures such as bulbils. It is obvious to biologists that the angiosperms must have evolved from gymnosperms, but beyond this the facts are obscure. They appear in the fossil record early in the Cretaceous Period as obvious angiosperms, without any hint of a connection to any particular group of gymnosperms. Many believe that among the gymnosperms the seed ferns provide the most likely ancestors. The Magnoliophyta consist of two large groups that have not been formally named: the eudicots and the magnolias. The eudicots are characterized by flowers that are highly organized in terms of the number and orientation of parts whereas the magnolias have many parts without any particular fixed patterns among the parts except for the monocots, in which the most developed groups, like the eudicots, exhibit developed flowers with highly organized patterns.

Class Magnoliopsida (Dicotyledons)

Dicotyledons or dicots are flowering plants whose seed contains two embryonic leaves or cotyledons. Dicotyledonous plant species are placed in the Class Magnoliopsida. Flowering plants, or *angiosperms,* that are not dicotyledonous are monocotyledonous: having one embryonic leaf. See How to distinguish a monocot from a dicot for other characteristics that separate these two groups. Traditionally the dicots have been treated as a class, originally called the Dicotyledoneae, but more recently called Class Magnoliopsida after the type genus Magnolia. Some botanists, however, divide the dicots into two or more classes in an effort to create monophyletic groups. The classification of dicots has undergone considerable revision as our understanding of their relationships has changed, and is still not entirely settled, though a general consensus is emerging.

Order Polemoniales

Order Polemoniaceae consists of Polemoniaceae; Solanaceae; Boraginaceae; Labiatae; Lentibulariaceae and Pedaliaceae; in some classifications includes the order Scrophulariales.

Family Solanaceae

The potato family, Solanaceae is one of the largest and economically important family of herbs or shrubs or trees often strongly scented and sometimes narcotic or poisonous; includes the genera Solanum, Atropa, Brugmansia, Capsicum, Datura, Hyoscyamus, Lycopersicon, Nicotiana, Petunia, Physalis, and Solandra.

Genus *Nicotiana*

The genus *Nicotiana* is one of the five large genera of *Solanaceae* and is represented by more than 60 species, which are grouped in three sub-groups, of which *N. tabacum* and *N. rustica* are the two commonly cultivated for producing commercial tobacco. *N. tabacum* is widely cultivated in most countries of the world while *N. rustica* is restricted to India, Russia and few other Asiatic countries. It is an herbaceous annual, however, *Nicotiana rustica* is found sometimes to behave like a perennial. Grow primarily for its aromatic leaves. The major characteristic features of some of important species are described as under.

N. tabacum L. (2n 48)

The plants of this species are usually taller attaining a height of 150-250 centimeter. The leaves are large but rather narrow. They may be sessile or petiolate. The colour of the flowers is reddish, pinkish or white. It is used extensively for smoking and chewing purposes. It is supposed to have been derived as an amphidiploid of a cross between *N. sylvestris* and - a member of the section *Tomentosae.* It has lanceolate leaves and pink flowers. Macrophylla is a very strong plant with large ovate leaves gradually tapering to a broad clasping base, with rounded but not prominent basal lobes. The flowers are deep red with a stout tube and an abruptly swollen, broad infundibulum; the limb is almost pentagonal, and at the base of every lobe is a whitish, triangular spot. The white tobacco is a small variety bearing a large number of small lanceolate and broadly decurrent leaves tapering towards the base on a relatively short stem. The corolla is relatively small, short, and cream colored.

N. rustica L. (2n 48)

In this species varieties humilis, Brasilia, and one which is apparently texana. The plants of this species are stocky, bushier in nature and also shorter in height, usually not more than 90 to 120 centimeters in height. The leaves are large and broad and ovate in shape and always possess a petiole. Flowers occur in cluster and are of dull greenish-yellow colour. It is used extensively for hookah, chewing and snuff purposes. It is supposed to have been derived as an amphidiploid from a cross between *N. undulata* and *N. paniculata.*

N. paniculata L. (2n 24)

N. paniculata-a long, slender corolla tube with short, symmetrical, greenish corolla lobes. The upper leaves were lanceolate, the basal leaves were ovate, slightly cordate, and with long petioles.

N. glutinosa L. (2n 24)

A small branched plant with small, cordate, long petiole, extremely glandular leaves. The flowers were short and cylindrical below, but suddenly swollen above into an irregular, obliquely one-sided funnel. The limb was bilabiate, the stigma and anthers being connivent just under the middle lobe of the upper lip. The colour was light yellow tinged with red.

N. glauca Grah. (2n 24)

This species is a tree tobacco with long, petioled, glaucous leaves. The flowers are pale yellow with tubes similar to those of paniculata. The stem and leaves are bluish green.

N. tomentosa R. et. P. (2n 24)

The two year old plants are not yet in flower, but their general character shows them to be true N. tomentosa (Leh- mannia tomentosa Spr.). Petunioides section.

N. Bigelovii WATS. (2n 48)

According to Setchell (1912) different varieties of N. Bigelovii exist. The plants used were small, their first leaves being slightly decurrent, nearly petioled, in fact, the later leaves becoming characteristically truncate, auriculate, and partly clasping at the base. The first leaves were more or less elongate deltoid, the later leaves more lance late. The corolla tube was narrow, and the corolla limb deeply divided into long pointed lobes. The corolla was pure white.

N. nudicaulis WATS. (2n 48)

N. nudicaulis is a very small plant with dark green ovate leaves narrowly decurrent. The corolla is shorter and broader than in N. *trigonophylla*. The tube is somewhat brownish, the limb cream. The corolla lobes overlap.

N. suaveolens Lehm. (2n 32)

The plants used were branched, with the lower leaves ovate, sessile, clasping at the base and pointed at the tip. The upper leaves were sharply lanceolate and clasping. The corollas were ivory white, salver shaped and with long slender, non-infundibular tubes. The corolla limb is large with broad, shallow rounded lobes.

N. sylvestris Speg. et Comes. (2n 24)

In general habit as in leaf shape this species of the Petunioides section is very similar to certain *Tabacum* varieties. Its corolla is white, and very long and slender, the tube being inflated in the middle. The corolla limb consists of 5 broadly triangular lobes. The flowers ordinarily droop.

N. acuminata Grah. (2n 24)

This is a branched plant with petioled leaves, cordate below and lanceolate above. The coarse, spine-like glandular hairs on the surface and on the margin are characteristic. The corolla is white with a cylindrical tube and shallow, rounded lobes on the limb.

N. trigonophylla Dun. (2n 24)

This is a very small plant with broadly lanceolate leaves, narrow at the base and then expanded into broad, partly clasping auricles. The corolla is very small with a short, cylindrical tube. The corolla limb consists of five broad, obtuse, and shallow lobes. The flowers are a deep cream tinged with green.

N. longiflora Cav. (2n 20)

This is quite a tall plant with broad, lanceolate or oblanceolate leaves, coarsely bullate and rugose above, the epidermis smooth except for coarse, spine-like, glandular hairs on the surface and margins. It early develops a characteristically compact rosette of large, coarse leaves which lie flat on the ground, the rosette persisting for a considerable time before the flowering stalk develops. The corolla tube is extremely long and slender. The broad, spreading limb is deeply divided into five moderately broad, blunt-pointed lobes.

N. plumbaginifolia Viv. (2n 20).

A species closely related to N. Longflora,

N. alata Lk. ET Otto. (2n 16)

They may have been segregates from crosses between N. *alata* and N. Forgetiana, since it is supposed that the different colored varieties of N. Sanderae have originated by crossing *N.alata* with *N. Forgetiana.*

N. Langsdorfii Weinm. (2n 16)

This is a small plant with sessile, elliptical lanceolate leaves, narrowed and decurrent at the base. The corolla is funnel-shaped below with a gibbous ring above and a concave, spreading limb slightly notched into five broad, shallow lobes, greenish yellow and pendent.

N. quadrivahis Pursh. (2n 48)

This is a small plant somewhat similar to Bigeloaii, except that the leaves are not elongate deltoid but broadly deltoid. The corolla is shaped somewhat similarly to that of Bigelovii, but is much larger. The capsule is spherical and twice as large as that of *Bigelovii.*

N. multivalvis Lindl. (2n 48)

This is a plant very similar to *quadrivalvis,* only larger, with very large corolla and capsules.

Botanical Description of Tobacco (*Nicotiana*)

Root System

Tobacco plant generally forms shallow branches root system. It has central tap root from which numerous short later roots extend. The majority of the root system is confined to the Upper.60-90 centimeter of the soil.

Stem

It has a simple cylindrical stem. The height of the plant and length of internodes vary in different types of the two species, *N. rustica* is short and more robust and densely growing plant than *N. tabacum.* Plant height ranges from 90 centimeter to 240 centimeter in *tabacum* and 50 centimeter to 140 centimeter in *rustica.* The inter-nodes may be very short presenting a rosette appearance or even as long as 20-24 centimeter. "The stems are usually round covered with glandular hairs with a tendency to produce branches near the base known as "Suckers". The plants of rustica are more robust and densely growing than that of tabacum. If there are few leaves, the plant has an open appearance, but if they are numerous, the plants have a bushy appearance.

Leaves

The leaves of tobacco vary greatly in shape, texture, and number, depending upon the variety, the environment, and the cultural practices. The leaves vary a great deal in number, shape and texture. In *N. tabacum* leaf shape ranges from broad ovate to linear or lanceolate. Leaf base can either be sessile or petiolate. In *N. rustica* leaves are always petiole with 4 to 9 centimeter long leaf stalk. The shape of leaf can be orbicular or ovate. The size of leaves may vary from 15 centimeter to 100 centimeter or even more in length. They may be narrow, long and drawn out like a whip, elliptical or broadly ovate with a marked variation even in the same plant. The leaf- angle varies from an upright to horizontal with dropping at the base, in the centre or at any point of the leaf. The leaf base may be sessile or petiole differing in the development of wings having a breadth of 5-6 centimeter. Sometimes in some spp.

Sometimes in some spp. the auricles are also observed. The lateral veins may be perpendicular to the midrib as in 'wrapper tobacco, or may make acute angle with various intergrades as in 'Natu'. The types to phyllotaxy normally observed are 1/3, 2/5, 3/8 and rarely 5/13. In general the texture of leaves of *N. tabacum* is finer than the *N. rustica,* which is more or less leathery and tough. Topping and suckering are generally practiced to promote growth of desirable leaves. The ability to accumulate nicotine in the leaves is a characteristic feature of the tobacco plant. Nicotine, an alkaloid having the formula $C_{10}H_{14}N_2$, is synthesized in the roots, but is found in all plant parts except the mature seed and is stored most abundantly in the leaves. The amount of nicotine, and a related alkaloid, nornicotine, differs greatly in different varieties and species of *Nicotiana.* The nicotine contents of leaves of *N. tabacum* ranges from 0.5 to 5.5 per cent while in *N. rustica* it varies from 3.5 to 8.0 per cent.

Tobacco plant at full vegetative stage

Flowers

The stem terminates in a raceme type of inflorescence, which may bear as many as 150 flowers. The tobacco raceme panicle may be compact to very lax with a short or long peduncle. The inflorescence in main axis always flowers first and the side branches thereafter in order from top towards bottom. Flowers of tobacco plant are complete and perfect and are usually 5 centimeter in length and vary in their colour from pink, yellow, purplish, or white. Those of *N. tabacum* are usually pinkish, whereas those of *N. rustica* are greenish white or yellow. The corolla contains five petals which are fused into a long tube and which terminate in five expanded lobes at the tip. The shape varies from

Chewing tobacco at flowering

funnel to flute. The development of green colour differs in its intensity both inside and outside the Corolla tube. The corolla may be wheel shaped with acute apices to pentagonal shape and obtuse tips. Calyx is usually tubular but in few types it is globular and inflated. The flowers bear five anthers, which are fused to the corolla tube. The pistil is compound and the stigma is generally sticky and pollen adheres to it readily.

Pollination

Tobacco is normally self-pollinated, although as much as 4 to 10 percent of cross-pollination occurs from pollen carried by insects. The extent of natural cross-pollination is such that it is desirable to bag seed heads to ensure self-pollination. Flowers that are open or those already pollinated should be removed before bagging. Dusting seed heads with an insecticide before bagging is desirable to prevent insects from feeding on valued capsule. Artificial cross-pollinations are easily made in tobacco if normal mature flowers are used. All, open flowers and seed pods are removed first leaving only flowers that have not shed pollen. The proper stage for emasculation may be identified by the presence of a pink colour in the tip of the petals of the unopened flowers. The nearly mature flowers are emasculated by tearing the petals apart and plucking off the anthers either by hand or with small tweezers. At the same time, flowers from the pollen parent are selected which have fully developed anthers that have not opened. With the point of a sharp knife, a slit is made in the mature anther and a small amount of pollen is transferred to the stigma of the emasculated flower on the point of the knife blade. After being pollinated, the flowers are marked and then bagged to keep out stray pollen.

Fruit

The pistil matures into a capsule type fruit. The capsules vary in size, shape and nature of apex, blunt or conical. The shape may be elliptical, ovoid or conical. A single flower may yield from 2,000 to 5,000 seeds and several hundred thousand seeds may be produced on a single plant of t*abacum*. Plants of *rustica* yield about one-fourth this numbers of seeds.

Seeds

Seeds are spherical or broadly elliptical with variation in degree of ridging on the seed coat. The coloure varies from dark brown to light brown. The seeds of tobacco are extremely small; a single fruit may contain as many as 8000 seeds. This tremendous capacity for seed production permits a rapid increase of new strains or varieties. Tobacco seeds are long-lived and if properly stored in a cool dry place may retain their viability for fifteen to twenty years.

Genetics

Many genetic studies have been made with common tobacco *(N. tabacum)* and other species of the genus *Nicotiana.* Hybridization of tobacco was accomplished by the German hybridizer, Koelreuter, as early as 1761 and by other early hybridizers long before the time of Mendel.During the past extensive studies of interspecific hybridization have been made by East, Good speed and others. Several reasons account for the vast genetic studies with the tobacco plant. Tobacco is normally self-pollinated but also easy to cross-pollinate. Large quantities of seeds are produced which remain viable for many years. Common tobacco is extremely variable and furnishes many different plant characters to study. Wide variations in characters are found in many closely related species which cross more or less readily with common tobacco. The extensive variability within the species *N. tabacum* has afforded an opportul1ity to study the inheritance of a large number of characters, such as flower colour and size, internode length, leaf shape and size, leaf texture, leaf base characteristics, and number and size of suckers. Some of these studies concern the practical breeding only indirectly, but those dealing with flowering time, disease resistance, leaf characteristics, leaf quality, nicotine content and similar character's are of agronomic importance and of direct interest to the breeder.

Chromosomal Taxonomy of Genus *Nicotiana*

Over 60 species and varieties of the genus *Nicotiana* have been under investigation, together with species of such related genera as *Petunia, Fabiana, Salpiglossis* and *Nierembergia*. Morphological, distributional and cytogenetic evidence indicates the desirability of recognizing among the races of *Nicotiana* under observation. The haploid chromosome numbers characteristic of these 40 species are 9, 10, 12, 16, 20, 22, 24 and 32, and in the following table the species are arranged under these various chromosome number classes. With the exception *of N. rustica and N. tabacum,* both of which have a chromosome complement of n=24, the rest of the species belonging to sub-genera *Rustica* and *Tabacum* have n= 12; in sub-genus *Petunioides* the haploid number is either 1:1 or 24 in 7 sections; there is an aneuploid series below 12 in one of the remaining sections and a similar series below 24 in the other. *N. tabacum* and *N. rustica* are two important cultivated species . Neither has been found growing wild. *N. tabacum* is grown and used most extensively for smoking and chewing tobacco although *N. rustica* is also grown for hookah, chewing and snuff as already related.

Table 13.1 The haploid chromosome number of various tobacco species

Haploid Chromosome numbers of *Nicotiana* species (n)							
8	**10**	**12**	**16**	**20**	**22**	**24**	**32**
alata bonarien sis Langsdo rffii Sanderae (4) **2n=16**	longiflo ra plumba gi Nifolia (2) **2n=20**	acuminata attenuata caudigera Cavanillesii cordifolia corymbosa glauca glutinosa Miersii pampasana paniculata Raimondii solanifolia sylvestris tomentosa tomentosifor mis trigonophyll a undulata Wigandioide s (19) **2n=24**	Suaveolen s and two unnamed species (4) **2n=32**	megalosip hon (1) **2n=40**	Rotundi olia (1) **2n=40**	Bigelovii Debneyi2 nesophila nudicaulis repanda rustica Stockton tabacum (8) **2n=48**	unnamed species (1) **2n=64**

Certain genetic groups characterized by similarity both in chromosome number and in certain basic morphological features have been recognized. The ***"alata group"*** includes *alata, Langsdorffii, Sanderae* and *bonariensis* (n = 8); *longiflora* and *plumbaginifolia* (n = 10); the ***"tomentosa group"*** (n = 12) consists of *tomentosa, tomentosiformis, glutinosa* and *wigandioides*; the ***"glauca group"*** (n = 12) of *glauca, Raimondii, paniculata, solanifolia, cordifolia* and *undulata*; the ***"corymbosa group"*** (n = 12) of *corymbosa* and *Miersii;* the ***"Cavanillesii group"*** (n = 12) of *Cavanillesii* and *pampasana*; the "***acuminata group***" (n = 12) of *acuminata, caudigera* and *attenuata*; the "***repanda group***" (n = 24) of *repanda, Stocktoni* and *nesophila*. The "***suaveolens group,***" consisting of *suaveolens, Debneyi, rotundifolia* and *megalosiphon*, is a natural one despite the distinctions in chromosome number involved. Thus *tabacum* (n = 24) may be traced to derivatives of hybrids between the "***tomentosa group***" (n = 12) and ***sylvestris*** (n = 12); *rustica* to derivatives of hybrids within the "***paniculata group***" (n = 12), Bigelovii obviously involves in its origin *attenuata* (n = 12) and probably trigonophylla (n = 12), while the origin of the "***repanda group***" and nudicaulis is not so clear, perhaps due to the absence of living descendants of their progenitors. Theoccurrence in our cultures of an amphidiploid between two species *diglutosa*-lends some weight to the postulated amphidiploid origins just mentioned.

□□□

14

Tobacco Genetic Resources and Varietal Improvement

A quality seed/planting material is mother of successful crop production. Germplasm has potential to produce the desired results. In the era of modern crop improvement and agriculture classical as well as new and innovative tools of breeding has got immense power to realize and break the yield barrier. The germplasm / genetic stock contain promising traits related to yield & yield attributing characters, quality characters and also resistance to various biotic and abiotic stresses. Now stage is set to reorient pre-breeding / breeding /advanced breeding works by utilized.

The flow of plant genetic resources from everywhere to India is slowing down gradually in the regime of strong intellectual property right (IPR). In order to satisfy our domestic need and to secure palace in world tobacco market there is need to improve our productivity and quality of tobacco produce, for the purpose, new cultivar/ varieties have to be breed out, with the help of new and desirable germplasms. India has diverse agro-climatic regions comprising tropical, sub-tropical and temperate regions, where we can grow various kinds of tobacco. There is a need to diversify the Indian tobacco, and for this purpose the new germplasms, having superior quality and high productivity should be introduced for their acclimatization/ adaptation in the country as well as for utilization crop improvement programme .India has introduced germplasm for this purpose, some of the important introductions are listed in the table below.

Table 14.1 Tobacco germplasm Introduction in India

EC Nos	Species Name	Source Country	Specific Traits
EC 444132-33	*Nicotiana sp*	USA	Burley Tobacco type & resistant to TMV
EC 444134	*Nicotiana sp*	USA	Dark Fire cured type
EC 444135 &137-38	*Nicotiana sp*	USA	Flue cured Virginia Type& resistant to TMV
EC 444136	*Nicotiana sp*	USA	Sun cured & resistant to TMV
EC 511325 -26	*Nicotiana longiflora*	USA	Resistant to TMV, root knot
EC 511341-42	*Nicotiana repanda*	USA	Resistant to tobacco Mosaic Virus, aphid and white fly
EC 511349	*Nicotiana stocktonii*	USA	Multiple resistant
EC 517324-25	*Nicotiana tabacum*	USA	Flue cured Virginia cigarette, type, mosaic resistant
EC 559244	*Nicotiana megalosiphon*	USA	Resistant to root knot nematode
EC 559245	*Nicotiana nesophila*	USA	Multiple resistant
EC439623	*Nicotiana sp*	USA	Natu type
EC439624-25	*Nicotiana sp*	USA	Aphid and caterpillar resistant
EC439626	*Nicotiana sp*	USA	Aphid tolerant
EC439638	*Nicotiana sp*	USA	Ornamental type, Free from aphids

Contd...

EC439663-64	*Nicotiana sp*	USA	Free from aphids
EC511310	*Nicotiana gossei*	Australia	Aphid resistant line
EC511320	*N. gossei*	Australia	Aphid tolerant
EC511325-26	*Nicotiana longiflora*	Australia	Resistant to tobacco Mosaic Virus
EC511325-26	*Nicotiana longiflora*	Australia	Resistant to tobacco Mosaic Virus
EC51134142	*Nicotiana repanda*	USA	Resistant to tobacco Mosaic Virus, aphid and white fly
EC511341-42	*Nicotiana repanda*	USA	Resistant to tobacco Mosaic Virus, aphid and white fly
EC516584-86	*Nicotiana tabacum*	USA	Flue cured Virginia hybrid Male sterile line
EC516584-86	*Nicotiana tabacum*	USA	Male sterile line
EC517324-25	*Nicotiana tabacum*	USA	Flue cured Virginia hybrid
EC519004-06	*Nicotiana tabacum*	Zimbabwe	Hybrid Burly type
EC519383-88 uniform font	*Nicotiana tabacum*	Brazil	High yielding with resistant to black shank
EC554858	*Nicotiana aeuminata*	USA	Aphid resistant line
EC554859	*Nicotiana africana*	USA	Aphid resistant line
EC554860-61	*Nicotiana arentsii*	Australia	Aphid resistant line
EC554862	*Nicotiana calyciflora*	Australia	Aphid resistant line
EC554863-66	*Nicotiana a eavieola*		Aphid resistant line
EC554867	*Nicotiana eavicola*	Australia	Aphid resistant line
EC554868-70	*Nicotiana cavicola*	USA	Aphid resistant line
EC554872	*Nicotiana goodspeedi*	USA	Aphid resistant line
EC554875	*Nicotiana goodspeedi*	USA	Aphid resistant line

Contd...

EC554876	*Nicotiana goodspeedi.*	USA	Aphid resistant line
EC554877	*Nicotiana gossei*	USA	Aphid resistant line
EC554878	*Nicotiana heteranflw*	Australia	Aphid resistant line
EC554879	*Nicotiana langsdorfii*	Australia	Aphid resistant line
EC554880	*Nicotiana linearis*	Australia	Aphid resistant line
EC554881 -86	*Nicotiana maritima*	Australia	Aphid resistant line
EC554887	*Nicotiana miersii*	Australia	Aphid resistant line
EC554888	*Nicotiana obrusijofia*	Australia	Aphid resistant line
EC554889	*Nicotiana occidentalic ssp obliqua*	Australia	Aphid resistant line
EC554890	*Nicotiana pauciflora*	Australia	Aphid resistant line
EC554890	*Nicotiana pauciflora*	Australia	Aphid resistant line
EC554891	*Nicotiana ramondii*	Australia	Aphid resistant line
EC554891	*Nicotiana ramondii*	Australia	Aphid resistant line
EC554892 -902	*Nicotiana rosulata*	Australia	Aphid resistant line
EC559238	*Nicotiana glauca*	USA	Black root rot resistant line
EC559239	*Nicotiana glutinosa*	USA	Tobacco Mosaic Virus resistant
EC559240	*Nicotiana gossei*	USA	Aphid resistant
EC559241	*Nicotiana hesperis*	USA	Root knot nematode resistant line
EC559242	*Nicotiana knighLiana*	USA	Root knot nematode resistant line
EC559243	*Nicotiana longiflora*	USA	Root knot nematode resistant line
EC559244	*Nicotiana megalosiphon*	USA	Root knot nematode resistant line
EC559245	*Nicotiana nesophila*	USA	Multiple disease resistant line
EC559245	*Nicotiana nesophila*	USA	Multiple disease resistant line

Contd...

EC559245	*Nicotiana nesophila*	USA	Multiple resistant line
EC559246	*Nicotiana noctiflora*	USA	Black shank resistant
EC559247	*Nicotiana nudicau/is*	USA	Black shank resistant
EC559248	*Nicotiana palmeri*	USA	Black shank resistant
EC559249	*Nicotiana plumbaginifolia*	USA	Black shank resistant
EC559250	*Nicotiana repanda*	USA	Multiple disease resistant
EC559251	*Nicotiana rustica*	USA	Insect resistant line
EC559252	*Nicotiana stenocarpa*	USA	Insect resistant line
EC559253	*Nicotiana stocktonii*	USA	Insect resistant line
EC559254	*Nicotiana trigonophylla*	USA	Insect resistant line
EC559255	*Nicotiana undulata*	USA	Insect resistant line
EC559256	*Nicotiana velutina*	USA	Insect resistant line
EC559257	*Nicotiana benthamianarepanda*	USA	Insect resistant line
EC559258	*Nicotiana umbraticanesophill*	USA	Insect resistant line
EC595359	*Nicotiana debreyi*	USA	Resistant to TMV
EC595360	*Nicotianagoodspeedii*	USA	Resistant to TMV
EC595361-75	*Nicotiana tabacum*	USA	Resistant to *Pernospora tabacina*
EC5l1349	*N stocktonii*	USA	Resistant to TMV and aphid
ECS54873	*Nicotiana ioodspeedi*	USA	Aphid resistant line

With the advent of new techniques in the field of molecular biology, the transfers of genes from wild to cultivated forms are now possible at will. Increasing cost of tobacco production due to highly susceptible nature of prevailing high yielding varieties, has led the scientists to look for varieties, which could be grown under low input conditions and could be resistant to various biotic and abiotic stresses. Wild relatives of tobacco plants represent a part of crops gene pool particularly resistant to biotic and abiotic stresses and have been the donors of many other useful traits. Therefore, National Bureau of Plant Genetic Resources continued its efforts to introduce important wild relatives of crop plants for their utilization in crop improvement programmes in the country.

Table 14.2 Wild *Nicotiana* introduced in India and utilized /to be utilized in tobacco improvement programme

Wild *Nicotiana*
N. acuminata, N. alata, N. arentsii, N. attenuata, N. benavidesii, N. benthamiana, N. bonariensis, N. calyciflora, N. cavicola, N. clevelandii, N. cordifolia, N. corymbosa, N. debneyi, N. foregetiana, N. gossei, N. heterantha N. kawakamii, N. knightiana, N. langsdorffii, N. linearis, N. longiflora, N. megalosiphon, N. miersii, N. noctiflora, N. nudicaulis, N. obtusifolia, N. occidentalis, N. occidentalis ssp. *hesperis, N. otophora, N. paniculata, N. pauciflora, N. petunioides, N. quadrivalvis, N. repanda, N. rosulata, N. rosulata* ssp. *ingulba, N. sanderae, N. setchellii, N. simulans, N. spegazzinii N. stocktonii, N. suaveo, N. sylvestris, N. thyrsiflora, N. tomentosa, N. tomentosiformis* and *N. velutina.*

Collection of Tobacco Bio-diversity

Biodiversity in tobacco species include species belonging to rainfed and irrigated ecosystem thriving in stress conditions such as cold and hot arid ecosystems, this vast diversity provides better choice for tobacco improvement with richness in tobacco gene pool, during recent time some of important germplasm collected by NBPGR and other organizations are given as in table13.3.

Table 14.3 Tobacco germplasm collected during recent past in India

IC Nos	Species name	Place of collection	Specific trait
IC 410309	*Nicotiana tabacum*	Senapati , Manipur	Profused branching
IC 415352	*Nicotiana tabacum*	Shimla, Himachal	Dark Brown in colour Pradesh
IC 415401	*Nicotiana tabacum*	Shimla, Himachal	Dark brown in colour Pradesh
IC 420433	*Nicotiana tabacum*	Mamit, Mizoram	Aphid tolerant line
IC 537534	*Nicotiana rustica*	Shimla, Himachal Pradesh	Temperate type, resistant to TMV
IC416658	*Nicotiana tabacum*	Anglon, Assam	Bitter in taste

Contd...

IC419728	*Nicotiana tabacum*	South Andaman Island	Disease resistant plant
IC528138	*Nicotiana plumbiifolia*	Dehradun, Uttaranchal	Wild tobacco, temperate type
IC537548	*Nicotiana rustica*	Shimla, Himachal Pradesh	Temperate type, resistant to TMV

Priorities for Introduction of Tobacco Genetic Resource in to India

Acquisition of more exotic germplasm as early as possible has become the priority due to latest development under Convention on Biological Diversity and other International treaties and laws, which makes flow of genetic resources somewhat restricted than it was before enforcement of new laws. The germplasm / genetic material contain promising traits related to yield & yield attributing characters, quality characters and also resistance to various biotic and abiotic stresses. Germplasm of tobacco crops were required for our National needs which will further distributed to various potential users (breeders/ crop improvement workers) in the country for its evaluation and further utilization in their ongoing/ensuing crop improvement works for our National security.

Table 14.4 Needs of trait specific germplams form specific country

Special feature	Source country
Flavored tobacco, low nicotine types and drought tolerant types.	Brazil Japan, USA Zimbabwe
Different germplasm resistant to important diseases like white fly, bud-worm, stem borer, leaf curl, TMV, CMV, black shank, brown spot, fusarium wilt, root knot nematode, aphid, broom rape, leaf eating caterpillar, root bug, white fly, stem borer	Australia, Brazil, USA, Zimbabwe
Nicotiana suaveolens	Australia

❑❑❑

15

Breeding Goals of Tobacco

Taste, aroma, nicotine content which differ greatly in different type of tobacco has necessitated development of breeding programmes for each of the various types of tobacco grown in India. Certain broad objectives which will apply to all classes of tobacco include yield, field and handling characteristics, disease and pest's resistance, and quality. Early maturity and, frost resistance are important for tobacco grown in certain areas of north India. Different quality characteristics are necessary for tobaccos used for different purposes.

The problems and prospects confronting the breeders of tobacco are multifold in this WTO regime. India needs broadly two traits to be enforced, first for its indigenous consumption and secondly for export purpose. Tobacco fetch handsome price for its quality viz., taste & aroma and physical appearance, which is again greatly differ for various tobacco type. The present breeding goals may be several as per specific needs, but increasing productivity improving quality and inducing resistance to biotic and abiotic stress is to name few, these may find place in any crop improvement programme.

Breeding for Yield

The yield of the tobacco plant is determined by number, size and body of the leaves. Varieties of different types differ greatly in these characteristics, but yields of these different types are never compared directly since the types are grown in different area and for different purposes. In general, yield has not been given first consideration by the tobacco breeder if it means a radical alteration in the characteristics of the variety already being grown. The market in each production area has been established largely on the basis of variety already in production.

Breeding for Physical Characters

As has been mentioned earlier, the qualities that determine a good tobacco, both from the view point of traders as well as the consumers, can be grouped as physical and chemical. At the initial purchase stage, the cured leaves are judged by their colour, resilience capacity and graininess of surface. Various improvements may be made which will improve the field and handling characteristics tobacco. These include toughness, so that the leaves will stand rough handling. Strong resistance, to prevent breakage in wet weather when plants are turgid. Scald resistance, to reduce the wilting and killing of leaf areas on hot days. Uniformity of ripening, to prevent lower leaves from falling off of deteriorating in quality before the upper leaves are ready to harvest. "Stand-up" types, which are easier to harvest and have less damage from leaves lying on the ground. Fewer suckers, small suckers, or slower-growing suckers to reduce the labour cost of their removal.

Breeding for Curing Properties

Curing properties of the leaf have received little attention, but could probably be improved by breeding. Curing properties are complicated by the fact that different varieties and types of tobacco have characteristics which respond differently to the various curing procedures.

Breeding for Quality

Breeding for quality has been equally important and our investigations are aimed at achieving some of the quality characters. Quality in tobacco is a complex characteristic which cannot be defined easily. Quality varies with the kind and variety of tobacco, the environment in which it is grown, the process employed in aging and curing the leaf, and the specific use of the tobacco. Quality cannot be measured with finality by simple mechanical or chemical means; it depends upon the desires of the manufacturer and the taste of the consumer. The tobacco breeder has done little to improve quality. His main concern has been to maintain the quality characteristics of the best tobacco types with which he is working. As genes for disease resistance have been brought into the common tobacco chromosome complex from other species of *Nicotiana,* it has become increasingly important to test a new variety for quality before it is distributed. Many of the species used as sources of disease resistance genes have few, if any, of the elements of quality desired in *N. tabacum,* so far as is known. Backcrossing to common tobacco to eliminate undesirable genus from alien species is resorted to in order to maintain satisfactory quality, as well as yield and plant type. Some of the components of quality that may be given consideration by the breeder are leaf characteristics, burning qualities, aroma and taste, sugar content and nicotine content. manufacturers are giving greater attention to chemical composition of the leaf than they did in the past.

Breeding for Leaf Characteristics

The size, shape; colour, thickness, and body of the leaf are characteristics affecting quality which vary with the variety. Although they are also modified by the environment, cultural methods, and position of the leaf on the stalk. Many of the so-called size characters of tobacco are of great commercial importance. For this reason their mode of inheritance is of much interest to the breeder. Extensive studies of inheritance of these size characters have been made.

It is more important that tobacco used in the manufacturing of cigars conform to specific requirements than tobaccos used for other purposes. For cigar, wrapper tobacco, types with short, thin leaves with less branched veins, fine texture, and elasticity are preferred. Thin leaves are also preferred for pipe-smoking blends, thicker leaves for cigarettes, and the thickest leaf type of chewing tobacco. Wide leaves are desirable in any variety to give a high stripping yield.

Breeding for Burning Qualities and Aroma

The burning quality, or combustibility, is an important consideration in quality of tobacco used for smoking. This characteristic is determined by fire holding properties, rate, evenness, and completeness of the burn and character of the residual ash. Burning qualities are affected by physical and chemical characteristics of the leaf, which vary with different varieties and soils. Aroma is developed with the process of curing and aging, but varies with the variety. Taste or flavour is important in the final product, but it is a difficult characteristic to evaluate.

Breeding for Low Nicotine Content

Recently, the trend in variety development has been towards milder, lower nicotine varieties that have high yield. It has long been known that the nicotine content is influenced by variety, environment in which the tobacco is grown, and cultural practices used in its production. Heavy fertilization with nitrogen to increase yields, topping or suckering, and other practices often result in excessive percentages of nicotine. Factors which control total alkaloid production (nicotine and nornicotine) are different, from the factors which affect the conversion bf nicotine to nornicotine or other products. Interest in breeding low nicotine varieties of cigarette tobacco has been stimulated by the desire to develop a variety which could be smoked without harmful effects. High nicotine content is preferred in bidi, hookah, and chewing tobacco. Breeding programme in future will keep a close watch on the level of nicotine and in view of the health hazard posed by cigarette smoking, the lesser the nicotine in the varieties, the better its acceptability by the trade. Nicotine content in tobacco is generally influenced by environment and a wet season tends to produce a tobacco low in nicotine. This aspect will be helpful when tobacco is grown as a monsoon crop in the light soil.

Breeding for Chlorine Tolerance

The relation between chlorine and the burn of tobacco leaf is well known. Higher chlorine content in the leaf retards good burning of the leaf in cigarettes. The evidence of chlorine in the leaf is seen well in the cured leaves by its pale whitish colour on the dorsal side of the midrib. Such leaves are priced low by trade. The chlorine effect is much felt in tobacco grown in saline soils and for this reason; tobacco has been discouraged from being grown in these areas. However, varieties may have differential uptake of such elements and it may be possible to breed varieties which can take in only negligible quantity of chlorine. If, suitable varieties are found, they may well be used as the breeding base for growing such chlorine tolerant tobacco in saline areas also.

Breeding for High Sugar Content

Some emphasis has been given to breeding for higher sugar content in the flue-cured tobacco leaf. With the increased use of the cigarette, tobacco companies have been seeking mild, thin leaf tobaccos lower in nicotine and higher in sugar. Most flue-cured varieties having average around 18 percent sugar, but it would be desirable if the sugar content could be increased to about 20 percent.

Breeding for Disease Resistance

There are numerous plant diseases which are very injurious. Few studies of the mode of inheritance of resistance and susceptibility have been made but resistant species, varieties, and strains have been observed in respect to several diseases. The reaction of 20 species of *Nicotiana* to wildfire caused by *Bacterium tabacum* All 41 varieties of *N. tabacum* which were studied proved somewhat susceptible although several varieties were somewhat less susceptible than others. Several species proved resistant and all varieties of *N. rutstica* proved resistant. Strains of White Burley resistant to root rot. Certain strains which are resistant to root rot are of the same type and quality as the original variety from which they were selected. Control of disease by chemicals, where such disease control is possible, interferes with quality in many cases; hence development of disease resistant varieties offers the primary solution of the disease problem. With some diseases in which practical control by chemicals as not available such as with root rotting organisms, breeding for resistance is necessary to maintain yield. The important diseases of tobacco in India are black shank, bacterial wilt, *Fusarium* wilt, powdery mildew, anthracnose, mosaic, leaf curl, and nematodes. These instances lead to the conclusion that breeding disease-resistant varieties will in some cases prove a valuable means of controlling certain tobacco diseases. In many areas of the world including India major emphasis in breeding tobacco has been given to the breeding of disease resistant varieties. Resistance to black shank and bacterial wilt is present in common tobacco. Adequate resistance to many diseases is found only, in wild species of *Nicotiana*. Intraspecific resistance is frequently polygenic; although interspecific resistance is often simply inherited. Transfer of high resistance or immunity from other species, usually results in the production of strains undesirable in plant type, yield or quality unless back crosses are made; to eliminate the undesirable genes introduced from the alien species. Desirable genes in other species are sometimes linked with undesirable genes. For example, genes for nematode resistance are linked with genes for narrow leaf shape.

Since several verity is under cultivation due to continuous and concentrated breeding work on disease resistance in tobacco in India, which is mentioned in the characteristics feature of leading and important varieties bread and release for general production in India. To know the historical background of disease resistance breeding examples are cited here the earlier work done in the U.S.A. with several diseases.

Black Shank *(Phytophthora parasitica* var *nicotianae)*

The black shank disease is characterized by blackened dead roots and the decay extends into the pith and cortex at the base of the stem. Infection later develops to the extent that the plants die the black shank disease was first identified in the U.S.A. about 1916. Florida 301, a resistant shade tobacco variety, was developed about 1930, by crossing and selection within local varieties of Big "Cuba and Little Cuba". Florida 301 was later used as the source of resistance in the breeding of other black shank resistant varieties. Resistance to black shank appears to be controlled by multiple factors and varieties differ in degree of resistance. In the moderately resistant varieties seedling plants may be killed by invasion of the black shank fungus, but in older plants only a portion of the root system is damaged and the tobacco yield is not greatly reduced. Some species of tobacco, *N. longiflora* and *N. plumbaginifolia,* are highly resistant or immune. Resistance genes from these species are being transferred to common tobacco.

Bacterial Wilt *(Pseudomonos solanacearum)*

Bacterial wilt was reported in the U.S.A. more than fifty years ago. It is a common disease in Sumatra, also, but is not so important in India. Roots of diseased plants decay, and the plants wilt in a manner similar to those infected with the black shank disease. Breeding work on resistance to bacterial wilt was started in 1931. The strain developed from the plant T.I. 448A has been used as a source of resistance in breeding the variety Dixic Bright 244, which has resistance to black shank and *Fusarium* wilt as well as bacterial wilt. Resistance to bacterial wilt is controlled by multiple recessive genes. Young plants are not as wilt resistant as adult plants. Many plants in a wilt resistant variety may appear to be infected early; however, most recover with only slight ill effects. Resistance has not been found in other species of *Nicotiana r*esistant lines have been developed in Sumatra.

Fusarium Wilt *(Fusarium oxisporum var. nicotianae)*

Fusarium wilt is an important disease in India. Tobacco plants infected with the *Fusarium* wilt turn yellow on one side of the plant because of the production of toxin by the *Fusarium* organism. After the tissue is broken down the plant wilts. Many flue-cured varieties-grown in the U.S.A. are resistant. Resistance to Fusarium wilt is inherited by a single factor pair. 50 Artificial inoculations to test for resistance may be made at the time plants are set in the field by dipping the plants into solutions carrying the wilt-producing organism.

Powdery Mildew *(Erysiphe cichoraceanun* var. *nicotianae)*

Powdery mildew (is an important disease in India. The infection spreads rapidly in the field destroying the crop. The affected leaves develop defects on curing, or even get scorched, rendering them useless for marketing. Immunity has been transferred to .*N. tabacum* varieties from N. *glutinosa* and other species.' Selections from these crosses are being used as sources of resistance in the East Indies. Crosses to transfer resistance of *N. glauco.* to *tabacum* have been made in India.

Anthracnose (*Colletootrichum tabacum*)

Anthracnose is a serious disease in tobacco nurseries in India. Several wild species, including *N. debneyi* and *N. longifera,* arc reported to be resistant. Resistance in *N. debneyi* is stated to be controlled by polygenic recessive genes.

Mosaic

Mosaic, caused by the tobacco mosaic virus, is recognized by the preset of leaves with a conspicuous dark green and yellow-green mottling which later may become puckered and deformed. The greatest mottling is usually found on the younger leaves. Tobacco mosaic may be transmitted to healthy plants merely by rubbing first a diseased and then a healthy plant, or by handling dried infected tobacco used for smoking or chewing and then handling living plants. As a result, the grower frequently spreads the disease in the field while handling his tobacco plants. Mosaic infected plants are reduced in yield and quality with losses ranging up to 60 percent, depending on how early in the life of the plant the disease is contracted. Sanitation and use of disease resistant varieties are the only known means of control. Resistance to tobacco mosaic has been observed in *N. glutinosa* (n = 12) and other species. The resistance of *N. glutinosa* to mosaic was observed in 1916 by H. A. Allard and was transferred later by Holmes to common tobacco. Resistance from *N. glutillosa* was used in producing resistant commercial varieties in the U.S.A. Mosaie resistance from *N. glutillosa* is inherited as a monogenic dominant character, one of the few tobacco diseases known to be inherited in so simple a manner. In resistant plants, nom1ally, leaf tissue is killed only in areas where the virus enters, and the virus is localized in these spots. However, if parenchyma tissue such as mid vein or cortex is infected, the virus is likely to spread and kill the plant.

Nematodes (*Meloidogyne sp.)*

Several species of nematodes cause injury to tobacco. The root knot nematode *(Meloidogyne sp.)* is the most important one. The root knot nematode causes the infected roots to become enlarged and knotted. As the roots decay, growth of the plant is retarded and the leaves are killed prematurely. Moderate resistance is found in the flue-cured varieties 400 and 401 in the U.S.A. and high resistance in the Central American variety T.I.706. Resistance of T.I.706 is polygenic. Most highly resistant selections from crosses with T.I. 706 have had small leaf size. Excellent resistance to root knot is found in *N. repanda, N. megalosiphon,* and other species.

Other Diseases

Other diseases that call for urgent attention in tobacco growing areas in India and Southeast Asia are the frog-eye leaf spot *(Cercospara nicaianae)* and the leaf curl disease. Not much information about sources of resistance is available.

□□□

16

Methods of Tobacco Improvement

The principal methods of tobacco breeding are introduction, selection and hybridization. Modern breeding work on tobacco was started about the beginning of the present century in the U.S.A., and other countries. Through the efforts of the Dutch scientists breeding work on tobacco was initiated in Indonesia which led to the development of some excellent strains of cigar wrapper tobacco for that area.

Progress in the breeding for yield and other characteristics is very much complicated by the fact that quality, which is of prime importance, is an elusive and complex trait that cannot be seen or measured quantitatively. It is expressed only through such qualitative features as taste and aroma, and then only after the tobacco has been properly aged and cured.

In India, improvement work on tobacco was initiated by the Howards at Pusa as early as 1906, and later continued at breeding work was started in several states of India. Tobacco processing and export companies were active in introducing and trying different foreign types from time to time. However, no comprehensive and coordinated programme of improvement was started until after the formation of the Indian Central Tobacco Committee in 1947. Since then a number of tobacco research stations have been developed where breeding of the various types of tobacco grown in India is conducted. The economic value of tobacco lies in the quantity and quality of the leaf. Leaf quality in particular is very complex depending mainly on chemical makeup, which cannot be seen or measured easily.

Methods of Improvement

Introduction

Introduction has played an important role .in the establishment of tobacco varieties in India. Tobacco was first introduced into India by the Portuguese in the beginning of the 17th century. The crop soon spread into the different regions of the country and various local or desi varieties became established. A large number of varieties from various countries especially from the U.S. A. have been introduced and tried from time to time. The flue-cured varieties, Harrison Special and Hicks, are introductions from the U.S.A. The "Delcrest" variety of flue-cured tobacco is an introduction' from southern Rhodesia. The wrapper tobacco variety, Rangpur Sumata, is a selection from an introduction from Indonesia. Other important introductions include White burley from the U.S.A. and Amarillo from South Africa. A genetic collection of nearly 500 strains being maintained at the Central Tobacco, Research Institute contains introductions from many areas of the world.

Selection

Pure line selection has been the main method of breeding in the establishment of most of the improved tobacco varieties in India. Selections have been made from some introduced varieties, such as Harrison Special-9 from Harrison Special and Chatham from an unselected cross made in Chatham, Virginia, U.S.A. The local tobaccos grown in India have a wide range of variability. The Lanka varieties, grown for many years on the Krishna and, Godavari tracts, have been found to be quite variable genetically. From collections of the latter area alone twenty distinct types have been established. The influence of environment is so great in tobacco that it was commonly believed that a variety could be induced to "break up" into many strains if moved into a new location, particularly from a semitropical to a temperate climate. It has been demonstrated, however, that much of these variations are due to segregation following natural cross-pollination and that true breeding lines could be established by continuous bagging of flowering heads to protect them from foreign pollen. In selection for local materials, bulks of the local materials are collected and salfed for two or three generations after which distinct types can usually be isolated by pure line selection. Improved varieties of all types of tobaccos in India have been established through selection procedures.

Hybridization

As in other crops, hybridization became more important in breeding tobacco as knowledge of genetics increased, for, recombination of plant characters could be obtained more or less to fit the breeder's design. Large numbers of improved varieties have been developed by hybridization, principally in the U.S.A., in which were combined genes for larger number of leaves, improved quality, and disease resistance. Hybridization has played a less important role in the development of new varieties in India but will become more important as present breeding programmes develop. Hybridization of introduced flue-cured varieties with local Indian varieties often' led to poor curing characteristics in. the leaves of the hybrid selections. It is quite possible that this will be remedied as additional crosses using more diverse genmplasm are made. Some synthetic strains have been built up in a multiple crossing programme involving superior varieties from many countries.

Interspecific Hybridization

Interspecific hybridization has been an important procedure in the breeding for disease resistance in the U.S.A. In many instates resistance genes for specific diseases could be found only in other species of *Nicotiana.* This necessitated the use of interspecific crosses as already described. With interspecific crosses, deleterious genes-are often added to the adapted variety along with the genes for disease resistance. In interspecific crosses, the removal of co-transferred genes creates a problem. However, if there is some recombination between the chromosomes of the donor and recipient parents the chances of successful transfer of the desired gene without any deleterious effects exist. To overcome these undesirable features and recover the plant type and quality characteristics of the adapted variety type, backcrossing to the common tobacco variety was practiced.

Backcrossing has been used also with intervarietal crosses, but the intensity of backcrossing is not generally as great as a with interspecific crosses. Even after extensive backcrossing, recombination and selection, along with the desirable genes, other genes are also introduced. When the gene transfer is between related forms of the same species, the hybrids may be viable, fertile and after sufficient backcrossing with the recipient variety, the genotype of the cultivated variety with a gene from the donor parent may be recovered. The large number of species of *Nicotiana* which have resistance to common destructive diseases of *tabacum* makes this a rich field for the breeder. In India, crosses involving *N. glauca* have been made to transfer genes for powdery mildew resistance to *tabacum.*

Utilization of Hybrid Vigour

Numerous studies have been made on the expression of hybrid vigour in tobacco. Hybrid seed production in tobacco does not present any problem as 2,000 seeds, can be obtained from a single hand pollination. About 300 flowers are sufficient to produce hybrid seed for one acre. The emasculation process in tobacco may be eliminated by the utilization of cytoplasmic male sterility. Cytoplasmic male sterile plants may be obtained by introduction of *tabacum* chromosomes into cytoplasm of *N. debneyi, N. megalosiphon,* and other species. By repeated backcrosses to male sterile plants varieties of *N. tabacum* may be sterilized. When the male sterile plants are supplied with fertile pollen, seed production is normal. Restorer genes are not needed since it is the leaves of the F_1 plants that are harvested and seed production in the F_1 is unnecessary. Although hybrid vigour has been reported for early maturity, height, leaf number, and other characters, the increase in yield has not been encouraging. In a comparison in North Carolina with variety crosses of flue-cured tobacco, 10 F_1 hybrids yielded about 14 percent above the average yield of the two parents. From these and other studies the commercial advantages of F_1 hybrids over available pure line varieties at present appear to be very slight. Studies need to be made of yields with crosses of diverse germplasms and also of quality and uniformity of hybrids as compared to standard varieties. The major problem in *Nicotiana* had been the poor fertility of the interspecific hybrids. The production of sesquidiploids, where the hybrids are diploid for the chromosome of one species but haploid for the other species, helped in improving the fertility of the hybrids.

Mutations in Tobacco

The sudden appearance of giant plants with abnormally high leaf number has been recorded in the Sumatra, Maryland, Cuban, and Connecticut Havana varieties of *N. tabacum*. These new forms under field conditions have it much longer period of vegetative vigor than the normal varieties. Consequently, blossoming does not take place under ordinary field conditions. Otherwise, the general habit of each of these new types is not very different from the normal variety from which it was obtained. Two of these new varieties of giant habit are of some commercial importance. A short account of their first recorded appearance together with their cultivation as commercial varieties will be given. A giant plant of tobacco was first time noted in 1912 in the Cuban variety which is grown under shade in the Connecticut Valley.

Normally the plant can attain the flowering and fruiting stages only when the length of day falls within certain limits, and, consequently, these stages of development ordinarily are reached only during certain seasons of the year. In this particular, some species and varieties respond to relatively long days, while others respond to short days, and still others are capable of responding to all lengths of the day which prevail in the latitude of Washington where the tests were made. In the absence of a favorable length of day for bringing into expression reproductive processes in certain species, vegetative development may continue and thus lead to the production of such varieties as Stewart Cuban and Maryland Mammoth which under ordinary conditions never reach the flowering stage. Thus, certain varieties or species may act as early or late maturing, depending simply on the length of day to which they happen to be exposed. The Stewart Cuban and Maryland Mammoth varieties of tobacco, as well as several other species were used in a discussing the effects of controlling light as a means of forcing flowering.

Use of Modern Breeding Techniques

In addition to the sexual methods of gene transfer through intervarietal crosses and wide hybridizations (interspecific and intergeneric crosses). the, use of radiated pollen, and Para sexual methods, such as somatic cell hybridizations, protoplast Culture and fusion, genetic modification of cells by uptake of DNA, and transfer through vectors have been attempted. Alien genetic materials provide a mean to introduce entirely new genes into the genome. As the aim is generally to transfer only a gene, the alien segment substituted should be as small as possible.

Creation of Variability and Diversity

Variability and diversity is source of any crop improvement which could be achieved by utilizing diver's germplasm, by hybridization or muataion etc. In addition to the sexual methods of gene transfer through intervarietal crosses and wide hybridizations (interspecific and intergeneric crosses). the, use of radiated pollen, and Para sexual methods, such as somatic cell hybridizations, protoplast Culture and fusion, genetic modification of cells by uptake of DNA, and transfer through vectors have been attempted. Alien genetic materials provide a mean to introduce entirely new genes into the genome. As the aim is generally to transfer only a gene, the alien segment substituted should be as small as possible. Even after extensive backcrossing, recombination and selection, along with the desirable genes, other genes are also introduced. When the gene transfer is between related forms of the same species, the hybrids may be viable, fertile and after sufficient backcrossing with the recipient variety, the genotype of the cultivated variety with a gene from the donor parent may be recovered.

Mutation Breeding

Use of irradiation or chemical mutagens for creating variability has already been demonstrated. Although creation of a large number of viable mutations is possible in tobacco due to its amphidiploid genotype, the quality requirements of the plant makes it difficult to obtain a desirable mutant directly. Hybridization may be necessary to transfer the desirable mutant characters to varieties. Progenies of each capsule should be advanced separately, following irradiation.

Polyploidy

Since cultivated tobacco is already a polyploid plant, achieving success by production of polyploids plant in tobacco may be difficult. Polyploidy is useful in making interspccifie crosses for the transfer of disease resistance or other desirable characters.

Hybridization

From the standpoint of the student of plant genetics the *Nicotiana* genus is especially favorable material. Some of the reasons are: Tobacco may be self-fertilized artificially with ease and the technique of crossing is very simple. Each plant produces a large number of seeds and the seed is viable for many years. There are a large number of varieties which are entirely fertile *inter se*. These furnish especially favorable material for a study of quantitative characters. The different species furnish very favorable material for a study of sterility. Different crosses furnish F_1 generations which differ from each other in sterility. The range extends from species crosses which give no viable seed and from completely sterile F_1 crosses, to entirely fertile ones.

Parthenogenesis

Parthenogenesis, meaning the production of viable seed without pollination, was shown by Goodspeed in 1915; occur in *N. tabacwn,* variety Cuba. Under normal conditions its occurrence is rare. Wellington several species as well as several commercial varieties of *N. tabacum* were used in this study. Parthenogenesis in *N. tabacum* does not occur under normal but may occur under abnormal field conditions, at Pusa, India. The occurrence of parthenocarpy and parthenogenesis has never been observed by Howard (1924) in any of the Indian types of *Nicotiana rustica.* No parthenogenesis was observed in India.

Sterility

Studies of crosses between *N. tabacum* varieties and *N. sylvestris,* which belong to the *Petunioides* section, have been made by Goodspeed and Clausen (1917). The F_1 generation proved to be nearly sterile, although a few apparently normal pollen grains were produced. A few normally maturing ovules capable of fertilization were produced by the F_1 plants. If the plants were kept under poor cultural conditions and the flowers pollinated by their respective parents, approximately 1 per cent of the number of seeds normally produced was obtained. If back crossed with the *sylvestris* parent, practically 10 per cent of the offspring of the seeds produced are pure *sylvestris.* When crossed with *tabacum,* parts of the plants from the seeds produced seem to be of normal tobacco type and are fertile; others resemble *tabacum* but are sterile. The F_1 plants closely resemble the particular variety of *N. tabacum* which is used as one of the parents.

Studies of self-sterility in tobacco crosses have been made, crosses between *N. jorgetiana* and *N. alala* which are self-sterile and, *N. langsdorffii,* a self- fertile species. *Alata* and *forgetiana* varieties sometimes produce seed late in the flowering season, although during periods of rapid growth they are entirely self sterile. The few seeds obtained under reduced cultural conditions from selfing these self-sterile species are spoken of as cases of pseudo-fertility. The self-fertile condition proved dominant in F_1 and a ratio of approximately 4 self-fertile: 1 self-sterile plant. The self sterile plants or F_2 proved self-sterile in later generations. East explained these results by a dominant factor, *F for* fertility and a subsidiary factor, D, for pseudo-fertility which exhibits itself only in the presence of the factors for sterility, *if.* This pseudo-fertility factor produces some fertility under certain conditions, thus tending to lower the number of self-sterile forms.

East has suggested that differences in the rate of pollen germination are largely responsible for the differences in the length of pollen tubes from compatible and incompatible pollinations. When self-sterile plants are self-pollinated the pollen grains germinate but the pollen tubes grow so slowly that abscission of the flower occurs before the pollen tube reaches the ovary. The progeny of a cross between two self-sterile individuals consists of small number of sterility classes, each .member of a given class being reciprocally sterile with every other member of that class; but reciprocally fertile with members of other classes. Three such classes were present in the populations studied. When members of two different classes were crossed, the progeny consisted of two of the three classes in equal numbers, the class of the female parent being always absent in the progeny. The results were explained by the hypothesis of the allelomorphic, sterility factors. The three classes were of the It was supposed that a plant could be fertilized only with pollen-bearing factors other than its own.

Environment Consideration and Tobacco Improvement

It is a matter of common knowledge that environmental conditions widely modify the expression of characters. The kind and quality of tobacco grown is greatly influenced by the soil and the climate. This is particularly noticeable in tobacco, where quality, size, and shape of leaf are of such marked importance. The tobacco leaf, which is the commercial product of the crop, develops its characteristic quality only under fairly precise environmental conditions. The change of environment usually the growing of southern grown seed in the north is the one of the most effective means of inducing variability. Statements of this nature have been used as evidence that environment modifies the characters of a pure line by inducing variability. A careful survey of experimental studies does not support this contention. The result was a disastrous failure, owing to a lack of knowledge of methods of handling and to the use of unselected seed. By further study of handling and through careful selection in which artificially self-pollinated seed was saved, the industry was placed on a firm foundation. Temperature and moisture are important climatic factors affecting the quality of the tobacco leaf. The two efficient means of inducing variability as a source of new types are change of environment and crossing. As a result tobacco production tends to be concentrated in small areas and the tobacco produced in a specific area will have certain peculiar quality characteristics which distinguishes it on the market from tobacco produced in other area. In India the crop is grown during the winter when the mean temperature for the crop growth is between 21^0C and 32^0C and the relative humidity is high.

□□□

17

Genetic Engineering for Stress Tolerance

Plant genetic engineering strategy has now been used to a great extent in several crops for developing tolerance to various stresses. Different approaches have been used to improve stress tolerance of several plants by foreign gene transfer. Genetic engineering of biosynthesis pathways associated with stress response has evolved as one of the most promising methods for improving stress tolerance in plants.

Mechanism of Plants Response to Biotic and Abiotic Stresses

Tobacco plant have to survive under various type of abiotic (draught, temperature, salinity, freezing Chilling, cold and Oxidative stress) and biotic (disease and insect pests) stress. It is crop of grown for its quality traits any stress has direct bearing on its productivity and quality. Although our understanding of plant tolerance mechanisms to stresses is incomplete, the plant hormone abscisic acid (ABA) is believed to be an essential mediator between environmental stimulus and plant responses. The plant hormone ABA levels increase when plants are subjected to various stresses viz., drought, salt, or temperature stress, and these increased levels have been shown to be essential for stress tolerance. ABA is a sesquiterpenoid hormone that plays an important role in many physiological processes, including seed development and dormancy, vegetative growth, and responses to environmental stress. Modulation of ABA levels prior to the onset of stress also provides plants with better stress tolerance. ABA is synthesized by the cleavage of a C40 carotenoid precursor followed by a two-step conversion of xanthoxin via an ABA-aldehyde Several Arabidopsis genes involved in ABA biosynthesis have been identified. One of these is NCED3, a gene encoding a 9-cis-epoxycarotenoid dioxygenase (NCED) that catalyzes production of xanthoxin, a key intermediate in ABA biosynthesis.

Genetics of Resistance in Tobacco

Stress resistance is a complex trait. An understanding of genetic basis of Stress resistance in crop plants is a pre-requisite for a geneticist to evolve superior genotype through either conventional breeding methodology or biotechnological approach. Various morphological (earliness, reduced leaf area, leaf rolling, wax content, efficient rooting system, awn, stability in yield and reduced branching), physiological (reduced transpiration, high water-use efficiency, stomatal closure and osmotic adjustment) and biochemical characters (accumulation of proline, polyamine, trehalose, etc., increased nitrate reductase activity and increased storage of carbohydrate) confer drought resistance. Morphological and physiological characters show different types of inheritance pattern (monogenic and polygenic) and gene action (additive and non-additive), whereas the genes responsible for biosynthesis of different compatible solutes have been identified and cloned from plants, yeast, mouse and human.

Very little is known about the genetic mechanisms that condition these characters. The identification of genes responsible for morphological and physiological traits and their location on chromosome have not been possible, but their inheritance pattern and nature of gene action have been reported. Polygenic inheritance of root characters is reported. The long root and high root numbers are controlled by dominant alleles and thick root tip by recessive alleles. However, leaf rolling and osmotic adjustment have shown monogenic inheritance.

Lack of multidisciplinary approach and precise screening techniques, incomplete knowledge about genetic basis of drought resistance, negative correlation of drought resistance traits with productivity and unavailability of appropriate genes to obtain transgenic plants are the main constraints for genetic improvement of drought resistance. Exploration of wide genetic variation of relevant characters, consideration of more genes at a time to transfer through breeding or genetic engineering method, application of antisense RNA technique, assessment of polypeptides induced under drought and multidisciplinary approach should be included in the future research programmes for drought resistance.

Breeding Approach

Different breeding approaches for resistance have emerged, with their merits and demerits. Efficient screening techniques are pre-requisite for success in selecting desirable genotype through any breeding programme.

Breeding for High Yield Under Optimum

The first is to breed for high yield under optimum (water-stress-free) condition. As the maximum genetic potential of yield is expected to be realized under optimum condition and a high positive correlation exists between performance in optimum and stress conditions, a genotype superior under optimum level will also yield relatively well under drought condition. This is the basic philosophy of this approach. However, the concept of expression of maximum genetic potential in optimum condition is debated as genotype environment interaction may restrict the high yielding genotype to perform well under stress.

Breeding Under Actual Stress Condition

The second approach suffers from the problem that the intensity of stress is highly variable from year to year and as a consequence environmental selection pressure on breeding materials changes drastically from generation to generation. This problem compounded with low heritability of yield makes for the complicated and slow breeding programme.

Breeding Under Incorporation of Stress Resistance

An alternative approach to the above two would be to improve stress resistance in high-yielding genotypes through incorporation of morphological and physiological mechanisms of drought resistance. However, transferring stress resistance in high-yielding genotypes is complicated due to lack of understanding of the physiological and genetic basis of adaptation in stress condition. In contrast, improving the yield potential of an already resistant material may be a more promising approach, provided there is genetic variation within such a material. Simultaneous selection in non-stress environment for yield and in various stress condition for stability may be done to achieve the desired goal of evolving various stress resistant genotype with high yield. As such, the breeding methodology to be applied for resistance in tobacco resistance is the same as that applied for other purposes. In general, pedigree and bulk method could be used for self-pollinated crops and recurrent selection for cross-pollinated crops. However, if transfer of few traits relating to various stress resistance to a high-yielding genotype is the aim, then back cross is the appropriate methodology? On the other hand, biparental mating (half sib and full sib) maintains the broad genetic base as well as provides the scope to evolve the desired genotype for various stress resistance. The success of any breeding programme depends on the availability of the screening technique, especially for stress resistance.

Screening Techniques

Any effort for genetic improvement in stress resistance utilizing the existing genetic variability requires an efficient screening technique, which should be rapid and capable of evaluating plant performance at the critical developmental stages and screening a large population using only a small sample of plant material. As stated earlier, stress resistance is the interactive result of different morphological, physiological and biochemical traits and thus, these different components could be used as selection criteria for screening appropriate plant ideotype. A combination of different traits of direct relevance, rather than a single trait, should be used as selection criteria. The importance of developing a reliable screening technique for stress resistance has been realized very early.

Genetic Engineering Approach

Biotechnology offers a promising array of tools that may be useful in achieving drought tolerance in plants. One such tool is the low input approach to crop production by which crops are modified to suit the environment in which they are growing, rather than modifying the environment to meet the needs of the crop. Improvement of crop plants for the tolerance of various stresses has been the priority area; however introgression of complex traits often limited due to their complex polygenic nature. Identification of numerous genes and putative functional transcripts involved with threshold tolerance. Various stresses have given new diversion to the rational engineering of tobacco plants against the various stresses. Genetic engineering of biosynthesis pathways associated with stress response has evolved as one of the most promising methods for improving stress tolerance in plants. Transgenic technology combined with the genome sequencing, functional genomics and proteomics has the potential to advance crop breeding beyond imagination. Several gene transfer approaches have been employed to improve the various stress tolerance in tobacco plant, which involve either product of an effected gene or a regulatory gene. However, tolerance developed in transgenic plants has been low in most of the cases. Thus there is a need of a comprehensive approach to pyramid a number of genes involved in the stress tolerance or simultaneous regulation of many stress responsive genes by using single gene encoding stress inducible transcription factor, so as to enhance tolerance towards more complex stresses including drought, salinity, temperature, and freezing for increasing the crop productivity.

The techniques for gene transformation of crop plants have been applied for identification of genes responsible for drought resistance and their transfer. Mainly two approaches, namely targeted and shotgun approach facilitate genetic engineering to obtain transgenic plants conferring drought resistance.

Targeted Approach

Metabolic pathways involving the synthesis of different metabolites such as polyamine, carbohydrate, *proline, glycine betaine and trehalose* have shown to be associated with various stresses resistance. This approach relies on the availability of relevant information on biochemical reaction for synthesis of these metabolites and utilizes the related genes to transfer them from different sources to crop plants. This approach is more precise and methodical, having a higher probability of success in comparison to the shotgun approach. In recent years, introduction of various stress induced genes involved in different biochemical pathways from different sources to sensitive plants has evolved as one of the promising methods.

Shotgun Approach

This approach to obtain the desired gene is indirect. A random analysis of stress-related alteration in cell process and gene expression is employed. Genes, which are expressed under drought and for which no specific role has yet been proven, are identified. Though the approach is less precise with low probability of success, the strategy can work even as there is no prior information about the gene or gene product. Thus, the shotgun approach appears to be better choice due to dearth of sufficient information on biochemical changes in the cell for drought resistance. Barley hva1 gene responsible for late mbryogenesis abundant (LEA) proteins has been transferred to rice to produce drought-resistant transgenic through a shotgun approach. Gene hva1 encodes for a group of three LEA (late embryogenesis abundant) proteins which gets accumulated in vegetative organs during drought condition. The tissue culture method has a potential to create somaclonal variation for drought resistance also, but difficulties in selection for desired variant restrict its use.

Marker-Assisted Selection

In most breeding programmes, the genetic improvement for drought resistance is accomplished through selection for yield and because of low heritability of yield under stress and the spatial as well as temporal variation in the field environment, conventional breeding approaches are slow. Whereas molecular markers such as restriction fragment length polymorphism (RFLP), random amplified polymorphic DNA (RAPD) and isozyme will facilitate to develop drought-resistant genotypes more effectively as their expressions are independent of environmental effects. After identification of the molecular markers associated with yield or other morphological traits related to drought resistance, those markers could be used as a selection criterion for drought resistance. The application of marker-assisted selection in evolving drought resistant genotypes is in an experimental stage; more specifically just identification of RFLP markers associated with osmotic adjustment, stay green, root traits has been achieved.

Transgenic Plants

Resurrection plants can tolerate stress better than the normal plant for the trait they are restructures. Transgenic technology combined with the genome sequencing, functional genomics and proteomics has the potential to advance crop breeding beyond imagination. The most consistently successful approach is the introduction of genes encoding enzyme that catalyze the conversion of naturally occurring substrate into product with osmo-protective properties. Other important genes encode membrane modifying enzymes, radical scavenging enzyme, stress induced proteins and hypoxia/anoxia reducing proteins. Several gene transfer approaches have been employed to improve the stress tolerance of crop plants, which involve either product of an effected gene or a regulatory gene. However, tolerance developed in transgenic plants has been low in most of the cases.

Genes imparting protection from stresses can be expressed in plants in two ways: they can be expressed all the time, whether or not the plant is under stress; or they can be engineered to express only when there is any stress. The second method is more favored, as it limits the side effects of the manipulations. The vectors provided herein can be used to transform plant cells and generate transgenic plants. Thus, transgenic plants containing the nucleic acids described herein also are provided, as are methods for making such transgenic plants. The polynucleotide's and/or recombinant vectors described herein can be introduced into the genome of a plant host using any of a number of known methods, including electroporation, microinjection, and biolistic methods. Alternatively, polynucleotides or vectors can be combined with suitable T-DNA flanking regions and introduced into a conventional Agrobacterium tumefaciens host vector. Other gene transfer and transformation techniques include protoplast transformation through calcium or PEG, electroporation-mediated uptake of naked DNA, electroporation of plant tissues, viral vector-mediated transformation, and micro projectile bombardment If a cell or tissue culture is used as the recipient tissue for transformation, plants can be regenerated from transformed cultures. A transgenic plant containing at least one exogenous nucleic acid, the at least one exogenous nucleic acid containing a nucleotide sequence that encodes a polypeptide having an amino acid sequence with at least 98.4 percent identity to the amino acid sequence set forth in SEQ ID NO:2.

Production of Transgenic Plants

Selection of desirable trait

↓

Identification of governing gene(s) that control the trait

↓

Cloning of the identified gene(s) in a desired vector

↓

Genetic transformation of plants

(*Agrobacterium* mediated or Direct bombardment)

↓

Molecular characterization of selected transgenic lines

↓

Transgenic crops

↓

Testing for agronomic superiority adaptability and for bio safety to humans, animals and environments

↓

Commercial release of the trait specific transgenic

Fig. 17.1 Flow diagram of genetic transformation and development of transgenic

□□□

18

Varietal Development

Advent of hybrid and high yielding varieties of tobacco and other crops of immense economic importance have necessitated in the maintenance and preservation in seeds, much of our success in increasing in tobacco production has been due to development of seed enterprises over past few decades. The development of seed enterprises is absolutely necessary in the context of modern tobacco cultivation.

Seed is the unique, cheapest and ultimate input to improve crop productivity and sustainability. It is root cause of any successful cropping; the history of crop production is based on seeds of new cultivar having potential for vigorous growth, resistant to insect pest and disease with wide adaptability to is surrounding environments. In the early days it was achieved through the cultivation of useful plants, selection of superior types from cultivated plants constituted next stage of progress. Later through use of well know techniques of selection and hybridization, the scientists made available many new better cultivar and varieties. The introduction, development and release of new varieties have helped to raise the sights as regards yield possibilities and consequently have stimulated interest among the tobacco growers. However to the farmers all this scientific research would be of little value unless he gets seeds of desired quality namely high germination percentage, high vigour, high purity, sound health *etc.*

During the early phase of varietal development i.e. up to 1960 introduction and acclimatization of exotics introduction in FCV tobacco and isolation of pure lines from local cultivars in non Virginia type took places. With changing requirements of the trade and consumer, suitable breeding programme were later under taken to evolve superior varieties with high yields, quality and resistant to disease, pests and drought. Keeping in view developing and producing tobacco seeds in desired quantity having required characteristic feature suitable for particular agro climatic and agro ecological region of India for various type of tobacco, efforts are continued science independence in crop commodity including tobacco, varieties developed and distributed in public domain is discusses here in this chapter with reference to chewing and hookah tobacco.

1. FCV tobacco Varieties

1.1 FCV Tobacco Varieties from Karnataka

FCV Special

The first Indian FCV variety known as FCV Special and famously known as Mysore FCV Special was developed by M/S ILTD, Rajahmundry in 1976. FCV Special is a double cross (Harrison Special x Virginia Bright) x (Yellow Chatam), has been an outstanding Karnataka for its leaf quality and is often taken as a standard reference of Mysore FCV Special variety has been predominantly under cultivation in Karnataka light - more than two decades. It is now outdated due to its low yield potential and susceptibility to root knot nematode and black shank disease and prompted the development of several new varieties.

Thrupthi

Thrupthi was released by CVRC and SVRC during the year 1998 for cultivation under southern transition zone of Karnataka. This is high yielding (1800 kg/ha) FCV line KST 19 developed at Shimoga centre. It has registered 43.8% increase in cured leaf yield and 49.9% increase in TGE over the check FCV special. It is tolerant to moisture stress and

Thrupthi
Courtesy by RARS Shimoga

moderately tolerant to black shank and root knot nematodes and popular in Shimoga area. Further, this variety is suitable for early and late plantings.

Swarna

Swarna is first variety developed at Hunsur by incorporating powdery mildew resistance from donor parent Kuofan to Kanakaprabha and released in 1984. It was recommended for KLS particularly the higher rainfall areas in Maysore, Coorg and Hasan districts, which are endemic to powdery mildew disease. Though its yield potential was relatively lower at 1450 kg/ha, but It has compatibly 14.0 % more cured leaf yield and 18.0% TBLE over the check Kutsaga E 2.

Swarna
Courtesy by CTRI
RS, Hunsur

Bhavya

Bhavya is derived from the parents FCH 6002 and FCH 6523, released in 1988 for cultivation in black shank sick areas of assured rainfall zones of KLS. It is result of continued breeding work which led to the development of a high yielding (2000 kg/ha) variety Bhavya. Bhavya is a resistance to black shank and tolerance to root-knot nematode as well as frog-eye spot disease. It has registered 39% increase in cured leaf yield and 43% increase in bright grades over the check variety, FCV Special in Station trials at Hunsur.

Bhauya
Courtesy by CTRI
RS, Hunsur

Kanchan (NLS 4)

An exotic introduction (Variety NLS 4), jointly released by CTRI and M/s ITC Ltd., ILTD Division as Kanchan (NLS 4) in 1999. This variety is recommended for cultivation in assured/heavy rainfall areas of KLS. It is a dark cast variety; with semi open plant habit and short internodal length. Leaves are long and narrow with wavy lamina. Cured leaf is bright lemon to

Kanchan (NLS4)
Courtesy by CTRI
Jeelugu milli

deep orange colour, heavy bodied, optimall spotted, pliable and has superior aroma. Kanchan yields more than 2000 kg cured leaf /ha with around 70% bright grades. Kanchan was superior to the control Bhavya in all yield parameters and recorded an increase of 10, 11, 16 and 13% of green, cured, bright grade and top grade equivalent yields, respectively over Bhavya. It is resistant to black shank disease and tolerant to root-knot and brown spot diseases.

Rathna

Rathna was released in 2001 for cultivation in for medium and low rainfall regions of KLS. It is a mutant selection of variety FCV Special developed at Hunsur as Rathna. It is a light cast variety with open plant habit and medium internodal length. The leaves are very broad and long with smooth surface. The variety has field resistance to black shank. The cured leaf yield potential is around 2000 kg/ha, with more than 75% bright grades, of which over 60% is first grade. Cured leaf is ripe, thin to medium bodied, pliable, with deep lemon to orange colour and optimum, uniform spots. Rathna registered 40 and 14% higher cured leaf yield and 43 and 13% higher top grade equivalent yield over FCV Special and Bhavya, respectively.

Rathna
Courtesy by CTRI RS, Hunsur

1.2 FCV tobacco varieties from Andhra Pradesh

Early phase

In early period of time (during 1950's and 1960's) Chatam, an exotic introduction from USA and Delcrest, an exotic introduction from Canada were extensively cultivation in TBS area of Andhra Pradesh. Kanakaprabha (Delcrest X DB 101) and Dhanadayi (Delcrest X Bisettes). Both varieties were green cast with low percentage of bright grades and both were released in 1971.Prior to the development of new varieties these two rule the area for long period.

CTRI Special

The first light cast FCV tobacco variety, CTRI Special (developed from the local variety SP Cross) was release in 1976. This variety was developed to improve percentage of bright grade leaf. It had a yield potential of 1365 kg/ha with high bright grade outturn from all the priming and registered 13.7% increase in TBLE yield over the check, Delcrest, in the black soils at Rajahmundry.

Jayasri and Jayasri (MR)

Jayasri
Courtesy by CTRI

Jayasri is a selection from a cross between a genetic mutant and structural mutant. The potential of the variety is of 1990 kg cured leaf/ha. It has registered 20.6% increase in cured leaf yield and 38.3% increase in bright leaf yield over the check, CTRI Special, in replicated trials at Rajahmundry. However, it was fast maturing and susceptible to TMV. To improve TMV resistance Jayasri (MR) was developed with incorporation of TMV resistance from TMVRR 3 to Jayasri. Maturity of Jayasri (MR) was medium and its yield potential was around 1503 kg/ha.

Gauthami and VT 1158

Gauthami
Courtesy by CTRI

Gauthami (L 617 X Delcrest) a high yielding (2000 kg/ha) varieties was released in 1992. Gauthami yielded higher than the check, Jayasri, by 21.8% in cured leaf yield and 19.5% in bright leaf and TBLE yield in replicated trials at Rajahmundry. It was released for cultivation in traditional black soils and southern light soils.

VT 1158 (L 617 X CTRI Special MR) was released in 1993.VT 1158 registered 8% increase in cured leaf yield and 10.2% increase in TBLE yield over the check, Jayasri, in bulk trials at Rajahmundry over four seasons (1987-1991). Variety VT 1158 is resistant to TMV and is popular in NBS and CBS.

Cy 135

CY135
Courtesy by CTRI

Cy 135 was developed by pedigree method of breeding with parents CM 16 and Gauthami. Cy 135 with a yield potential of 2860 kg/ha. Cy 135 yielded higher than the better check, VT 1158, by registered 24% increase in cured leaf yield and 20.8% increase in grade index over the better check, Hema, in coordinated trials at Guntur.

Hema

Hena
Courtesy by CTRI, Guntur

Hema, developed from cross between two vigorous plants found in black soils of West Godavari and light soils of Jeddangi in East Godavari district. Hema variety which has wide adoptability, was released in 1987 for black soil belt of Andhra Pradesh Yield potential of Hema is around 1560 kg/ha. It has registered 20.8% increase in cure leaf yield over the check, Jayasri, in replicated trials at Guntur. It is still a prominent variety in SBS and SLS.

Hemadri

Hemadri
Courtesy by CTRI, Guntur

Hemadri, develop from the parents line 754-2#1 and 16/103 is recommended for traditional black soil zone of Andhra Pradesh. Hemadri is medium cast and high yielding (2240 kg/ha) variety, Hemadri yielded higher than the check, Hema, by 31 % in cured lea yield and 27% in grade index.

MC Nair 12

MC Nair 12
Courtesy by CTRI, Guntur

To overcome the black shank disease problem led to the release of its first variety, MC Nair 12, for NL Sarea, in 1986. MC Nair 12 is an exotic introduction from USA. and has a yield potential of 1880 kg/ha. MC Nair 12 gave higher yields over the check, 16/103, by 8.4% in cured leaf and 21.5% in grade index in coordinated trials at Jeelugumilli.

CM 12 (KA)

CM 12 (KA)
Courtesy by CTRI, Guntur

CM 12 (KA) is a chemical mutant of MC Nair 12. It is in order to increase the yield potential in NLS to 2000 kg/ha level; breeding efforts were made for successful release of CM 12 (KA). This variety is resistant to black shank disease. It has yielded 2021 kg cured leaf and 1359 kg grade index/ha, producing 11.3% more cured leaf and 13.4% more grade index over the check, 16/103. It also registered 20.5% increase in cured leaf yield and 2.6% increase in grade index over the check, Mc Nair 12.

Kanchan

Kanchan is perhaps first variety which is developed for its high yielding, bodied and flavourful tobacco and released in 1998, for NLS area, which has a yield potential of more than 2000 kg/ha. It has registered 33-36% increase in cured leaf yield and 60-78% increase in grade index over the checks (CM 1 2 (KA) and 1 6/1 03) in coordinated trials in NLS. With improved production practices, the variety is yielding around 2500 kg/ha and presently it is occupying more than 95% area in NLS.

Kanchan
Courtesy by CTRI, RS, Hunsur

Kanthi

Kanthi, a suitable variety for SLS area is derived from the parents Jayasri and L 61 7. It is recommended for general cultivation in SLS and SBS areas. It can withstand moisture stress to certain extent and has a yield potential of around 1600 - 2000 kg/ha. Kanthi registered 23% increase in cured leaf yield and 21 % increase in grade index over the check, Jayasri MR, in coordinated trials at Kandukur. It also produced significantly higher cured leaf yield and grade index over the check, Kanakaprabha, in coordinated trials at Guntur.

Kanthi
Courtesy by CTRI, Kandukur

2. Bidi tobacco Varieties

2.1 Bidi Tobacco Varieties of Gujarat

Gujarat is leading bidi tobacco producer in the country. With the increase in canal and lift irrigation facilities major bidi tobacco growing area in state was brought under irrigation, However, some pockets still remained under rainfed cultivation and tobacco grown under these pockets always fetched higher price due to it quality bidi tobacco produces.

A 119 and A 2

Anand-2
Courtesy by BTRS, Anand

Bidi tobacco varieties A 119 and A 2 developed at Anand prior to the establishment of All India Coordinated Research Project on Tobacco and released in 1969 by State Varietals Release Committee for Gujarat. These two varieties were popular not only in Gujarat but also in Karnataka, Maharashtra and Andhra Pradesh, because of their high adaptability coupled with better yield and quality these two varieties were released as National varieties by CVRC in 1984.

Gujarat Tobacco (GT4)

Gujarat Tobacco (GT 4)
Courtesy by BTRS, Anand

Gujarat Tobacco (GT4) was developed at Anand centre, was released in 1976 for rainfed bidi tobacco growing areas of Gujarat. This is outcome of breeding for drought resistance, to cater the needs of farmers of these areas. It has a yield potential of 2605 kg/ha under rainfed conditions and 2841 kg/ha under normal conditions. It is a dwarf type with short internodes, tolerant to moisture stress and leaf burn disease.

Gujarat Tobacco (GT 5)

Gujarat Tobacco (GT 5)
Courtesy by BTRS, Anand

Gujarat Tobacco (GT 5) was released in 1985 developed for irrigated tobacco with high yield potential. GT 5 possesses high degree of tolerance to root-knot nematodes and wet foot. It has shy suckering habit. It has a yield potential of 3301 kg/ha. It also possesses high nicotine content and smooth smoke characteristics.

Gujarat Tobacco (GT 7)

Gujarat Tobacco (GT 7)
Courtesy by BTRS, Anand

Gujarat Tobacco (GT 7) was developed and released in 1993, especially for the rainfed tobacco belt to further increase in productivity of the crop. It is a tall variety and has a yield potential of 2535 kg/ha.

Gujarat Tobacco Hybrid (GTH 1)

GTH-1
Courtesy by BTRS, Anand

Gujarat Tobacco Hybrid (GTH 1) first of its kind in India, for irrigated bidi tobacco growing areas of Gujarat. This is resultant of breeding programme undertaken to develop high yielding hybrids. Hybridization work was done with CMS lines of popular varieties resulted in release of GTH 1 in 1995, It is tolerant to root-knot nematodes, has high nicotine content with long and broad leaves. Its yield potential is around 3600 kg/ha, this was achieved through exploitation of hybrid vigour and boon for the changed agricultural scenario in middle Gujarat where competitive crops like banana and vegetables are grown and there was a need to improve tobacco yields further.

Gujarat Tobacco 9 (ABO 64)

Gujarat Tobacco 9
Courtesy by BTRS, Anand

ABO 64 was released in 2001 as Gujarat Tobacco 9 (GT 9), which is resistant to tobacco mosaic virus (TMV), is one of the three serious virus diseases of bidi tobacco in Gujarat. Gujarat Tobacco 9 (GT 9), yielded 36% higher than the popular variety A 119. Its yield potential is around 3100 kg/ha.

2.2 Bidi Tobacco Varieties of Karnataka

In early time bidi tobacco variety, A 119, released by Anand centre in 1969 was also popular in Karnataka. Karnataka first bidi tobacco was release of its first variety NPN190 in 1979 with a yield potential of 1964 kg/ha. Its parentage is 320 X (169-19 X Olar 121-2-1). It was released for cultivation in bidi tobacco growing areas of Belgaum in Karnataka and Kolhapur and Sangli districts in Maharashtra state.

NPN 22 (Bhavyasree)

Bhavyasree (NPN22)
Courtesy by ARS, Nipani

NPN 22 (Bhavyasree) was released as Bhavyasree for general cultivation in Karnataka during the year 1997. This Bidi tobacco variety is derivative of a cross between NPN 190 and PL 5, developed at Nipani, This gives about 30% higher yield than the popular variety, A 119. It is tolerant to frog-eye leaf spot disease with shy suckering habit. This variety gives an average yield of 2000 kg/ha under rainfed condition and 2500 to 3000 kg/ha under irrigated condition.

NBD 43

NBD 43 was developed using PL 5 and NPN 23 as parents. Karnataka State Variety Release Committee in 2005 for consideration of its release for northern transitional zone of Karnataka state. It is high yielding (18% higher yield) than A 119 with better smoking qualities and tolerance to frog-eye leaf spot.

NB 043
Courtesy by ARS, Nipani

3 Natu Tobacco Varieties

3.1 Natu Tobacco Varieties of Andhra Pradesh

Guntur centre developed four high yielding varieties namely Prabhat (1500 kg /ha), Vishwanath (2429 kg/ha) Natu special (1600 kg/ha) and Bhairavi (2600 kg/ha).

Prabhat

The first variety of Natu tobacco desi Prabhat was released in 1977. Its parent is Tokaku and Harrison Special. It is suitable for cigarette natu areas with a yield of 1500 kg/ha and recommended for natu tobacco tracts of and Kurnool districts in Andhra Pradesh and area s around Sehore in Madhya Pradesh. It was acceptable to trade due to its quality traits.

Prabhat
Courtesy by CTRS, Guntur

Viswanath

Viswanath, a desi type, developed from (1 - 452 Karuvazhai X Rayala) released in 1986. It has a yield potential of 2429 kg/ha with high filling value, good bodied leaf and aroma and found suitable for natu tobacco growing tract of Guntur district in Andhra Pradesh. It has registered 13.5% increase in yield over the better check, DG 3 at Guntur.

Viswanath
Courtesy by CTRS, Guntur

Natu Special

Natu Special variety released in 1992, It is derived from Botcha Vithanam (Addanki) X KVT (Sattenapalli) X Mutant (Tokaku) with a yield potential of 1600 kg/ha quality natu tobacco variety, suitable for sun-curing having registered 22.2 and 7.4% increase in cured leaf yield over WAF and Prabhat, respectively at Guntur. Natu special also produced 14.9 to 51.9% more yield than WAF. Physical and chemical properties of leaf are in the desirable range and the quality of produce is acceptable to traders & manufacturers. It is recommended for natu tobacco growing areas of Guntur, Kurnool, Mahaboobnagar, Nalgonda and Khammam districts of Andhra Pradesh.

Natu Special
Courtesy by CTRS, Guntur

Bhairavi

Bhairavi a high yielding natu tobacco variety developed with crosses between Ash Fri II (WAF) and Viswanath, which gives an average yield of 2000 kg/ha under normal conditions. It gave 27.0 and 29.7% higher cured leaf yield over WAF at Guntur and Nandyal centres, respectively. It has got good lusture and deep colour in addition to the high yielding ability. It is found suitable and recommended for black soil zones of Guntur, Kurnool, Prakasam and Mahaboobnagar districts of Andhra Pradesh.

Bhairavp
Courtesy by CTRI, RS, Guntur

3.2 Natu Tobacco Varieties of Orissa

Gajapati

Gajapati a pikka variety released in 1999 by Orissa SVRC for cultivation in Orissa. Gajapati has a yield potential of 1774 kg/ha. Gajapati has registered 3.4 to 45.9% increase in yield over the check JJP Local lover the period 1989-98 at Berhampur centre. It has moderate tolerance to damping-off; frog-eye pot, black shank, TMV and leaf curl diseases.

Gajapati

4. Chewing & *Rustica* Tobacco Varieties

Chewing and Hookah (*Rustica*) is commercially cultivated in at least in five states (Uttar Pradesh, Bihar, West Bengal, Gujarat and Tamil Nadu) in India. Several local types of poor yielding ability, low input responsive and inferior in quality are still prevalent in several parts in India and Bihar. Since systematic breeding work started in India and regular release of varieties by central varietals release committee (CVRC) or concerned states varietals release committee (SVRC) has made available the quality seed to the farmers.

4.1 Chewing Tobacco Varieties of Bihar

The revolution in the varietals development was achieved only during sixty in case of chewing and hookah tobacco in Bihar. Since inception of Hookah and Chewing Tobacco Research Station in 1950 seven high yielding better quality varieties of chewing tobacco (*Nicotiana tabacum*) viz. DP 401 (1961), Gandak Bahar (1976), Sona (1977), Prabha (1981), PT-76 (1990), Vaishali Special (1993) and Lichichavi (2001) has developed and released by this Research Station. In case of hookah tobacco (*Nicotiana rustica*) three superior cultivars of viz. R-12, NP-220 and NP-22 have also been isolated and recommended for general cultivation in Bihar and adjoining areas. In general the yield potential of chewing tobacco and hookah tobacco varieties ranges from 20 to 35 q/ha and 15 to 25 q/ha respectively under normal condition. The distinguishing features of different varieties recommended for North Bihar are given hereunder.

DP 401

First improved variety of chewing tobacco was developed through selection by CTRI, RS, and Pusa during 1961. The typical genotypic characters are its long broad, sessile and elliptical leaves. The leaf lamina droops downwards on either side. The cured leaf is dark brown in colour. It is a pure line selection form Bori Bharao-10 a local and indigenous material of chewing tobacco. It gives a total cured leaf yield of 1670 kg and first grade leaf yield of 1050 kg per ha.

DP 401

Gandak Bahar

Second variety chewing tobacco was developed and recommended for cultivation in north Bihar is Gandak Bahar in the year 1976. This variety was evolved from the cross NP 35 X Bori Rampur Rohua. The plant type is erect; leaves are big, long and broad, sessile, closely set with crinkled leaf surface and dark green in colour. The cured leaf is brown in colour, it produces 1850 kg total cured leaf and 1210 kg first grade leaf yield per ha and matures in 140-145 days. Apart from its use for chewing purpose, this variety is best suited for the manufacture of black zarda and extraction of quiwam both taken along with paan.

Gandak Bahar

Sona

Cultivar was developed by crossing Bori Malinagar Thahra and DP 401 fallowed by selection and released in1977. Plant type is erect with big, long and broad, elliptical, semi open leaves which are sessile, closely set, well spangled and puckered. The cured leaves are dark brown in colour with white incrustation and pungent in taste. It matures in 130-135 days and gives 2321 kg per ha total cured leaf yield and 1496 kg first grade out turn per ha.

Sona

Prabha (HP-67-5)

Prabha is outcome of the crosses between Desila Mandhata X Hingla Kiratpur released in 1981. This is an early variety mature in 120-125days and fits well in multiple cropping (one year cycle) system. It has erect plant type, semi open, well spangled and puckered, thick, broad, lanceolate leaves. The cured leaf is dark brown in colour and with pleasing chewing taste, it gives on an average 2300 kg per ha total cured leaf yield (Sinha *et al.*,1984)

Variety Prabha (HP-67-5)

Pusa Toabcoo-76

New high yielding variety of chewing tobacco was released in 1990 s for general cultivation in Bihar. It gives 2633 kg total cured leaf and 1728 kg first grade leaf yield per ha. A comprehensive intervarietal recombination breeding programme led to the evolution of this variety as a result of the multiple cross and interse cross between [(Sona X HP 66-1)] X (SonaX HP 66-1)]X [(B.K.Rajwa X Gandak Bahar) X Gandak Bahar]. The main characterstic feature of this variety is semi open, long and broad light green leaves with pointed tips. It has greater plant height and internodal length which favours higher photosynthesis and thereby registering higher yield and better leaf quality. The spangling and puckering quality of leaf at green crop state as well as its colour, elasticity, white incrustation after curing and chewing taste are excellent.

Pusa Toabcoo-76

Vaishali Special

It is first variety that has been bred by back crossing and selection from a cross between (Sona×Bandi) X Sona. It gives 2599 kg total cured leaf and 1630 kg first grade leaf yields per ha. It has been found suitable for cultivation in Vaishali district or local "Bandi" area. It gives about 26% more first grade out turn over the local variety Bandi. It is a medium tall variety with semi open, long and broad leaves with vigorous growth habit. The leaves are well puckered and well spangled. Leaf curability is good with high first grade out turn; cured leaves are dark brown to reddish in colour, elastic with heavy deposition of white incrustation. Chewability is exceptionally good. In general incidence of diseases is low and also tolerant to black shank disease. It was released in 1993 for general cultivation in Bihar and areas identical in agroclimatic condition.

Vaishali Special

Lichchivi

New chewing tobacco variety Lichchivi has been released in 2001for cultivation in all chewing tobacco growing areas of Bihar and U.P. It has been developed by crossing variety Sona and elite germplasm Bori Tepari followed

Lichchivi

by pedigree selection method of breeding. Its specific feature is its thick bodied cured leaf. Its total average yield is 28 q/ha with 15 q/ha first grade leaf outturn. Two important physical leaf quality parameters e.g. spangling (4.2) and puckering (4.2) scores superior in this variety than the earlier released and established variety e.g. Pusa Tobacco-76 (3.8 & 3. 7) and Vaishali Special (3.6 & 3.7).

Nicotine content in the cured leaves of this variety 3.53% being at par to PT-76 (3.54%) but superior to Vaishali Special (3.50%). Highest cured leaf thickness of 11.00mm has been observed in "Lichchavi" as compared to Vaishali Special (11.00mm) and PT-76 (8.50mm). The chewing test evaluation scores of this variety was highest (77.44) than PT-76 (68.7) and Vaishali Special (61.2).

4.2 Chewing Tobacco Varieties of Gujarat

In Gujarat state chewing tobacco is grown in an area of 15,000 to 20,000 ha in Anand district of Gujarat, characterized by method of curing (Lal or Kala chopadia), and mainly used for chewing, snuff and hookah paste preparations. Prior to the start of AICRP on tobacco, variety A 45 was popular in Gujarat.

Gujarat Tobacco 6 and 8

Chewing tobacco variety, Gujarat Tobacco 6, was released in 1985 with high yielding (2700 kg/ha) potential for dual purpose *(Lal* and *Kala chopadia).* But this variety could not become popular because of its medium body and erect plant type.

An improved variety Gujarat Tobacco 8 suitable for Kala chopadia is released in 1998 by Tobacco Research Station, Dharmaj. It has a yield ptential of 4469 kg/ha.

Gujarat Tobacco-6
Courtesy by BTRS, Anand

4.3 *Rustica* Tobacco Varieties of Gujarat

Rustica tobacco, known as Calcutti tobacco in Gujarat, grown in an area of 15,000 to 20,000 ha in north and middle Gujarat as a *rabi* crop. The produce of which either lamina flakes *(Bhuka* or *Chura)* or leaf bundles *(Chopadia)* are mainly used for chewing, *hookah,* snuff, and blending with bidi tobacco.

Gujarat Calcutti Tobacco 1 (GCT 1)

An improved variety of *rustica* tobacco Gujarat Calcutti 1 (GC 1) was released in 1981 for *rustica tobacco* growing areas of Gujarat. Gujarat Calcutti 1 is a result of intensive selection from the local material at Anand. It is dark green foliage, suitable for *Lal* and *Kala chopadia* as well as *Bhuka* with a yield potential of 2693 kg/ ha. It is still popular among the farmers.

Gujarat Calcutti Tobacco-1
Courtesy by BTRS, Anand

Gujarat Calcutti Tobacco 2 (GCT 2)

The promising hybrid derivative of parents GC 1 and Cocker 1 known as culture AR 26, exhibiting its superiority in yield (3512 kg/ha) and acceptable quality over GC 1 was released in 1994 as Gujarat Calcutti Tobacco 2 (GCT 2), for general cultivation in *rustica* tobacco growing areas of middle Gujarat.

Gujarat Calcutti Tobacco-2
Courtesy by BTRS, Anand

Gujarat Calcutti Tobacco 3 (GCT 3)

Gujarat Calcutti Tobacco 3 with a yield potential of 3285 kg/ha in middle Gujarat and 5986 kg/ha in north Gujarat N as developed and released by Ladol in 2001 for cultivation in north Gujarat area. The produce of the variety is suitable for Lal chopadia and has light green colour with moderate degree of spangling.

Gujarat Calcutti Tobacco-3
Courtesy by BTRS, Anand

4.4 *Rustica* Tobacco Varieties of West Bengal

Dinhata centre developed three high yielding varieties (DD 437, Sonar Motihari and Dharla) for cultivation in Motihari tobacco growing areas of West Bengal.

DD 437

DD 437 is a single plant selection from bulk crop of Hemti. It matures in 120 to 130

DD 437
Courtesy by CTRS, Dinhata

days. Its average yield is 2200 kg/ha. It is released in 1977. It is still popular in Motihari tobacco growin areas of West Bengal.

Sonar Motihari

Sonar Motihari is also released in 1977, an early maturing variety (100-110 days) derived from parents BVM 5 X Sel. 47-33. It has leathery and dark brown leaves with pungen aroma. Its average yield is 1690 kg/ha.

Dharla

Dharla a late maturing and high yielding variety, released by Tobacco Variety Release Committee of West Bengal released it for sandy loam and sandy clay soil region in Motihari tract of West Bengal in 2001. It matures in 120-130 days and yields 2700 kg/ha. It registered 20-22% more yield than the check DD 437.

❑❑❑

PART-IV
Tobacco Trade

19

Economic Significance of Tobacco

There is no other alternate to replace tobacco in India, as in many other countries, it yields higher net returns per unit of land than most other cash crops, and substantially more than food crops. Tobacco being one of the major carcinogenic agents because it contains nicotine and some other compound, is used by masses in various form and its serious impact on health is well established hence in the interest of public health; India is signatory and party to the WHO Framework Convention on Tobacco Control (FCTC). India is the world's third-largest producer of leaf tobacco. It is also a very large consumer of tobacco products. Tobacco is one of the important cash crops in the country, and makes a significant contribution to the Indian economy in terms of employment, income and government revenue. It generates nearly Rs 20 billion of income per annum. The economic importance of the crop can be considered at three levels: farm households engaged in tobacco growing and processing; major tobacco producing States; and central government level.

There are at least two important characteristics of the agricultural industry that makes it distinct from all others. One is the cyclical nature of production caused primarily by physical and biological factors. The other is price instability resulting from the effects of changes both physically caused and from within the market for agricultural products. The cyclical nature of production can be seen in the cattle market for instance, where there is a long lag between the times when increasing beef prices motivate producers to increase production and when the results of their decisions show up at the meat counter of the local grocery store.

Farm Level

There are an estimated 8,50,000 growers of tobacco in the country, characterized by small family farms, with farmers owning less than 2 ha of land forming about half of all tobacco growers. However, based on field surveys carried out by Gujarat Agricultural University, Anand, in selected tobacco growing districts in Gujarat and Karnataka, the small-scale producers account for about a quarter of the tobacco area. In total, nearly 6 million farmers and workers depend on this sector for their livelihood. In addition, the tobacco sector provides direct and indirect employment to a large number of people in many related industries.

State Level

The tobacco sector contributes to the States economies through crop production and to the exchequer through excise duty, as well as from their share in the net central excise duties. Tobacco cultivation is mainly concentrated in three states: Andhra Pradesh, Gujarat and Karnataka. The additional excise duties on tobacco, which is distributed only among tobacco growing states, increased from Rs 3,477 million in 1979/80 to Rs 18,532 million in 1999/2000. During the fiscal years 1995 to 1998, the State Governments received annually 47.5 percent of the net proceeds of union excise duty levied on a number of commodities, including on tobacco products. It amounted to around Rs 2,40,000 million annually, of which the share from the tobacco sector is estimated to have been around 9 percent. These are an important source of revenue for the States.

Central Government Revenue

In India, excise duty is imposed on the entire range of manufactured tobacco products. In 1998- 99, tobacco contributed about

Rs 59,400 million to the central government's revenue or 10.6 percent of total excise collection. Tobacco contributed Rs 7,790 million to export earnings in 1998-99, which was around 5 percent of the foreign exchange earnings from agricultural products. In addition, the central government also realized on average around Rs 2 000 million per annum from tobacco enterprises in the form of corporate tax during the last three tax years. However, raw tobacco was exempted from excise with effect from 1979-80, primarily because the administration and collection of excise duty on raw tobacco was expensive as well as cumbersome, and control was ineffective. The consequent revenue loss from the exemption was more than compensated for by steep increases in tax rates on manufactured items. Although most manufactured tobacco products attract excise duties, tobacco products from cottage industries (bidis, etc.) are taxed at a much lower rate than those from the organized sector (*i.e.* cigarettes). Bidi manufacturers producing less than 2 million pieces annually do not have to pay any excise duty. Bidis (other than paper rolled) produced without the aid of machines pay Rs 5 per thousand pieces. Other bidi manufacturers currently pay Rs 15.5 per thousand pieces. Pan masala is taxed 40 percent ad valorem (24 percent basic duty plus a special duty of 16 percent). Chewing tobacco and snuff with a brand name attract 50 percent excise duty (ad valorem).

Indian Tobacco Industry

The Bidi Industry

Bidi is tobacco rolled in a tendu leaf and tied by a string. Tendu leaf accounts for 74 percent by weight of bidi. Dark and sun-dried tobacco varieties are used in bidi production. Almost 80 percent of bidi tobacco comes from Gujarat, and the rest comes from Karnataka. Bidis account for over 50 per cent of total tobacco use, compared with less than 20 percent by the cigarette segment. There are an estimated 290 000 growers of bidi tobacco. The collection of tendu leaf that is used to wrap bidis forms an important link for the bidi industry. Tendu leaf is almost wholly grown on government-owned forestland, with around 62 per cent of tendu leaf being grown in Madhya Pradesh.

Annual production of tendu leaf in 1994-95 had an estimated value of Rs 14 700 million. About 2 million people are engaged in leaf collection, while another 4.4 million people are employed directly for bidi rolling. Bidi rolling is concentrated in the states of Madhya Pradesh,

Andhra Pradesh, Tamil Nadu, Uttar Pradesh and West Bengal. Bidis are manufactured largely in the independent small-scale and cottage industry sector. There are a few large manufacturers of branded bidis, which tend to be closely-held, family-run businesses. The bidi industry is estimated to have used 268 000 tonnes of tobacco in 1998-99, 54.4 percent of the total apparent tobacco use.

Cigarette Industry

Currently, there four major cigarette manufacturers in India: ITC Limited (formerly Imperial Tobacco Co.); VST Industries Limited (formerly Vazir Sultan Tobacco Co.); Godfrey Philips India Ltd; and GTC Industries Limited (formerly Golden Tobacco Co., Ltd.). There are a couple of smaller-sized cigarette companies with manufacturing facilities. As they lack the necessary marketing infrastructure, they produce cigarettes for the large cigarette companies on a sub-contractual basis. Production of cigarettes reached a peak of 96.1 billion pieces in 1984-85 and then declined. It recovered again in the 1990s. The Indian cigarette market was reportedly worth Rs 66 billion in 1997. The average price for a pack of ten cigarettes increased from Rs 4 in 1990-91 to Rs 6 in 1998/99. Currently, the retail price of a pack of 10 cheap cigarettes is Rs 6, against Rs 3 for a bundle of 25 bidis. Bidis largely escape tax as most are produced in cottage industries across the country. Attempts to raise the taxes on bidis are often interpreted as an attack on the poor, and therefore regarded as politically inexpedient.

Economic Analysis of Tobacco Manufacturing Enterprises

It is estimated that over 2.3 million persons depended on this sector for their livelihood. The annual wage bill in these enterprises averaged Rs 4 300 million, and annual wages per worker varied from Rs 8 400 in bidi factories to Rs 55 730 in cigarette, cigar and cheroot factories. The total net value added by all enterprises averaged Rs 15 000 million per annum, of which bidi factories contributed 41.2 percent, and cigarette and allied industries 34.3 percent. The total annual wage bill in the cigarette and allied industries, despite wages per worker being substantially higher, was only 4 percent of its gross value of output, compared to 16 percent in the bidi factories, because bidi manufacturing is more labour intensive. Bidi manufacturing is estimated to provide employment to more than 4.4 million workers, a large number of whom are women and children. If the forward and backward economic linkages are taken into account, bidis generated 1 310 million workdays, whereas cigarettes generated 340 million workdays. The bidi industry provides employment to a large number of workers - some 4.4 million workers employed in bidi rolling alone. Around 22 percent of these workers depend upon bidi rolling as their sole source of income. The employment figures do not include the employment created in tendu leaf picking (in the case of bidis) or in the supplier industries for the cigarettes industry, such as the manufacture of paper, packaging and machinery used in cigarette manufacture.

Impact of Introduction of New Processing Technologies and Equipment

The cigarette manufacturing companies have been modernizing their processing and manufacturing facilities, with new or upgraded machines introduced for cigarette making. The impact of the advanced technology has been to improve processing, reduce component prices and improve quality and efficiency. It has also reduced cigarette damage and waste; reduced tar and nicotine levels through filterization; reduced usage of tobacco per cigarette; and improved cigarette paper and filter design. The share of filter-tipped cigarettes increased from less than 30 percent in 1970 to 66 percent in 1998. The reduction in tobacco used per cigarette has been significant. According to industry sources, currently about 750 g of processed tobacco is used to manufacture 1 000 cigarettes, against the 1 000 g used three decades ago. However, the cigarette industry in India has generally not kept pace technologically with the developed countries. For instance, the speed of cigarette makers and packers in India is 2 000 to 3 000 cigarettes per minute (cpm), compared to 7 000-10 000 cpm abroad. The plants abroad have very high levels of automation in primary processing, material handling systems and secondary technology.

Shifting from Tobacco to Other Crops

Based on the research conducted by respective research organization only limited number of specialized crops in various areas that provide similar incomes, but it is estimated that these crops would not remain remunerative if total production increased. The economics of alternative crops is generally based on experiments carried out on a limited area at research stations under optimal conditions. More detailed research work is needed on a wider scale at farmers' field level before firm recommendations can be made about them. In general, under farmers' field conditions, most other alternative crops, as discussed below, are currently not as remunerative as tobacco. If tobacco farmers go for crop diversification with other crops, they are likely to not get more or even equal income as in case of tobacco and may suffer economic loss.

More importantly alternative crops also require high levels of inputs, assured irrigation and market support. Whereas in some part of country, tobacco is preferred due to its drought resistance and suitability for growing under rainfed conditions. Other problems associated with substitution by other crops include the capital invested in specialized facilities created for tobacco processing, which cannot be used for other crops; the difficulties of finding substitute crops for rainfed areas; and the dependency of millions of people on bidi rolling and tendu leaf collection. Moreover, with an assured market and prompt payment of sale proceeds through the Tobacco Board, it will be difficult to replace FCV tobacco as a crop.

It is worthwhile comparing rates of return from tobacco with competing crops such as groundnut or cotton. For this, Karnataka is taken as a representative state for FCV tobacco. The data show that while tobacco typically yields high profits, it can also show severe losses at times. Of the three crops, groundnut consistently showed reasonable profits over cost of cultivation. In contrast, cotton and tobacco showed fluctuations in profits. Though data are not available for comparable years, it may be assumed that the trend observed for several years for these crops will not have changed much in recent years.

In the light of oversupply of FCV tobacco in recent years, some comparisons have been carried out on profits from growing tobacco versus other crops in FCV-growing areas. The comparison indicates that some other crops can be grown in that soil, and that some of the crops can be more remunerative than tobacco. For instance, the Central Research Institute found that combinations of hybrid maize and black gram yielded promising levels of net returns per hectare in black cotton soils of Andhra Pradesh, as did hybrid maize and soybean.

Table 19.1 Net return from different crops in Karnataka (Rs/quintal)

Crop year	Groundnut	Cotton	Tobacco
1982-83	71.11	46.77	313.71
1983-84	121.61	103.13	146.94
1985-86	77.27	-57.71	N.A.
1986-87	119.15	-17.72	-1 069.31
1987-88	N.A.	N.A.	-432.25
1991-92	145.50	N.A.	1 110.00

Notes: N.A.= not available. 1 quintal = 100 kg.
Source: Directorate of Economics and Statistics. Cost of Cultivation of Principal Crops in India. (various issues).

Bidi tobacco is generally less remunerative to farmers compare to VFC tobacco. Some studies have shown other crops, such as chili or cotton, or a combination of soybean and rabi sorghum, as well as groundnut and rabi sorghum, could give higher returns than a sole crop of tobacco. However, other studies. Studies conducted by the Department of Agricultural Economics, University of Agricultural Sciences, Dharward, on Economics of Bidi Tobacco in Nippani Area in Belgaum District; and have shown different results, with bidi tobacco yielding higher net return per hectare than soybean, sorghum, cotton or groundnut, with only sugar cane being more profitable than tobacco. Sugar cane could be the most favoured crop in the region wherever irrigation is available. Moreover, the extensive research programme carried out by CTRI show that currently no alternative crop tested under monocropping gives returns comparable to tobacco. Intercropping or double-crop returns were equal to monocropped tobacco. It is important to state that tobacco is generally raised as a sole crop, except in areas where ample and assured irrigation facilities allow a second crop.

Micro-Level Information on Shifting from Tobacco to Other Crops

A field studies carried out by the National Council of Applied Economic Research (NCAER, 1994) and by CMDR showed a number of major socio-economic factors encouraging tobacco growing:

- Richer farmers tend to prefer tobacco to other crops.
- Small-scale farmers take to tobacco cultivation as something inevitable in the absence of a suitable alternative.

- Tobacco as a crop gives superior net economic returns compared with alternative crops.
- Tobacco is preferred due to its drought resistance and suitability for growing under rainfed conditions. Due to tobacco's soil preferences, cultivation is concentrated in certain states, and even within major tobacco growing states, the crop is grown in specific districts.
- A widespread belief prevails among farmers, especially in bidi growing areas, that no other crops should be grown in the same land where tobacco is cultivated, as it will lower the quality of the subsequent crops. However, this is contrary to scientific recommendation that tobacco should be grown alternate years.
- The prevalent practice of growing only tobacco every year is reinforced by bidi manufacturers through their agents, who may refuse to purchase tobacco if any other crop has been grown on the same plot. Marketing of non-FCV tobacco has been a major problem and there have been allegations of agents exploiting farmers.
- A well organized marketing system for FCV tobacco through the Tobacco Board assures prompt payment to farmers, which is not the case for many other crops.
- Farmers are reluctant to give up tobacco cultivation because of heavy investment in irrigation equipment and barns.
- A change in cropping is practicable only when some assured irrigation is available. For example, the coming online of Nagarjuna Sagar Dam led to a radical change in cropping pattern - from tobacco to sugar cane.
- Failure of other crops raised in the past.

The farmers favour tobacco cultivation as it generally yields higher returns than other crops. However, the cost of cultivation of tobacco is also much higher. It seems that farmers often do not consider the full economic implications - both costs and returns - of tobacco cultivation. The Tobacco Board declared a complete crop holiday for FCV tobacco in Andhra Pradesh and lowered production targets for Karnataka from 40000 tonnes in 1999/2000 to 25000 tonnes for the 2000-01 season. As a result, some of the progressive farmers devoted a part of their tobacco areas to other crops, such as sugar cane, chili, groundnut or cotton. The CMDR team surveyed 74 non-tobacco growers in the tobacco region of Karnataka, to learn about reasons for not cultivating tobacco. It seems that the majority of such farmers were convinced of the economic difficulties of tobacco cultivation. It should be stressed that more detailed studies of these aspects are needed to derive further insights useful for policy formulation.

Economics of Inter-cropping Systems as an Approach to Agricultural Diversification

Field experiments on inter-cropping conducted by CTRI's station in Pusa, Bihar, from 1990 to 1997, showed that mixed cropping was more profitable than tobacco monocropping. Tobacco plus garlic; tobacco plus red kidney bean; and tobacco plus potato all showed potential to provide better returns than tobacco alone. Such inter-cropping systems may be the first step to moving away from tobacco. A package of mixed cropping, shifting to other crops with suitable crop insurance facilities, adequate farm inputs for the alternative crops, adequate marketing facilities, etc., would be needed to ensure the success of a policy of gradually shifting from tobacco.

Economic Implications of Diversification on Tobacco Land

An attempt has been made to assess the direct implications for farmers and agricultural labourers of shifting land from tobacco to alternative crops. CTRI identified certain best crops for different areas in a number of tobaccos producing States. Net returns per hectare from all of these crops were generally lower than from tobacco alone. Thus, even if the farmers can grow these alternative crops, they are likely to lose rather than gain. Moreover, given the wide variations in quality and other specifications of tobacco and other alternative crops, the determining factors will not only be prices but also the capacity to expand production of alternative crops with requisite quality and to put in place support infrastructure. The results of the CTRI's studies are used here to assess the impact on farmer income of substitution of tobacco by the next best alternative crops, while the official data on Cost of Cultivation of Principal Crops in India has been used to analyse implications for employment. The assessment has been worked out on a per hectare basis where such data are available. The reduction in net income that might result from a shift to the next most profitable crop is estimated to average about 23 percent. This would amount to a considerable loss, especially for the marginal and small-scale farmers that constitute the bulk of the farmers in most States. Bigger-scale farmers tended to adopt tobacco cultivation more than others. In rural areas, a large number of households depend largely on tobacco for their cash income, especially in the three major tobacco growing states. Most of the marginal and small-scale farmers practice tobacco monoculture and rely mainly on this crop for their livelihood. They have to buy staple grains and other essential items for family consumption.

A study on the Tobacco Industry in Andhra Pradesh, commissioned by the Tobacco Institute of India, assessed the expenditure of tobacco farmers on baskets of household goods and services. The marginal and small-scale farmers depended on tobacco as the main source of their family income and livelihood. The overall average annual income of the tobacco farmer in the State came to about Rs 15,000. The largest share of household consumption expenditure was devoted to food items, some 67 percent. In the absence of detailed information in the study on consumption patterns, results of the survey conducted by the National Sample Survey of Household Consumer Expenditure for 1993-94 for Rural Andhra Pradesh (NSSO, 1996) can serve as a proxy. Among food items, cereals and pulses accounted for 24 percent and 4 percent respectively, with annual cereal consumption per household about 800 kg. Other major food items included milk and milk products (9.7 percent); fruit and vegetables (7.7 percent); and meat, eggs and fish (3.3 per cent). Of non-food items, fuel and light accounted for 7.3 percent, followed by clothing (5.3 percent), while cigarettes and bidi were 2 percent of total household expenditure.

To assess the impact on the food security of the tobacco-growing farmers if they were to shift their tobacco land to other crops, the overall reduction in income per hectare of 23 percent from the alternative crops has been used. With the loss of income from the tobacco crop, the farmer would need to readjust the family consumption expenditure pattern, either by devoting a higher share of family expenditure to food or by reducing food consumption, especially among the marginal and small-scale farming households. In the event of a total switch from tobacco production, with a consequent estimated average drop of 23 percent in annual household income, cereal consumption would contract by around 10 percent. Average family cereal consumption could decline from 800 to 720 kg.

This would have an adverse affect on the health and well-being of marginal and small-farmers families, who are already undernourished.

A shift away from the tobacco crop would be likely to reduce the demand for hired labour in most states, as the hired labour required for alternative crops is substantially lower. Total labour requirements for alternative crops, including family labour, would be some 35 percent lower. Any reduction in employment of hired labour in rural areas would seriously affect the landless labourers, who depend for their livelihood by working mainly in the agricultural sector, as well as for the marginal and small-scale farmers who supplement their earnings by working on other large farms as hired labourers. Therefore, the reduction in employment for hired labour in areas where the tobacco crop is replaced with other crops would result in increased economic and food insecurity, not only among a large section of the landless in rural areas but also among the marginal and small-scale farming families.

Costs of Smoking

A task force of the Indian Council of Medical Research conducted a research study from 1990 to 1996 on the Cost of Tobacco-Related Diseases in India. The average economic cost of major diseases due to tobacco use in India in 1999 was estimated at Rs 3,50,000 for cancer, Rs 29,000 for Coronary Artery Disease (CAD) and over Rs. 23,000 for Chronic Obstructive Lung Disease (COLD). Total losses from these tobacco-related diseases in 1999 were about Rs 277.6 billion, equivalent to US$6.5 billion. The medical experts, however, consider the estimated total loss to be an underestimate as it is derived from only a small sample.

□□□

20

Tobacco Consumption and Production

The study reviews the status of the tobacco sector in India, both raw tobacco and its manufactured products. An attempt has been made to identify major economic and social factors affecting tobacco production and consumption, export and to explore the economic implications of government policy measures for tobacco control. Although people have used tobacco for centuries, cigarettes did not appear in mass-manufactured form until the 19th century. Since then, the practice of cigarette smoking has spread worldwide on a massive scale.

Today, about one in three adults, or 1.1 billion people, consumes tobacco. Of these, about 80 percent live in low-and middle-income countries. Partly because of growth in the adult population and partly because of increased consumption, the total number of smokers is expected to reach about 1.6 billion by 2025. In the past, tobacco was often chewed, or smoked in various kinds of pipes. While these practices persist, they are declining. Manufactured cigarettes and various types of hand-rolled cigarette such as bidis-common in Southeast Asia and India-now account for up to 85 percent of all tobacco consumed worldwide.

Regional Patterns of Consumption

Data on the number of smokers in each region have been compiled by the World Health Organization using more than 80 separate studies. For the purpose of this report, these data have been used to estimate the prevalence of smoking in each of the seven World Bank country groupings. As table 20.1 shows, there are wide variations between regions and, in particular, in the prevalence of smoking among women in different regions. For example, in Eastern Europe and Central Asia (mainly the former socialist economies), 59 percent of men and 26 percent of women smoked in 1995, more than in any other region. Yet in East Asia and the Pacific, where the prevalence of male smoking is equally high, at 59 percent, just 4 percent of women were smokers. The populations of the low- and middle-income countries have been increasing their cigarette consumption since about 1970. The per capita consumption in these countries climbed steadily between 1970 and 1990, although the upward trend may have slowed a little since the early 1990s. While the practice of smoking has become more prevalent among men in low- and middle-income countries, it has been in overall decline among men in the high-income countries during the same period. For example, more than 55 percent of men in the United States smoked at the peak of consumption in the mid-20th century, but the proportion had fallen to 28 percent by the mid-1990s. Per capita consumption for the populations of the high-income countries as a whole also has dropped. However, among certain groups in these countries, such as teenagers and young women, the proportion who smoke has grown in the 1990s. Overall, then, the smoking epidemic is spreading from its original focus, among men in high-income countries, to women in high-in-come countries and men in low-income regions. In recent years, international trade agreements have liberalized global trade in many goods and services. Cigarettes are no exception. The removal of trade barriers tends to introduce greater competition that results in lower prices, greater advertising and promotion, and other activities that stimulate demand. One study concluded that, in four Asian economies that opened their markets in response to U.S. trade pressure during the 1980s-Japan, South Korea, Taiwan and Thailand-consumption of cigarettes per person was almost 10 percent higher in 1991 than it would have been if these markets had remained closed. An econometric model developed for this report concludes that increased trade liberalization contributed significantly to increases in cigarette consumption, particularly in the low and middle-income countries.

Table 20.1 Estimated smoking prevalence by gender and number of smokers in population aged 15 or more, by World Bank region, 1995.

World Bank Region	Smoking prevalence (%) Males	Females	Overall	Total smoker (millions)	(% of all smokers)
East Asia and Pacific	59	4	32	401	35
Eastern Europe & Central Asia	59	26	41	148	13
Latin America & Caribbean	40	21	30	95	8
Middle East & North Africa	44	5	25	40	3
South Asia (cigarettes)	20	1	11	86	8
South Asia (bidis)	20	3	12	96	8
Sub-Saharan Africa	33	10	21	67	6
Low/Middle Income	49	9	29	933	82
High Income	39	22	30	209	18
World	47	12	29	1,142	100

Note: Numbers have been rounded.
Source: Author's calculations based on World Health Organization. 1997. Tobacco or health: a Global Status Report. Geneva, Switzerland

Smoking and Socioeconomic Status

Historically, as incomes rose within populations, the number of people who smoked rose too. In the earlier decades of the smoking epidemic in high-in-come countries, smokers were more likely to be affluent than poor. But in the past three to four decades, this pattern appears to have been reversed at least among men, for whom data are widely available.2 Affluent men in the high-income countries have increasingly abandoned tobacco, whereas poorer men have not done so. For example, in Norway, the percentage of men with high incomes who smoked fell from 75 percent in 1955 to 28 percent in 1990. Over the same period, the proportion of men on low incomes who smoked declined much less steeply from 60 percent in 1955 to 48 percent in 1990. Today, in most high-income countries, there are significant differences in the prevalence of smoking between different socioeconomic groups. In the United Kingdom, for instance, only 10 percent of women and 12 percent of men in the highest socioeconomic group are smokers; in the lowest socioeconomic groups the corresponding figures are threefold greater 35 percent and 40 percent. The same inverse relationship is found between education levels a marker for socioeconomic status and smoking. In general, individuals who have received little or no education are more likely to smoke than those who are more educated.

Trend in Tobacco Consumption in India

Total consumption of tobacco products has been increasing in India, as in many other developing countries, notwithstanding the increased awareness of negative health effects. Total tobacco consumption increased from around 3,00,000 tonnes in 1971-72 to 4,50,000 tonnes in 1998/99 - an increase of 1.6 percent annually. Only about 20 percent of the total tobacco consumed in India (by weight) is in the form of cigarettes. Bidis account for about 40 percent of tobacco consumption (about 950 billion bidis), with the rest divided among chewing tobacco, pan masala, snuff, hookah, zarada and other mixtures. Cigarette smoking is essentially an urban phenomenon 80 percent of cigarette smokers are in urban areas, while 80 percent of bidi smokers are in rural areas. Per capita consumption of cigarettes remains very low by international standards. This reflects the lower level of incomes and the substantial use of cheaper alternative tobacco products resulting from traditional habits that are supported by the very low or non-existent taxes on these products. Manufactured cigarettes, on which heavy excise duty is levied, remain unaffordable for many. Per capita consumption of cigarettes declined from 99 pieces in 1990 to 87 pieces in 1993. However, the reduction in taxation on small non-filter cigarettes in 1994 attracted more smokers to switch from bidis, and also encouraged downgrading, both of which have helped increase per capita consumption quite substantially in the latter half of the 1990s. Thus, per capita consumption had reached 108 pieces/ year by 1997.

Annual average consumption of cigarettes per adult (aged 15+ years) increased from 170 pieces during 1970-72 to 180 pieces during 1980-82. However, it declined to 150 pieces during the next decade, but recovered to 170 pieces during the last three years. Annual average consumption per adult of bidi increased from 840 to 1350 during 1996-98. The increase in consumption in tobacco products in general has been due to population increase, especially the smoking population, higher per capita income in real terms, (which increased from Rs. 1650 in 1970-71 to Rs. 3780 in 1998- 99), changing tastes accompanying increased income, changes in prices of various products over time, and, last but not least, government taxation policy. Data on the smoking population are not available except for a survey conducted by NSSO (1996), which provided information on tobacco users in terms of proportion of population. Based on this information, the tobacco using population is estimated to have increased from 201 million in

1986-87 to 203.7 million in 1993-94, or an increase of around half a million annually.

Table 20.2 Annual consumption of cigarettes and bidis by adults in India

Period	Annual average number consumed per adult (aged 15+)		
	Cigarettes	Bidis	Total
1970-72	170	840	1 010
1980-82	180	1 130	1.310
1990-92	150	1 220	1 370
1996-98	170	1 350	1 520

Sources: WHO, 2000; ERC Statistics International, 1998.

The consistent growth of the bidi segment in the smoking market has been partly due to taxation being lower then cigarettes, which has given bidis a considerable price advantage. Other factors supporting rapid growth in demand for bidis include the traditional habit of bidi smoking in the family, which is passed on to the children; the relatively low income level of a large part of the population, especially among the rural masses; and increased use of bidis by women in rural areas in certain states, as there is no inhibition to their smoking. By contrast, the government taxation policy seems to have restricted growth of the cigarette market. From 1970-71 to 1997-98, cigarette taxes increased almost 15-fold (from Rs 31 to Rs 439 per 1 000 cigarettes). This resulted in smokers using more bidis. The Indian tobacco market is highly price sensitive. The price elasticity for cigarette consumption was estimated at -0.66 between 1967-68 and 1992-93, i.e. a 10 percent increase in price would result in a reduction in consumption of 6.6 percent. While increased prices have a constraining affect, the wide use of sponsorship of sport events, publicity and advertisement by the cigarette companies have helped introduce large numbers of young people to cigarette smoking.

Based on the NSSO (1998) survey data, the total population using tobacco was estimated at 203.7 million in 1993-94. Among these over 53 percent were smokers and the remaining 47 percent were using non-smoking tobacco products. Chewing tobacco was used by 34 percent of the tobacco using population. Around 12 percent used more than one product. Bidi smokers formed 78 percent of the smoking population. Total annual private final consumption expenditure

(PFCE), in current terms, on tobacco and tobacco products increased from Rs 8 billion in 1970-71 to Rs 236.7 billion in 1998-99. In real terms, based on official sources and industry estimates, it increased from nearly Rs 19 billion to Rs 54 billion. The share of tobacco and tobacco products rose from 2.6 percent of PFCE in 1970-71 to 2.9 percent. The cigarette segment accounted for 37 percent of total consumer spending on tobacco, while the bidi segment contributed 30 percent. The corresponding figure for the non-smoking sector was 33 percent. However, while for cigarettes as much as 61 percent of the total consumer spending went towards government excise revenue, the corresponding figure for the bidi segment was 5 percent. Moreover, the figure for the non-smoking sector, which has the highest share in consumer expenditure, was also only 4 percent in 1998-99. The annual spending per tobacco user was Rs 1,150. Annual expenditure per smoker is estimated to have been Rs 2,750 for cigarettes and Rs 835 for bidis, while for non-smoking tobacco users it was Rs 906. Government policy regarding consumption of tobacco products applies mainly to its taxation measures and a limited control on use of tobacco products. The motivating rationale for tobacco taxation seems to be fiscal (i.e. to maximize the excise revenue). Measures to control the use of tobacco on health grounds seem to have had only a small impact, partly because only a few control measures are in place, and those are not fully implemented. These measures include bans on advertising cigarettes, but applied only on television and commercial radio stations; a single health warning on cigarette packets; and bans on smoking in public places, but in only three States.

Age and the Uptake of Smoking

It is unlikely that individuals who avoid starting to smoke in adolescence or young adulthood will ever become smokers. Nowadays, the overwhelming majority of smokers start before age 25, often in childhood or adolescence, in the high-income countries, eight out of 10 begin in their teens. In middle-income and low-income countries for which data are available, it appears that most smokers start by the early twenties, but the trend is toward younger ages. For example, in China between 1984 and 1996, there was a significant increase in the number of young men aged be-tween 15 and 19 years who took up smoking. A similar decline in the age of starting has been observed in the high-income countries.

Global Patterns of Quitting

While there is evidence that smoking begins in youth worldwide, the proportion of smokers who quit appears to vary sharply between high-income countries and the rest of the world, at least to date. In environments of steadily increased knowledge about the health effects of tobacco, the prevalence of smoking has gradually fallen, and a significant number of former smokers have accumulated over the decades. In most high-income countries, about 30 percent of the male population is former smokers. In contrast, only 2 percent of Chinese men had quit in 1993, only 5 percent of Indian males at around the same period, and only 10 percent of Vietnamese males had quit in 1997. With these assumptions, one may calculated that the number of children and young people taking up smoking ranges from 14,000 to 15,000 per day in the high-income countries as a whole. For middle- and low-income countries, the estimated numbers range from 68,000 to 84,000. This means that every day, worldwide, there are between 82,000 and 99,000 young people starting to smoke and risking rapid addiction to nicotine. These figures are consistent with existing estimates for individual high-income countries.

Tobacco Production

The increasing yield of VFC tobacco during the 1990s reflected the boost from government policy, through the Tobacco Board, for this variety. The improved technology and the cultural practices recommended by the Central Tobacco Research Institute (CTRI) and other institutions also helped to boost production. The widespread adoption of improved varieties released by CTRI and other research stations, combined with the adoption of improved cultural practices, have improved unit yields. Substantial increases in the use of fertilizers and insecticides for tobacco have also played an important role. Data on total quantities of these inputs used on the tobacco crop are not available for the country as whole, but research indicates that tobacco farmers in India use dosages of these inputs that are substantially higher than those recommended by research stations. However, in Andhra Pradesh, according to estimates fertilizer application per hectare on tobacco increased almost 250 percent in the five years from 1990- 91 to 1994-95, while the use of insecticides doubled. Similarly, increased use of irrigation, which gives higher yields compared with rainfed production, has also made an important contribution. All of these inputs are supplied at subsidized prices. Finally, price increases also seem to have stimulated higher output. Farm harvest prices for tobacco have increased three- to ten-fold in the last three decades, depending on region. Moreover, the increase in wholesale prices for tobacco has been higher than for cereals or other alternative crops, such as cotton, pulses, chili or groundnut.

A wide variety of tobaccos are grown in 16 states in India under diverse agroclimatic conditions. However, most of the varieties grown (other than Virginia, Burley and Oriental) are of non-cigarette types. These include natu, bidi, chewing, hooka (hookah), cigar and cheroot tobaccos and account for about 77 percent of the total output . Cultivation of FCV tobacco was initially confined to the traditional black soil areas of Andhra Pradesh. However, to suit the quality requirements in internal and export markets, cultivation of FCV was encouraged in light soils in Karnataka and Andhra Pradesh. In the initial years, the varieties grown were limited to Havana tobacco used in cigars, and Lanka tobacco used in the manufacture of snuff and bidis. Subsequently, other forms, like FCV, were introduced.

Impact of Government Policies on Production of Tobacco

Even though tobacco comes under state jurisdiction, the Government of India plays an important role in the growth and development of the tobacco industry. In fact, at least six ministries of the Union Government - Agriculture, Commerce, Finance, Industry, Labour, and Rural Development - deal with one or another specified aspects of the industry. Following the increasing health concern about tobacco consumption, the central Ministry of Agriculture has not launched any development scheme for the crop since the completion of the Seventh Five-Year Plan (1985-90). However, in general, government policy has been to promote production, improve quality and ensure remunerative prices for growers.

□□□

21

Tobacco Marketing

Marketing is art as well as science in general; tobacco marketing in India is a complex event right from raw producer (growers) to middleman at village level to district, state as well as National level. Different channels (system) of marketing are prevailed for various kind of tobacco. In this chapter a brief account has been given tobacco marketing channels in India with background information tobacco business.

The introduction of the auction system by the Tobacco Board brought an element of competition to the tobacco leaf market and freed the market from pricing and grading anomalies. Farmers intending to grow Virginia tobacco are required to register with the Board every year. Production quotas are fixed. Nevertheless, the Tobacco Board, which has responsibility for regulating production, marketing and exports of FCV tobacco grown in the states of Andhra Pradesh, Karnataka and Mahaarshtra, has not been completely successful in controlling the area under tobacco. In fact, controlling the area to be planted is an ineffective instrument for controlling production since price incentives, climatic conditions and the expertise of individual growers can greatly change the yield per hectare on individual farms from season to season. The Board's attempts to reduce fluctuations in auction prices have met with mixed success. Moreover, cultivation and trade of non-FCV varieties, constituting over 75 percent of tobacco production, is still outside the purview of the Tobacco Board. The Directorate of Tobacco Development handles marketing of non-FCV tobacco. The Central Ministry of Agriculture continues its efforts to streamline the marketing of non-FCV tobacco in consultation with the State Departments of Agriculture. They provide limited facilities, such as market yards, and their impact on prices, etc., is restricted. In this sector, traders and manufacturers are more active and they take advantage of weak bargaining positions, especially in cases where they provide loans and or inputs on credit to farmers for raising the tobacco crop.

Tobacco Business in India-Major Points

- India ranks second in the production of tobacco in the world
- India is even one of the largest consumers of tobacco and tobacco based products
- The yearly production of tobacco in India is 5,80,000 metric tonnes
- Nearly 12 lakh people are involved in the cultivation of tobacco in India
- Tobacco and tobacco products consists of cigarettes, bidis, zarda, cigars, etc
- Bidi manufacturing is the largest cottage industry in India
- Nearly 5 million people are involved in the commercial operation pertaining to tobacco
- Tobacco is consumed by smoking of cigars, hukka, bidis, cigarettes, by chewing of zarda, pan masala, tobacco leaves, etc.

Tobacco Business in India-Revenue

- The contribution of the excise duty on cigarettes to the total revenue is 8.8 %
- The excise duty on cigarettes was levied at a rate 10%, which was later reduced to 8.8 %
- The contribution of the excise duty on other tobacco products such as bidis, cigars, etc, to the total revenue is 3.3%

Tobacco Business in India-Exports

- The domestic consumption of tobacco is very high
- The export of tobacco depends on the demand for the same in the country
- The highest export earnings pertaining to tobacco comes from tobacco leaves
- The earnings from the export of cigarettes has declined by 76%
- The earnings from the export of tobacco leaves has increased by 99%.

The Tobacco Board, established in 1976, provides marketing services for FCV tobacco, through its compulsory auction system. Virginia tobacco at the primary level is sold through auctions conducted by the Board. Exporters of tobacco, manufacturers of cigarettes and dealers in tobacco wishing to participate in the auctions have to register as a buyer for each auction floor, wherever they intend to operate. However, the auction system has not been successful in reducing fluctuations in the levels of auction and export prices. Table 4.4 compares price levels for selected representative grades at both the production and export levels. These grades include F2 at the farm level and its equivalent export grade, namely L2. CTRI is an apex research body for tobacco in India, and has been successful through its multidisciplinary programmes in evolving a number of high yielding cultivars of tobacco and in improving quality. In addition to the CTRI programmes, six research stations - located in Andhra Pradesh, Gujarat, Karnataka, Orissa and Uttar Pradesh-under the All India Coordinated Research Project on Tobacco have evolved high yielding varieties for different types of tobacco for various states. They have also developed improved crop management and crop rotation systems.

Government Interventions in Support of the Industry can Broadly be classified into

(i) Institutional and regulatory support;
(ii) Price and market support;
(iii) Export promotion;
(iv) Research and development (R&D); and
(v) Direct fertilizer and credit subsidies.

All these interventions involve explicit or implicit subsidies for the tobacco industry.

Marketing Channels and Price Determination of Cigarettes

In 1984, a compulsory auction system was introduced for the Virginia varieties normally used in cigarette manufacture and sold in overseas markets. Generally, the farmer delivers bales of leaf to the auction platform. The market for bidi tobacco, in contrast, is largely unregulated. As marketing of bidi tobacco is not under the control of any government agency, bidi farmers do not get prices as remunerative as in the case of FCV tobacco. Bidi tobacco is sold to traders at negotiated prices. In most cases the agent buys the bidi tobacco crop based on the smoke of leaf, leaf spangle and nicotine content. In most cases, the trading community finances the farmers, and so price setting power lies with the trader, and in most cases the farmer is not paid until almost a year after the sale occurs.

Processing of bidi tobacco is not a technology-intensive process. After the processing of the tobacco into flakes, the agent stores the tobacco until it is dispatched to the manufacturer. The processor also blends the tobacco according to the manufacturer's requirements. The purchase of bidi tobacco from the farmer begins in February and March. After processing, the tobacco is stored for a period of 6-12 months for ageing. The marketing and distribution of bidi tobacco continues to be the domain of the private sector and the industry is totally free. Currently, no effective institutional arrangements for the regulation or marketing of bidi tobacco exist, due to strong opposition of bidi tobacco traders. Marketing channels for leaf tobacco, bidi and cigarettes are shown in Figure 18.1. All the marketing agencies are in the private sector, except for the Tobacco Board and the State Trading Corporation. The cigarette market is oligopolistic, with four large manufacturing companies. The Indian market for cigarettes and other tobacco products is highly price sensitive. Following the reduction in excise duty and consequently prices in 1994, there was an explosion in demand for micros (cigarettes shorter than 60 mm). Trade sources estimated that consumption of micros rose from 300 million pieces in 1993 to around 4 000 million pieces in 1994, 18 000 million pieces in 1995 and over 19 000 million in 1996. However, with the increased excise duty on these cigarettes since the 1996-97 budgets, demand has declined, and led to a drift back towards small filter products by some smokers and towards bidis by others.

Prices of different types and size of cigarettes depend inter alia on the level of and changes in excise duty imposed by the central government in its annual fiscal budget. Any more than moderate increase in excise duty (say over 3 percent) effectively raises prices of cigarettes. At times, a modest increase in taxation has helped to maintain retail prices. Owing to the variations in prices as a result of differential changes of taxation on different types and size of cigarettes, their share in total sales has changed considerably.

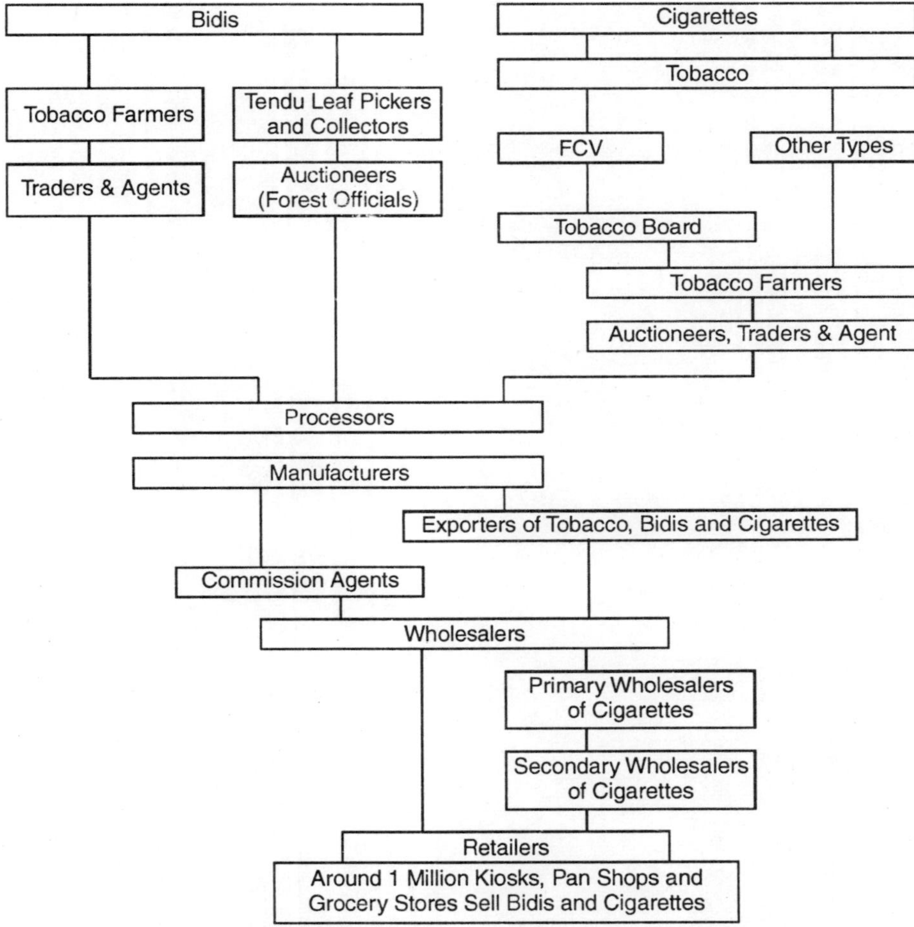

Fig. 21.1 Marketing channels for bidis and cigarettes

❑❑❑

22

Indian Tobacco in World Market

India is the one of the largest producer and exporter of tobacco in the World. Total tobacco production in India is about 700 million Kilograms annually. Rich and varied Indian geographic and agro-climatic conditions foster consistent availability of wide range of tobaccos for export all through the year. Indian tobacco, by virtue of its qualities, sheer volumes and diversity, is progressing gracefully to occupy its rightful place in the world tobacco market.

In Indian as per conservative estimate about 250 million Indians use tobacco, and the market's already worth a massive $5 billion. And they don't only smoke it. Many Indians chew tobacco, mixed with lime, spices and other substances - some of them possible carcinogens. The tobacco companies are keen to expand their share of one of the biggest global markets and they are deploying their marketing skills to the full to achieve their ends. As well as television commercials, they use sports sponsorship and sell sports equipment. It's more than sponsorship, says Suhel Seth, who used to be in marketing with the Indian Tobacco Company (ITC). "It's very careful target marketing where they're looking at young people who watch the sport, who are almost fanatic about it see their icons being associated with a cigarette brand. Cricket is the only common religion in India."

Continuing with a policy to control smoking, the tax measures became harsher in 2007, with an increase of 6% in excise and an additional 12.5% in VAT, bringing the total effective tax to about 20% (although the most populous state, Uttar Pradesh, imposed 32.5% VAT). As expected, volume sales declined due to the subsequent price rises. ITC Group, the leading company, is estimated to have taken price hikes as high as 20% on some brands. The industry expected a volume decline of 5%-10%, but the overall industry declines have been lower than expected, reflecting consumer resilience to price rises. The government has put severe regulatory restrictions of all types on cigarettes to restrict sales and reduce smoking prevalence. Apart from consistent rises in taxes each year on cigarettes, there is also a total ban on advertising and promotion, a ban on smoking in public, and also in films, a ban on sale to minors, and there are impending graphic health warnings on packs in 2008. Moreover, the domestic industry also faces threats from imports and smuggled cigarettes. Although imports constitute only a negligible proportion of the domestic pie, their rising share poses a threat to players. Tobacco Business in India is important for the economic development of the country. The tobacco contributes to the major portion of gross domestic product of India. It also helps in the rural economic development. Tobacco and tobacco products are exported to different countries and helps in earning stronger foreign currencies.

Exports of Tobacco

Exports of tobacco, along with other agricultural items, receive various incentives. For example, export incomes are exempted from income taxation. Also, credit at subsidized interest rates is available for tobacco exports. Moreover, tobacco exporters can import necessary inputs duty-free, while they can import capital goods at concessional duty rates provided they are able to fulfil an export obligation, which may vary from four to six times the value of the capital goods. To promote exports of various agricultural products, including tobacco, the government also extends financial assistance in the form of grants for certain export promotion activities. Raw leaf accounts for around 80 percent of the total volume and value of tobacco exports. Of raw leaf, FCV, which forms less than 25 percent of total production in the country, accounts for the bulk of exports, in terms of both volume and value, although its share has declined during the last three decades.

India accounts for only 4 percent of the world exports of raw tobacco. Total export earnings from both leaf tobacco and manufactured products in India have increased almost 25-fold, from Rs 326 million at the beginning of the 1970s to about Rs 8 100 million in 1998-99. This was partly due to the general increase in tobacco prices at the world level. During the last decade, tobacco export earnings increased from Rs 2 600 million in 1990-91 to Rs 10 600 million in 1997-98, but declined in the next year. Earnings from raw tobacco during this period increased almost three-fold, from Rs 1 900 million to Rs 6 300 million. However, for manufactured tobacco products they rose by only 145 percent, from Rs 699 million to Rs 1 700 million. Among tobacco products, exports of hookah paste accounts for around 80 percent in volume but only 40 percent in value terms. The share of cigarettes has declined since 1993-94, following the reduced demand from the former Soviet Union, with which India had a bilateral trade agreement.

Nearly two-thirds of exports of raw tobacco were to European countries, with Russia being a leading importer. Saudi Arabia is the single largest importer of hookah paste. Bahrain, Oman, Singapore and the United Arab Emirates import bidis from India. Russia has been the largest importer of Indian-made cigarettes. Globally, most tobacco is consumed in the form of cigarettes. In India, however, the bulk of India's consumption of tobacco is in non-cigarette forms. This mismatch between India's product mix and world demand has hampered India's export growth. To emerge as one of the leading players in the global tobacco market, India has to grow more cigarette-type tobacco (Virginia, Burley and Oriental) to integrate its economy with the world market.

FCV tobacco accounts for the bulk of the volume and value of exports and generates around 80 percent of revenues from exports of raw tobacco. With an average cost of production of Rs 25.60 per kg in 1997/98, Indian FCV tobacco has been perhaps the cheapest in the world market. Further, by virtue of absence of any governmental protection for this crop, unlike in many tobacco-producing countries, Indian FCV tobacco acquires an added edge in the competitive export market. Judged from the internationally accepted parameters of levels of protection, Indian FCV tobacco is the least protected and therefore has more favourable terms of trade. The contribution of cigarettes to foreign exchange earnings is substantial. Cigarette exports touched nearly Rs. 490 million in 1993-94 and accounted for 61 percent of the total earnings from manufactured tobacco products. The absence of an internationally acceptable brand name, non-conformity with international standards, and overseas marketing limitations imposed by foreign collaborators seem to be some of the reasons for India's limited exports of cigarettes. Bidis are exported to several West Asian countries. If India could reframe its tobacco production strategy and maximize the marketable surplus in the export-oriented varieties, so as to raise its production levels - especially of exportable varieties -it could boost its exports. The present pattern of exports, where raw tobacco forms the bulk of exports in both volume and value, needs to be reversed and value-added manufactured products (cigarettes) should form a greater part of export earning from the tobacco sector as a whole by the end of the current (9th) Five-Year Plan.

Imports

India imports small quantities of raw tobacco required for blending purposes for manufacturing international-brand cigarettes in the country. In recent years, imports of raw tobacco totalled around 300 tonnes per annum. To be effective, measures to control the use of tobacco would need to address all forms of consumption, not only cigarettes. In particular, this would mean dealing with the economic and political difficulties of taxing, or otherwise controlling, bidis and other non-cigarette forms of tobacco consumption.

Table 22.1 Tobacco trade (Import and Export)

S No.	Tobacco group	Name of product	No of company engaged in production
1.	Chewing Tobacco	31	13
2.	Flavored Tobacco	26	75
3.	Mouth Freshener	48	135
4.	Tobacco	53	111
5.	Others	88	153
6.	Total	246	487

In India to conduct smooth and efficient tread in tobacco and tobacco related product, a number of company is involved presently more than 450 companies are actively participating in the as many as 246 product under five different catogry of tobacco product as mentioned in table . Out of 450 plus company 79 companies are treated as tread leaders. Import Export Trade Leads for chewing smoking tobacco, scented and flavoured tobacco beedis, cigarettes, snuff, supari, pan masala, zafrani zarda, betalnut leaves, pan parag, gutkha, pipe tobacco, filter bidi, tendu leaves and other tobacco products.

Table 22.2 Trade leader in tobacco and tobacco product

S. No.	Name of Company	Product trade	Lead product treads
1	bluebell Fragrances India	All tobacco product	All types of tobacco products such as gutkha, khaini, zafrani zarda, sweet supari and saffron blend fennel seeds.
2	Ningbo Shining Electronic Technology Co., Ltd. [China]	Electronic Cigarette	Product description e-cigarette: it is a elv electronic atomizer, adopts the advanced microelectronic technology and super critical physical atomized technology, to atomize the high-purity nicotine diluent..
3	Mohan Food Products [India]	Tobacco, Flavored Supari	Chewing tobacco, smoking tobacco, snuffing tobacco, tobacco bidi, flavored tobacco, sweet supari, flavored supari, saffron supari, scented supari and elaichi mix supari.
4	Mohan Food Products [India]	Mouth Freshener	Mouth freshener and flavored mouth freshener such as dry dates, fennels, melon seeds, coriander seeds, silver coated cardamom seeds, saffron, coconut, natural sweetener and herbs. Also supplier of namkeen like plain namkeen, salty...
5	Mohan Food Products [India]	Flavored Tobacco	Offer to selling tobacco product like flavored tobacco, chewing tobacco, smoking tobacco, flavored shisha tobacco, flavored smoking tobacco, snuffing tobacco and tobacco bidi
6	Mak Fiber Tech Private Limited [India]	Ship Bidis	Beedi
7	Tej Ram Dharam Paul [India]	Flavoured Tobacco	Flavored tobacco with selected natural ingredients like Spices, Flavors & Fragrances. - GANESH 701 Khaini - Ganesh 701 Filter Khaini - GANESH 701 Zipper
8	Golden Tobacco Limited [India]	Style Cigarettes	We are offering diverse range of cigarettes and tobacco products like: - Loe Tabac Cigarettes - Panama Cigarettes

Table 22.3 Important tobacco manufacturers in India

S. No.	Name and Address of the company	Registration No.
1	Godfrey Phillips (India) Ltd. Chakala, Andheri (East),MUMBAI – 400 099,Maharashtra , Ph: 022-28367301/8	2006/01
2	G.T.C. Industries Ltd. Tobacco House,S.V.Road, Vile Parle (West), MUMBAI– 400 056, Maharashtra Tel:26713951	2006/02
3	I.T.C. Limited – I.L.T.D.Division Virginia House, 37, J L Nehru Road,KOLKATA – 700 071, West Bengal, Phone:033-22889371,Website: www.itcportal.com, www.itc-iltd.com	2006/03
4	International Tobacco Co. Ltd., Kermani Bldg, Dr.D.Naroji Road,MUMBAI – 400 001, Maharashtra Ph:022-2788235/240	2006/04
5	V.S.T. Industries Ltd. Post Box No. 1804,Azamabad HYDERABAD – 500 020, Rangareddy Dist., A.P. Ph: 91-040-27610460, Fax:91-040-27615336	2006/07
6	New Tobacco Company Lessee : RDB Industries Ltd.,Bikanir Bldg., 8/1, Lalbasar Street, KOLKATA – 700 001,West Bengal, Fax:91 33 22420588, Ph : 91 33 22485146/22486941	2006/13
7	Hilton Tobaccos Pvt. Ltd. H-1, Samrat complex, Saifabad, HYDERABAD – 500 004. Rangareddy Dist., A.P. Fax No.040-23240198. Ph: 040-23233981, 23240006, E-mail: hilton@bbmbommidala.com	2006/15
8	The Hyderabad Deccan Cig. Factory Private Limited 1-7-140, Musheerabad, HYDERABAD – 500 048, Rangareddy Dist., A.P., Fax:No: 040-27612760, Ph:040-27616658, 27616167, E-mail: hyd2_golconda@sancharnet.in	2006/20
9	Western Tobacco Ltd., A-6, Koh-e-Fiza, Indore Road,BHOPAL – 462 001, Madhya Pradesh, Fax No:0755- 2739332, Ph: 0755-2739713, 5235331, 5266601, E-mail: Westobit@sancharnet.in	2006/23
10	Asia Tobacco Co. Ltd., Guna Complex, 3rd Floor, 304/305, Annesalai, CHENNAI-600 018 ,Tamil Nadu ,Tel No:24364179, 24364224, Fax No: 24341068,E-mail: atcltd@vsnl.com	2006/26
11	Rajam Extractions Limited Penubaka, RAJAM – 532 127, Srikakulam Dist., A.P., Ph: (O) 08941-252308	2006/48 (Solanesol)
12	Aster Naturals Pvt. Ltd., Plot No.92,Vijaypur Colony, ECIL Post SECUNDERABAD – 500 062, Rangareddy Dist., A.P., Tel No. 040- 30917224, 30949514, Fax: 040 – 27110535	2006/61 (Solanesol)

Contd...

13	Neutra Specialities Pvt. Ltd., 26, Vandalur-Kelambakkam Road, Keelakottaiyur, Kanchipuram Post, CHENNAI - 600 048, Tamil Nadu , Ph: 04114-279293/94, Website: www.neutrasplty.com	2006/65 (Solanesol)
14	Premier Tobacco Packers Pvt. Ltd., P.B.No.19, Mangalagiri Road, Bommidala House, GUNTUR - 522 001, Guntur Dist., A.P., Ph : 0863-2356772, Fax :0863-2356771, Email: premire@bbmbommidala.com	2006/66 (Solanesol)

Table 22.4 Others Exporters for Tobacco Products

S No.	Name and Address of the company	Tobacco product	Registration No.
1	Bharat Beedi Works Pvt. Ltd., Kadri Road, P.B.No.730, Bharat Bagh, MANGALORE – 575 003 Karnataka Fax: 91-824-2444530Ph 91-824-2443230, 2443330, 2412830Website: www.30beedi.co.in, Email: info@30beedi.co.in	Beedies	2006/12
2	Western Tobacco Ltd., A-6, Koh-e-Fiza, Indore Road, BHOPAL - 462 001, Madhya Pradesh , Telex No: 705 426 MEPI IN, Fax: 91-755-2739332, Ph: 91-755-2739713, 5235331, Email: westoblt@sancharnet.in	Cigarettes, Cut Tobacco	2006/15
3	Flavour Jurak Traders P.B.No:523, Faiz Ganj Road,MORADABAD - 244 001 , Uttar Pradesh, Fax: 0591-2470644, Ph: 0591-2743450	HTP	2006/20
4	Gobind Tobacco Mfg. Co., Bawa ganga puri Road, PANIPAT - 132 103., Haryana, Phone: (O)0180-2666881, 2666180,R) 0180-2662081, 2661748, Fax: 0180-2666180	Chewing Tobacco (Khaini)	2006/23
5	Godfray Phillips India Ltd., Chakala, Andheri (East), MUMBAI – 400 029,Tel:28367301/08	Cigarettes	2006/24
6	Khalil Shahi Zarda Factory RISIA , Bahraich Dist., Uttar Pradesh, Ph: 222339	Chewing Tobacco	2006/38
7	International Tobacco Co. Ltd., Kermani Bldg, Dr.D.Naroji Road, MUMBAI – 400 001 Maharashtra ,Ph:022-2788235/240	Cut Tobacco	2006/45
8	New Tobacco Company RDB Industries Ltd., Bikanir Bldg., 8/1, Lalbasar Street, KOLKATA – 700 001 West Bengal, Fax:91 33 22420588, Ph : 91 33 22485146/22486941	Cigarettes, Smoking Mixture	2006/56
9	Nimex Trading Corporation (Tobacco Products) 2nd Floor, R.No.17,Topiwala Mansion,93-97, Mohamad Ali Road, MUMBAI - 400 003 Maharashtra,Telex: 752306 BIDI IN Fax No.91-22-23420826, Ph: 91-22-23421441, 3420420. Email: nimex@bom5.vsnl.net.in	Beedies, Snuff, Chewing Tobacco, Quiwam	2006/59

Contd...

10	Renown Commercial Corporation Snuff, 35, Kalbadevi Road, P.B.No.2471, Kanta Terrace,2nd Fllor, MUMBAI – 400 002.Maharashtra, Ph: 22014907, 55748825 Fax: 22089572,Email: renown_cc@vsnl.net	Snuff, Quiwam	2006/78
11	Prabhat zarada Factory (I) Pvt. Ltd., New Area, Sikandarpur, MUZAFFARPUR – 842 001, Bihar, Ph: 0621-2217974, 2217959, 2217933, Fax:0621 - 2213380	Zafrani Patti, Quiwam	2006/69
12	Esquire Exporters Plot No. 39 A/2, Sindhi Society, Chembur, MUMBAI – 400 071., Maharashtra, Fax.No.(91-22)-2524 4525, Ph. (91-22)-2529 8275	Chewing Tobacco, Beedies	2006/82
13	Veera Prabhu Exports House 71, Veer Vithaldas Chandan Street, Vadgadi, MUMBAI – 400 003, Maharashtra, Phone: 022-23425101, 2344 1314, Fax: 022-2341 1158, Email: jenil@mtnl.net.in, Website: www.veerprabhu.com	Beedies, Snuff, Panparag, Gutkha	2006/90

Economic Policies for Tobacco

Introduction

The tobacco companies are always ready with economic arguments against the control of tobacco. Tobacco is vital to our economy, they argue, because:

- Tobacco produces tax revenues
- Tobacco exports bring hard currency income
- Tobacco creates jobs
- Tobacco is a vital farm crop

They argue that action to reduce smoking will bring great harm to the economy. Jobs will be lost; tax revenues will fall; exports will suffer; farmers will be devastated. But they greatly exaggerate the economic losses which tobacco control measures will cause and they never mention the economic costs which tobacco inflicts upon every country.

Tobacco Companies Argument: excise tax increases will discourage smoking and thereby reduce government revenues.

1. Raising tobacco taxes, no matter how large the increase, has never once led to a decrease in cigarette tax revenues.
2. When cigarette taxes are increased, fewer young people take up smoking; light and moderate smokers tend to quit; but most heavy

smokers continue to smoke and to buy cigarettes. This keeps tax revenues high.

3. Some of the money saved by quitters will be spent on other goods which are also taxed.

4. Raising tobacco taxes is not an unpopular move. Studies from several countries show that most people-smokers and non-smokers-support increased tobacco tax as a means of preventing young people from taking up the habit.

Tobacco Companies' Argument: tobacco exports furnish critical hard currency income.

1. Much tobacco is consumed in the country of production. But in many countries much of the proceeds from cigarette sales leave the country as transnational tobacco companies' profits, directors' fees, payment for such costs as imported equipment, licences, advertising of foreign origin and packaging.

2. Two-thirds of developing countries spend more on importing cigarettes than they gain from exporting tobacco. Thus, reducing smoking will reduce loss of foreign currency, as well as government revenues lost to smuggling.

Tobacco Companies Argument: the tobacco industry provides thousands of jobs which will be lost if tobacco control policies are adopted.

1. If all tobacco control policies were adopted, tobacco consumption would fall gradually over time. Most cigarette-related jobs would remain for many years, and the gradual reduction of jobs would allow workers to make an orderly transition.

2. The children of tobacco farmers will go into other occupations; most tobacco farmers will not need to.

3. In future, the introduction of mechanization in tobacco farming and manufacturing will mean less employment anyhow.

4. Money not spent on cigarettes will still be spent, but for other goods and services, thereby creating **new** jobs. The tobacco industry itself admits: "If the (tobacco) industry would vanish tomorrow, most would find alternative work".

5. There will be less work for lung surgeons, heart disease specialists, nurses and firemen.

6. But is a "balance sheet mentality" really appropriate? The health profession earns its living from caring for the sick and car repair workers from mending crashed cars. Is this an argument against preventing disease or promoting traffic safety? International drug dealers are certainly well paid but no one suggests that this is an argument against trying to control their activities. So should it be with tobacco.

Tobacco Companies' Argument: Tobacco farmers benefit from the cultivation of tobacco, and other farmers benefit from the new skills and technology which the tobacco companies introduce. Therefore, tobacco cultivation and consumption should be encouraged.

1. "The lion's share of profits goes to these (tobacco) companies and not to producers" states the World Health Organization. (11) In Malaysia, for example, the profit margin for farmers is only 2%, but for manufacturers, 79%.

2. As tobacco control programmes throughout the world take root, tobacco will be in oversupply and the price will fall.

3. Other crop programmes supported by the World Bank and other development agencies could provide the same skills and technology without these harms. Alternative crops are being ntroduced (such as wheat, sweet potatoes, coriander and yellow evening primrose) but these have only been on a very small scale.

Costs to Government, Business, the Individual and the Environment

The tobacco companies never talk about the great economic costs that tobacco inflicts upon governments, upon business and industry, upon individuals and their families, and upon the environment.

The Costs to Government

1. Where governments provide medical care, smoking-related illness results in added medical care costs.

2. Cigarette smoking strikes down smokers in their working years, thus bringing added welfare costs for supporting families which have lost their wage earners.

3. In developing countries, where educated, middle-class citizens tend to be the heaviest smokers, death from smoking-caused disease reduces the countries' leadership at the very age when they may be making their greatest contribution to society.

4. Developing countries produce 73% of the world's output of tobacco. While some land used to grow tobacco is unsuitable for other crops, tobacco usurps the place of food crops that could feed an estimated 10-20 million people. This may lead to the government having to bear the higher costs of imported food.

5. Cigarettes cause at least one-quarter of all fires. The government must therefore pay more for fire-fighting services. Also, it must pay for the replacement of government-owned buildings, forests, and other property lost to smoking-related fires.

6. Government must pay more for maintaining the cleanliness of government buildings and public areas littered with cigarette butts, discarded packages, matchboxes and matches.

The Costs to Business

1. Smokers are less productive workers because they are sicker: they are absent from work more often and also die sooner than non-smokers. New workers have to be trained to replace them.

2. Smoking workers cause more accidents and fires than non-smokers.

3. Workers interrupt their work to smoke, taking on average 7-10 minutes to smoke each cigarette-time often not spent working.

4. Employers who provide health care, or health and life insurance for their workers, must now or soon pay higher insurance rates.

5. Smokers force businesses to spend more money to clean (or replace) ventilating systems, smoke-stained walls and ceilings, and on cleaning up ashtrays and other smokers' litter.

6. Businesses which permit smoking bear an increasing risk that courts or other authorities will rule that such employers are financially liable for illness in non-smokers exposed to passive smoking at work.

The Costs to the Individual Smoker and to the Smoker's Families

1. In developing countries, the cost of buying cigarettes can rise to over 25% of an individual's disposable income. This takes money that could otherwise be spent on food, clothing and shelter.
2. Many smokers lose wages from work days lost from smoking-related illnesses.
3. Since smoking often kills smokers in their working years, smoking deprives the smoker's family of many years of income. Following a smoker's premature death, a partner, children or elderly parents may even be left destitute.
4. Smoking raises the costs of health care necessitated by smoking-related illnesses suffered both by the smoker and the smoker's family who have been exposed to passive smoke in the home.
5. Smoking affects smokers' family members in the time needed to look after the smoker at home or to take him/her to hospital (the latter may sometimes be measured in days in developing countries).

The Costs to the Environment

1. Tobacco is mostly flue-cured by wood in developing countries. It requires one acre of forest to cure one acre of tobacco. Already in Tanzania, 12% of all trees felled annually are used for tobacco curing. (15) Except where replanting programmes exist (virtually all of which are totally inadequate), this destruction of forests brings floods and threatens water supplies.
2. Cigarette wrapping, packaging and advertising involve the heavy use of paper. A single cigarette manufacturing machine can use 4 miles of paper per hour.
3. Other environmental aspects include smokers litter, the use of pesticides and herbicides on the tobacco crop, the environmental impact of passive smoking and fires. One of the world's worst forest fires, that swept north-east China in 1987, ravaged 1.3 million hectares of land, killed 300 people and made 5,000 people homeless.

However, the value of human life is beyond costing. While economic arguments are important, and must be answered, the rising death toll from tobacco is an epidemic. No civilized country asks the economic costs before it decides to fight an epidemic.

Strategies and Opportunity

Tobacco & Tobacco Products

India is currently the world's 2nd largest producer of tobacco next to China. Indian tobacco is exported to more than 80 countries spread over all the continents. The tobacco industry provides employment to 25 million people. Of the total exports of tobacco items from the country, unmanufactured tobacco accounts for 80% to 85% and manufactured tobacco products account for 20% to 25%. Of the unmanufactured tobacco exports Flue Cured Virginia tobacco is the single largest item accounting for 75% to 80% of the tobacco exports. The other varieties exported are Burley, HDBRG, Natu, DWFC, Top Leaf & Jutty-all are cigarette tobaccos. Non-cigarette tobaccos exported are Lalchopadia, Judi, Rustica for chewing purposes and Bidi tobacco in small quantities. The export of non cigarette tobaccos are at levels of 8-10%.Contrary to international trend, India's tobacco production is dominated by non-cigarette tobaccos. Indian tobacco exports have not registered a steady growth. Our flue cured tobacco exports, the major exportable variety have also declined significantly while the competitors could excel. Currently the situation in the global tobacco market is transforming into a favourable market environment for tobacco exports due to several reasons. Brazilian export prices have almost levelled the most expensive American tobacco prices. Zimbabwean farm prices have also seen higher than Brazilian prices. Increasing cost of production and the significant amount of export cesses imposed on tobacco exports by the governments of Zimbabwe and Malawi recently made their tobaccos further expensive. The newly opened East European and CIS markets are not in a position either to absorb high cost tobaccos or cigarettes made with high conversion cost. Phasing out of farm subsidies by European Union will further enhance price competitiveness of Indian Tobacco. Under these circumstances, India can become a major player in the Indian tobacco market if it can harness the emerging opportunities and enhance its exports to the major import markets. The advantages for Indian tobacco are low unit production cost; average export prices of Indian FCV tobacco are more competitive than that of Brazil, USA, Zimbabwe; low conversion costs of tobacco into cigarettes of 0.80 US$/1000 pieces in 1999 in India compared to 3.50 in UK and 4.0 in USA; low to medium nicotines to suit the current requirement of world markets; anticipated decline in production in China, USA, Zimbabwe, EU in the next five years due to declining consumption in USA & EU, government controls to restrict production in China, phasing out of Agri subsidies by EU and Land invasions and Land acquisition in zimbabwe i.e, change of tobacco farms from White farmers to Native farmers; and phasing out of Agricultural subsidies in European Union and Argentina; etc.

There is a need to examine the adoption of the following strategies:

i) Production Front

- Expanding production of quality tobacco (semi flavourful/ flavourful tobaccos) of FCV and Burley types in the soil regions of AP and Karnataka.
- Identifying high productivity areas in the central and northern block cotton soil region and sustaining cultivation to meet the demand from the price sensitive/price responsive importers from CIS countries.
- Release of new varieties with higher yield potential and with low to medium nicotine levels to suit the changing demand pattern in the International markets.
- Allowing FDI through joint ventures with multinational leaf tobacco merchants/cigarette manufactures to go for contract farming in new areas to generate additional exports.
- To undertake special projects of financing the cost of digging / building small water ponds in the Southern light soil region of AP (where per hectare yields are very poor) with a view to give life saving irrigation with the stored water to enhance productivity and quality of the leaf schemes for supplying inputs like fertilizer, coal briquettes at subsidized cost to the tobacco farmers, schemes to bring down curing cost and transportation, development programs in the production of light soil Burley/ oriental/Dark western Fire cured tobacco, schemes to help in the use of Malathi on for control of mosquitoes in the light soil burley areas to avoid DD Tresidue problems, initiating efforts for semi mechanization of tobacco cultivation in AP & Karnataka through introduction of leaf stitching machines, tobacco planters, etc.

ii) Production presented in the Auction Floors

- To introduce standardized packing material for wrapping tobacco in bales as in Zimbabwe.

- Offering well-graded product on the floors to ensure speedy sales, reduce post auction expenses & handling losses and reduce shipping delays through enforcing compulsory grading of tobacco by farmers in the licensed commercial/community grading centers.

iii) Export Front

- Tobacco trade delegations and publicizing Indian tobacco and tobacco products through International tobacco exhibition.
- Allowing FDI in 100% EOUs or special economic zones for export of cigarettes.
- Pursuing with the Government of USA at highest level for allocation of separate quota for India under its Tariff Rate Quota (TRQ) regime.
- Negotiations at the highest political level with Japanese counterparts for the re-entry of Japanese tobacco companies into Indian tobacco market
- Negotiating with China for importing Indian tobacco regularly.
- Permitting exporters to pay admissible commissions (at least up to 5%) for their rupee exports to Russia in hard currency.
- Entering into long term contract with Russia for import of 45,000 tons to 50,000 tons of tobacco every year for next 5 years under Debt Repayment Route.
- Undertaking propaganda for Indian tobacco and obtaining market situation report of Russian market and financial status of each cigarette factory by entering into a cooperation agreement with TABKPROM, Moscow, an association of Russian cigarette Factories.
- In Exim Bank credit line, inclusions of tobacco also as an item for export to countries like Egypt, Iraq, and Algeria.
- Encouraging Indian cigarette manufactures to set up joint ventures in overseas markets viz., East Europe, CIS countries, Myanmar, Vietnam, for manufacture of cigarettes which would result in increased tobacco export to these countries.
- Pursuing with the Governments of Tunisia and Morocco to enhance their tobacco imports from India highlighting their huge trade balances in their favour.

- Pursuing with the Governments of Romania, Bulgaria to extend further tariff concession under Globalised System of Trade Preferences (GSTP) for Indian tobacco.
- Tying up tobacco exports to Egypt with imports of crude Fertilizers, cotton etc.
- Helping cigarette exporters to undertake brand promotion by extending financial assistance.

□□□

PART-V
Issue and Concern

23

Tobacco Cultivation Issue and Concern

Tobacco has got some issues and concern related to its cultivation practices like green tobacco sickness (GTS), a common disease of persons involve in tobacco harvesting, or farmers are not getting good quality seed. In this present chapter light has been focused on such type of issue related to production of tobacco crop.

Green Tobacco Sickness (GTS)

Tobacco cultivation is a seasonal work and hazardous cultivation practices last for 2-3 months. Workers engaged in tobacco cultivation particularly in non flue cured tobacco e.g. chewing and hookah tobacco suffers from occupational illness known as green tobacco sickness. Green tobacco sickness (GTS) is a result of dermal exposure to dissolved nicotine from wet tobacco leaves. Worker engaged in various processes gets cuts and abrasions on their palms and the skin around their nails get peeled off also facilitates nicotine absorption. Nicotine being alkaloid is easily absorbed through the skin to cause acute toxicity and repeated small dose may cause chronic toxicity also. Green tobacco sickness (GTS) was first reported among tobacco workers in Florida in bay back in 1970. India is third country to report GST as green symptom among tobacco harvester. During its onset, early symptoms often include headache and nausea followed by vomiting we is potentially preventable through the use of protective clothing, worker should use seamless knitted nylon gloves while come in to contact to tobacco.

Alternative Uses of Tobacco

Tobacco is one of the best sources of secondary metabolites; of course nicotine is one of them, but other phytochemicals viz. solanesol, organic acid nicotine sulphate, protein and oil as well as other bio compound like enzymes and vitamins. These entire products are derived from leaf or leaf waste. Now in the era of high health consciousness it is high time to develop cultivar for phytochemicals having high flavors and lower levels of harmful substances in leaf to the tune of international standard. Breeding suitable variety for value added end product and increasing yield of these high end products by adopting specific agronomic practices for each product.

Alternative Crops to Tobacco

Tobacco in India, as in many other countries, yields higher net returns per unit of land than most other cash crops, and substantially more than food crops. Currently, there are a few specialized crops in various areas that provide similar incomes, but it is estimated that these crops would not remain remunerative if total production increased. The economics of alternative crops is generally based on experiments carried out on a limited area at research stations under optimal conditions. More detailed research work is needed on a wider scale at farmers' field level before firm recommendations can be made about them. In general, under farmers' field conditions, most other alternative crops, as discussed below, are currently not as remunerative as tobacco. Should tobacco farmers need to diversify into other crops, they are likely to suffer economic hardship. Our country has enacted a comprehensive central legislation. There is need for agricultural diversification and crop alternative to tobacco. Accordingly a study was conducted by the Indian Council of Agricultural Research, Central Tobacco research Institute has outlined that no single crop is as economically viable as tobacco. In general tobacco grower do not suffer from dept bondage hence improve their socio economic conditions of farmers and farm worker. It is important to compelling tobacco grower to give up tobacco cultivation and go for alternative crops might be proving fatal to Indian economy too. There is no single remunerative and alternative crop to tobacco means that the much higher net returns per unit area is much higher than other crops beside provides employment maximum as compare to other enterprise. However one could go for high value vegetables, fruits, spices and condiments may be the some alternative to tobacco. New crops like pawpaw's (*Asimina triloba*), and maypops (*Passiflora incarnata*) as potential new crops. Both can be adapted to the climates and soils of those areas. Both bear delicious fruits, which have been gathered from the wild and eaten by local people for centuries.

Future Thrust in Respect to Tobacco

In the era of global village in the regime of WTO we should developed varieties/ hybrids and equipped with proper agro and plant protection technique to suit International market. Our product must be possesses low levels of pesticide residue and other harmful substance like TNSA. We are the only country where many di?erent types of tobacco, viz., fue-cured Virginia (FCV), burley, natu, cigar fller, cigar wrapper, cheroot, hookah, bidi and chewing are grown under different agro-climatic conditions. Our product must be free from insect pest moulds genetically modified organism, non tobacco related materials and low level of nicotine contents. The uniqueness of tobacco makes it an very good source of phytochemicals which are very much beneficial to human economic welfare. Considerable efforts have been made to develop technologies for extraction of protein, nicotine, seed oil and solanesol from various tobacco types' andsuccess has been achieved in case of bidi tobacco.

Among them FCV and burley are exportable tobacco types and are mainly used in cigarette manufacturing. Indian FCV tobacco is famous in the world market for its ?lling qualityin cigarettes, and burley tobacco for cigarette blending. Non FCV tobacco i.e. chewing, hookah, zarda, bidi etc contributes major share to the Indian tobacco. Now its perfect time to concentrate our research efforts towards improving the quality productivity and reducing harmful constituents to match up with international requirements. Bihar is leader in Non FCV tobacco viz., chewing hookah, paste, zarda, bidi production. Some of our Non FCV type product like hookah, paste, zarda, bidi have already got god export demand .Under this emerging situation, we could earn more foreign exchange than ever before by taking suitable correctives measures and ensuring product quality to the tune of international markets hence quality related issues need to be address properly on top priority.

Tobacco Seed Production

Seeds, the root cause of good crop production. Tobacco seeds are so minute and delicate to handle and only one institute are engaged in tobacco research and seed production. Tobacco being often self pollinated crop cross pollination take place only to the account of 5-7 per cent. Proper isolation distance of 50 meter must be maintained to avoid any chance of getting mixed with other tobacco varieties. Seed production of tobacco is done exactly in the same manner as the leaf crop is taken. In case of seed crop roughing should be done twice to maintain purity topping and desuckering operation is avoided to enable the plant to come in to the flowering. Due to non synchronized maturity picking is done and 2-3 picking is sufficient to harvest fully. Proper plant protection measures should be taken for aphids, capsule borer and capsule drop.

Field view of tobacco seed production

Proper moisture should be maintained during flowering stage. Low application of nitrogenous fertilizer and balance application of phosphorus and potassium is utmost required. In case of capsule drop and cracking application of micro nutrient viz. boron, zinc and manganese will be beneficial. A good crop seed tobacco can produce 10-12 q/ha. Depending upon variety and management practices.

Tips for Farmers to Produce Seed

Tobacco growers are advised that they should take every care while raising /producing / multiplying tobacco crops for seed purposes. Points to be taken on priority are mentioned bellow

- Ensure the quality of seed before sowing for that purpose they should always use breeder, certified or truthfully leveled seed.
- Seed should be procured from reliable source/ agency every time.

- Farmers can produced seeds for their self by selecting little best plant and left it to come in to flowering.
- If there is more than one variety in vicinity after selfing the flower, it should be protected by selfing bag to avoid contamination and mixing.
- At flowering stage proper care especially should be taken water and capsule borer management.

Tobacco Seed Production

Quality seed is mother of good harvest hence; every effort should be taken to develop new varieties, having desired traits, as per International demand and national needs, so that our produce can compete in world market under in the era of WTO etc. To achieve this goal there are urgent needs to developed novel models of seed production chain, so that seeds of promising varieties of chewing and hookah tobacco may reach to the farmers instantly.

□□□

24

Ethnobotany and Indigenous Technical Knowledge (ITKS)

It seems that use of tobacco for various purpose is old age practices, However its initially used purely for treatment of several disease as miracle medicine A brief account has been given on its ethonobotany and indigenous technical knowledge(ITK) in this chapter.

Current commercial uses of tobacco are not only a great health risk to our society, but also pose a threat to the ceremonial (traditional) uses of tobacco. According to some First Nations, Aboriginal and Métis tradition, tobacco is a sacred plant. The plant symbolizes cleansing, decision-making, communication with spirits, sacred beings and the Creator. In the cases of traditional tobacco being burned either in a pipe or in a fire, the smoke is generally not inhaled and is left to rise up and carry prayers or words to the Creator. The sharing and passing of the pipe is done to create a spiritual connection within the group of people that are sharing it. Ceremonial pipes are frequently used in many different rituals and are themselves of considerable symbolic significance and accordingly, are treated as sacred objects. Pipes are the route that the smoke travels through to communicate with the creator and allows for the spiritual message to be shared from person to person in a ceremonial circle.

1. Traditional Tobacco Use

According to some First Nations, Aboriginal and Métis tradition, tobacco is a sacred plant. The plant symbolizes cleansing, decision making, communication with spirits, sacred beings and the Creator. In the cases of traditional tobacco being burned either in a pipe or in a fire, the smoke is generally not inhaled and is left to rise up and carry prayers or words to the Creator. The sharing and passing of the pipe is done to create a spiritual connection within the group of people that are sharing it. Ceremonial pipes are frequently used in many different rituals and are themselves of considerable symbolic significance and accordingly, are treated as sacred objects. Pipes are the route that the smoke travels through to communicate with the creator and allows for the spiritual message to be shared from person to person in a ceremonial circle.

Tobacco, probably mixed with lime or chalk, appears to have been used in these Native American populations as toothpaste to whiten the teeth, as observed by Nino and Guerra in 1500.

2. Ceremonial Use of Tobacco

The ceremonial use of tobacco occurred once a day, or in special circumstances two or three times at most. Those who smoke, or dip several times a day are not using tobacco for its sacred purpose.

Tobacco is used in Ceremonies For:

- Communication with the spirits (smoke).
- Thanking the Creator.
- Praying for a good harvest or better fish catch.
- Rites of passage ceremonies (birth, weddings, funerals).
- Sealing the peace with enemies (peace pipe).
- Thanks to Mother Earth or other Spirits.

Tobacco is not always burned. It is sometimes placed on the ground as an offering to the earth, or on water, or is left in sacred places. It is offered in the spirit of "you give something to get something." It can also be spread over the individual or object as a blessing.

Traditionally, when plants or animals or other objects are taken for use, particularly sacred use, by Aboriginal peoples, tobacco is left in its place to acknowledge the gift from Mother Earth and as a sign of respect and gratitude.

3. Medicinal Use of Tobacco

Some nations have a "pipe carrier", a special medicine person who uses tobacco the traditional way to communicate with the Creator. Today, traditional tobacco is still regarded as sacred by most Aboriginal groups and retains its spiritual and, in some cases, medicinal value.

Successful' Uses of Tobacco as Identified by Stewart

Tobacco Administered Externally

Bites of poisonous reptiles and insects; hysteria; pain, neuralgia; laryngeal spasm; gout; growth of hair; tetanus; ringworm ; rodent ulcer; ulcers; wounds; respiratory stimulant.

Tobacco Administered by Rectum

Constipation; haemorrhoidal bleeding

Tobacco Administered by Mouth

Strangulated hernia (smoke by mouth); malaria or intermittent fever; dislodging obstructive material from esophagus by inducing vomiting.

Tobacco Administered by Inhalation

Nasal polyps.

Tobacco is a beautiful plant that has both spiritual and medicinal purposes. Cultivating this plant in mass quantities for use in cigarettes and chewing tobacco removes its spiritual and medicinal purpose and strips the beauty from the plants.

Vernacular Names of Tobaccos

The vernacular name is used to display common name equivalents to proper botanical names. The Vernacular name displays additional information about a common name such as its country of origin and whether or not it is a non-English language name. Common names are associated with entries in the specimen list and can be seen when the Identification View is active.

Table 24.1 Vernacular Names in differecountries

Species	Vernacular Names in different countries
Nicotiana spp.	Pacharo, rume (Peru)
N. alata Link & Otto	Jasmine tobacco, flowering tobacco, winged tobacco (English)
N. clevelandii A. Gray	Tobacco de perro (Mexico)
N. glauca R.C. Graham	Wild tobacco, shrub tobacco, Mexican tobacco, tree tobacco, canelon, mustard tree, San Juan tree, tobacco plant, tobacco bush, grey blue tobacco (English), Dokkhane (Arabic), Guang yan cao(China), Helado, karalawa, karallanta (Bolivia), and Makahala, paka (Hiawaii)
N. goodspeedii H	Small-flower tobacco, small-flowered tobacco, smooth-flowered tobacco (English)
N. gossei Domin	Gosse's tobacco, native tobacco tobacco (English)
N. knightiana Goodspeed	Tabaquilla (Peru)
N. longiflora Cav.	Long-flowered tobacco Japan: naga-bana-tabako (English)
N. maritima H.	Coast tobacco (English)
N. megalosiphon	long-flowered tobacco (English)
N. otophora Griseb.	Tabacachi, tabaquilla (Bolivia)
N. paniculata L	Ccamasayri, kkuru, koles-koles, sairi, tabaco cimarron ani (Peru)
N. plumbaginifolia Vivi	Tabaquillo (Mexico)
N. rustica L.	Wild tobacco, tobacco, Aztec tobacco (English), Dokhan akhdar, dokhan, dokhan soufi (Arabic),Huang hua yan cao (China) tabaco, tabaquillo, andumucua, macuche, picietl, yeti, k'uts, tabaco macuche, teneshii, tenapete, tabaco pequeno (Mexico) and Petum, piciete(Peru)
N. suaveolens Lehm	Auseral tobacco (English)
N. tabacum L.	Tobacco, common tobacco (English), tabac (French), Tabacco (Italian), Paka (Hawaii), iyaa (people Nya Hon),(South Laos), Yan cao, yen tsao, jen tsao, yu yen tsao (China), Tamrakuta, tamak, tamaku, tamakhu, tambaku, tanbak, pogaku, pugaiyilai, pugaielai, pugere, pukayila, pukayil. hogesoppu, pokala, dhurapan Malaya: tembakau (India), Tabako, tabaco, tobacco (The Philippines), iyatl, ju'uikill, me-e, otzi, picietl, ro-hu, ru-u, uipa, ya, yana, hapis copxot (Mexico), Sidl (Paraguay), Chiri, chiri tseri, iri, yiri, shahuano, pori, petima, tsaang, rume, tabaco, romu, sairi, seri, sheri, shifia, ssina, tsifia, yemats (Peru), Tabaco, furno, petum (Brazil) and Fumu, mbuli (Congo)
N. tomentosa	Jarato (Peru)
N. undulata	Asnak-kcora, ccjama-sairi, ccjamasairi (Peru)
N velutina H. Wheeler	Velvet tobacco (English)
N. wigandioides	Kura kura, kuro kuro(Bolivia)

Tobacco Based Pesticide

Tobacco was one of the first material used as an insecticide as early as 1763. The use of finely powder tobacco mix with water was recommended to destroy aphids without injury to the foliage. In 1828, an alkaloid namely nicotine was recognized which was synthesized by Picted and Rotschy in 1904. Nicotine occurs in at least 15 species of *Nicotiana* but the commercially important species are *Nicotiana tobacum,- and N. rustica.* Nicotine commonly occurs in *N. tobacum* plant to the extent of 2-5% in the leaves and 5 to 14 % in *N. rustica.*

The alkaloid nicotine when freshly distilled is a colourless nearly odourless liquid boiling point 247°C, specific gravity 1.009 at 20°C and vapors pressure 0.0425 mm Hg. at 25°C. When exposed to air, its darkness changes from red to brown to black becomes more viscous and develops specific odour, it is soluble in alcohol, ether or petroleum ether and is miscible with water due to of its basic nature. Nicotine usually forms salts with acids and dibasic salt with many metals and acids. The nicotine sulphate $(C_{10}H_{10}N_2)$ 2. H_2SO_4 is widely used as an insecticide as it is less toxic to warm blooded animals than free nicotine and is less volatile, therefore, more stable. When alkaline water or soap solution is added to nicotine sulphate, the free nicotine is liberated and shows higher insecticidal activity than the sulphate.

The nicotine enters in the body through the spiracles and acts on the ganglia of the nervous system. The less amount of nicotine produces excitement on the insect while more amounts produces paralysis due to which death takes place. It can be used by various ways as follows:-

(i) *As a dust* : The concentrated nicotine extract is mixed with talc, gypsum and sulphur to prepare the dust. The type of carrier has an important effect upon the dust.

(ii) *As a liquid extract* : A decoction of tobacco leaves with soap is prepared by boiling 1 kg of Ieaves, stalks etc. of tobacco in 10 lit. of water, One kg of soft soap is added after the water has boiled. This is continuously stirred and after a thick dark coloured liquid is obtained, strained. This stock solution should be diluted 10 to 15 times with water before spraying. A standardized 40% nicotine sulphate is also available to use as spray which remain as toxic residue for a short time after application. The nicotine sprays are diluted so as to contain 0.05% nicotine. It has been successfully used in the control of sucking insect viz., aphids, thrips, psyllids, leaf minor and jassids.

(iii) *As Fumigants* : It is also used as fumigant either by vapourising the concentrated extract with heat or by burning papers or powders soaked in the extract. Nicotine is compatible with' many of the common insecticides and fungicides, including sulphur, lime sulphur and copper preparations.

Nicotine- Chemical struture

Nicotine formulation

Indigenous Technical Knowledge Related to Tobacco

Tobacco is crop of multiple uses being one of the oldest know organic pesticide it has its own history of indigenous uses in curing human livestock's and crops and plant of economic uses. Indigenous technical knowledge (ITKs) has been developed over the years of experiences. ITK covers a wide spectrum-soil, water and nutrient management pasture and fodder management; crop cultivation; plant protection; animal rearing and health care; animal management; fisheries and fish preservation; and ethnic food management. Tobacco leaves are used for treating foot and mouth disease of livestock's. Different parts of this plant are being used by local man to control of various insect pests in maize and pigeon pea, whitefly in cotton and leaf spot disease in banana. Some of tested and well documented indigenous technical knowledge (ITKs) are listed here.

1. Growing *of Trap Crop to Control Pests*

Trap crop controls pests in tobacco as well as in other crops effectively. *Tagetes* spp. are cultivated as trap crops along with the main crop during rainy *(kharif)* season to control pest attack. Trap crop is grown between and also along the sides. Insect pests lay their eggs on yellow flowers of the trap crop, which cause death due to lack of food marigold, flowers also serve as a source of additional income.

2. Use of Tobacco Decoction to Control Larvae of *Helicoverpa armigera* in Soybean

About 1.5-2 kg green leaves of tobacco are boiled till the solution becomes dark brown. It is filtered and kept for 6-7 hours for cooling.

Then 4-5 liters decoction is mixed with 70-80 liters of water. The solution is sprayed over the affected crop. The practice is repeated at an interval of 5-6 days this controls *Helicoverpa armigera* in soybean.

3. Incorporation of Tobacco Residues in Soil for Controlling Termite

Farmers incorporate sticks and rootlets of tobacco in to the soil, using l disc harrow, which controls termite in the field due to nicotine content in the tobacco debris.

4. Control of Pests in Maize by Tobacco

Mix 1 kg tobacco powder with 5 kg sand. The mixture is made moist by sprinkling -250 ml water 20 days after sowing the crop, the mixture is slowly put in the crown of the crop. Six kg of this mixture is sufficient for 1 ha for controlling stem-borer. Similar treatment is given on the 40th day (from sowing). This technique is very cost effective (Rs 25/ha).

□□□

25

Medicinal Uses of Tobacco

Pharmaceutical companies are developing nicotine like drugs that target chemical imbalances in the brain that are thought to cause anxiety, schizophrenia, attention deficient hyperactivity disorder, Alzheimer's disease and Parkinson's disease. The study also suggests that people prone to depression may need extra help in order to quit smoking, such as nicotine replacement therapy, the scientists said. Despite the positive effects of nicotine discovered in their study, the researchers emphasize that it is not currently appropriate for treatment of any medical disorder outside of nicotine dependence.

When tobacco smoke is inhaled, the nicotine passes quickly to every organ of the body. The brain and nervous system are stimulated by small doses and depressed by larger ones. Nicotine increases the heart rate and the blood pressure, and may contribute directly to the excess of thrombosis and atheroma in smokers. Tobacco has long been removed from pharmacopoeias and from medical practice. First, too much was expected of tobacco. In medieval times, most herbs would be used only for a few conditions in which it was deemed effective—

not for a vast range of disorders from head lice to haemorrhoids, from hysteria to tetanus, as happened with tobacco. Secondly, writings on this subject commonly imply that nicotine is the only active medicinal constituent, yet the various species of *Nicotiana* contain many other alkaloids. Thirdly, the leaves and juice were much used for skin disorders, possibly including basal cell cancer. Might tobacco leaves contain an anticancer agent, as proved to be the case with periwinkle (*vinca alkaloids*)? Fourthly, in therapeutic applications of tobacco, dosage was largely uncontrolled. With any useful agent, excess dosage will do harm. We should set aside the prejudices generated by the ill-effects of tobacco smoking and examine the leaves systematically for substances of therapeutic value.

Uses in Before Eighteenth Century

It has been remarked that the first introduction of tobacco into the several countries of the world has always been in the use of medicine. As early as 15 October 1492 Columbus noted that dried leaves were carried by a man in a canoe near the island of Ferdinandina because they were esteemed for their healthfulness. Tobacco, probably mixed with lime or chalk, appears to have been used in these Native American populations as a tooth paste to whiten the teeth, as observed by Nino and Guerra in 1500 and by Vespucci at about the same time in Venezuela. This practice continues today in India, where powdered tobacco, or *masheri,* is rubbed on the teeth for this purpose and tobacco toothpaste is marketed commercially. It was perhaps in 1500 that the notion of tobacco as a panacea became prevalent. In that year, a Portuguese explorer, Pedro Alvarez Cabral, in Brazil, reported the use of the herb *betum* for treating ulcerated abscesses, fistulas, sores, inveterate polyps and many other ailments, and said it was called the holy herb because of its powerful virtue in desperate cases. Also, reports on medicinal use of tobacco by Native American populations continued to emerge in quantity. In 1529 Spanish missionary priest, Bernadino de Sahagun, collected information from four Mexican physicians about use of tobacco for medicinal purposes. He recorded that breathing the odour of fresh green leaves of the plant relieved persistent headaches. For colds and catarrh, green or powdered leaves should be rubbed around inside the mouth. Diseases of glands in the neck could be cured by cutting out the root of the lesion and placing on it crushed tobacco plant hot and mixed with salt, on the same spot. In the days when treatments for many diseases were being sought and herbs of all kinds were considered worth trying, the news of an unfamiliar herb with reputed therapeutic efficacy generated much enthusiasm. So great was the excitement that Nicolas Monardes, the Spanish physician-botanist, included it in a work originally published in the 1570s and later rendered into English as *Joyful Newes out of the New-Found World*? . Herbals at this time described not only the plants but also their medicinal applications. A notable example is the illustration from about 1555 by the Franciscan monk André Thevet in Brazil, of smoke being blown at a man from a primitive cigar. The most interesting, and perhaps the most convincing, indication was in the treatment of *Noli-me-tangere*. This name was given to slow-spreading ulcerating lesions of the skin. Later publications suggest that the condition embraced such conditions as lupus and syphilis but that the most frequent cause was probably basal cell cancer (rodent ulcer). In about 1560, according to Monardes, the French ambassador to Lisbon, Jean Nicot, was presented with a herb by the keeper of a prison he was visiting. It was described as a strange plant brought from Florida. The ambassador had it planted in his garden where 'it grew and multiplied marvelously'. One of Nicot's pages had a *Noli-me-tangere* on his cheek which was beginning 'to take root already in the gristles of the nose', and had himself been applying bruised tobacco leaves and juice to it. Hearing of this, Nicot ordered that the tobacco treatment should be continued for eight or ten days, and at the end of this time 'this said *Noli me tangere* was utterly extinguished and healed'. Throughout the treatment Nicot had the patient's progress monitored by a respected physician to the King of Portugal, who certified the happy outcome. So pleased was Nicot with this cure that, when he heard of two ladies in France who had carcinomas for which no cure could be found, he sent the herb to King Francis II, the Queen Mother and many Lords of Court. Nicot was so liberal and generous with tobacco that it became known as the ambassador's herb or *nicotiane* – the origin of the name by which we now know it. Nicot used it to treat the father of one of his pages for an ulcerated leg of two years' duration, and healing was reported after ten to twelve days. Similarly, complete healing was described after eight or ten days' treatment in 'a woman that had her face covered with a Ringworme rooted, as though she had a visor on her face', and a captain's son was cured of the 'king's euill' (scrofula). When a cook in Nicot's household nearly cut off his thumb with a chopping knife, the steward ran for the tobacco plant and bound the thumb back on; after five or six dressings of the same sort, the wound healed. All these uses involved external application of tobacco leaf and its juice, and various recipes are described. Monardes, for instance, specifies that the leaves must be stamped in a clean mortar and both the juice and the leaves applied to the lesion. To 'cleanse, incarnate, and knit together all manner of wounds'. 'Take a pound of the fresher Leaves of the said herb, stampe them and mingle them with a newel Wax, Rosine, common oyle, of each three ounces, let them bole altogether, until the juice of the Nicotiane be consumed, then add thereto three ounces of Venise Turpentine, straine the same through a linen cloth, and keep it in Potts to your use.

Uses in Nineteenth Century

After the isolation of nicotine from tobacco leaves in 1828, the medical world became yet more mistrustful of tobacco as a general treatment, now aware that the plant contained a dangerous alkaloid. Nicotine began to be used alone and more effort was made to measure doses. For example, a preparation of nicotine salicylate as a 0.1% salve replaced an infusion of leaf tobacco boiled in water as a treatment for scabies. However, even tobacco smoke *per rectum* was still being advocated, for conditions as varied as strychnine poisoning, constipation, strangulated hernia, tetanus, hydrophobia and worms. In a 1958 paper Silvette and co-worker scanned the medical press for case studies of tobacco treatments published between 1785 and 1860 and provided an overview of treatment outcomes for a range of conditions.

Uses in Twentieth Century

Even in the twentieth century, the therapeutic use of tobacco did not completely lapse. For example, in 1924, a salve made of burned tobacco leaves mixed with lanolin was said to be desiccant, stimulant and antiseptic for pruritus, ringworm, athlete's foot, superficial ulcers and wounds (it was also said to be good as a metal polish). Moreover, its disinfectant properties continued to generate debate. We have seen how, in the New World discovered by Columbus, tobacco smoke was used to ward off disease, and the sixteenth century doctors applied the leaves or a tobacco ointment or poultice to infected wounds. During the London plague of 1665 children were instructed to smoke in their schoolrooms, and in 1882, in a Bolton outbreak of smallpox, tobacco was actually issued to all the residents of a workhouse, However, claims for such protective effects did not go undisputed. For example, in 1889 an anonymous article in the *British Medical Journal*, whilst acknowledging the experimental evidence that the pyridine in smoke kills germs and the evidence that smokers appeared to be at lower risk of diphtheria and typhus, concluded that people who can tolerate tobacco are likely to be robust in other ways and thus able to resist infection; non-smokers, the article concluded, would be ill advised to take up smoking, which would make them more vulnerable. An anonymous article in *The Lancet* in 1913 discusses the 'pyridine' content of tobacco smoke and describes experiments showing that tobacco smoke destroys the comma bacillus of cholera; but again it warns that tobacco smoking can 'give rise to constitutional effects which diminish the resisting power of the body to disease'. Later in the twentieth century, attention switched to diseases affecting the brain and nervous system. In 1926, Moll reported that, when thirteen patients with post-encephalitic Parkinsonism were treated with subcutaneous injections of nicotine, nine showed immediate improvement in muscular movement. He concluded that, although the benefit was only temporary, 'the immediate results were indisputable'. A kindred observation is that, in at least three case-control studies, the relative risk of Parkinson's disease was lower in smokers than in non-smokers, though other factors could be operating to produce this apparent effect. Case-control studies also suggest a possible inverse association between cigarette smoking, Alzheimer's disease and Tourette's syndrome, but the same reservations apply.

Contemporary and Recent Development

Alzheimer's and Parkinson's Diseases

"When you look at people who smoke, and people who don't smoke...you find those who smoke cigarettes are about half as likely to get Parkinson's disease".

Researchers have long been aware that fewer smokers get Alzheimer's and Parkinson's diseases than non-smokers. A reduced risk for smokers and only four found no difference between smokers and non-smokers. Similar findings have been published on the effect of smoking and Parkinson's disease. In an article in The Times of London Dr. James Le Fanu provided an examination of the research on smoking and its apparent protective effect for certain diseases. Dr. Le Fanu stated unequivocally: "Smokers have a 50 per cent reduced risk of developing Alzheimer's—and the more smoked the greater the protection." He also noted that emerging research points to a similar effect of smoking on Parkinson's disease. So striking was the apparent protective effect of smoking on Alzheimer's and Parkinson's that increasingly biomedical researchers are experimenting with nicotine to treat the symptoms of these dread disease in-patients who have been diagnosed as having them. Results from these experiments have all showed promise in alleviating the symptoms of these diseases with the administration of nicotine. The mechanism by which the nicotine in tobacco works to protect smokers is that it increases the number of so-called "nicotinic" receptors in the brain, which in turn influence the production and release of the neurotransmitter acetylcholine. Those who come down with Alzheimer's show a marked loss of "nicotinic" receptors in their brains and thus have reduced levels of acetylcholine, which is necessary for memory and other brain functions. Research has shown that tobacco smoke (and the nicotine therein) inhibits the activity of monoamine oxidase B (MAOB), were resistant to the neurodegenerative effects of MPTP, a toxin that induces a condition reminiscent of Parkinson's disease.

Other Diseases

Other diseases for which smoking and nicotine appear to be protective are ulcerative colitis, Tourette's Syndrome, and possibly rheumatoid arthritis and colorectal cancer.

Ulcerative Colitis

The influence of smoking on the risk of developing ulcerative colitis is well documented. Compared with lifetime nonsmokers, the risk is reduced in smokers. When association between cigarette smoking and UC [ulcerative colitis] are examined, never-smokers are approximately three times more likely to develop UC than smokers. A consistent finding from study to study is that quitters have a mildly increased risk of developing UC which suggests that cigarette smoking may have a protective effect. It is beyond doubt that smokers are protected against ulcerative colitis, and the more that is smoked the greater the protection—so those on 25 cigarettes a day or more have a risk as little as one-tenth that of non-smokers. The finding does not mean that people with depression should smoke or even start using a nicotine patch, the researchers caution it may be possible to manipulate nicotine's effects to safely reap its potential medical benefits, according to the researchers. As an example of the drug's potential, they said, pharmaceutical companies already are developing compounds for treating other brain disorders by mimicking the beneficial properties of nicotine while avoiding its addictive nature.

Depression

In the study, the researchers recruited 11 people who did not smoke but who were experiencing symptoms of depression. Participants were randomly assigned to wear either a nicotine patch or a placebo patch that did not contain any nicotine. The researchers used a standardized method, a 20-item questionnaire called the Center for Epidemiological Studies Depression scale, to measure depression symptoms among the study participants. The team found that participants who wore the nicotine patch for at least eight days experienced a significant decline in their depression-assessment rating scores; this finding indicates that the drug led to an improvement in depression symptoms. As a possible explanation for how nicotine exerts its beneficial effect, the same areas of the brain that are stimulated by nicotine appear to be involved in the regulation of mood. Nicotine stimulates the release of specific neurotransmitters, including serotonin, dopamine, and nor epinephrine, which carry messages between nerves cells. Depression has been linked to chemical imbalances of this neurotransmitter. The researchers say the nicotine molecule can be manipulated to remove its addictive effect while maintaining its ability to increase levels of the various brain chemicals that can alter mood.

Cigarettes 'Keep you Sharp After Dark

The results showed that smokers were subsequently able to perform various tests of reaction, memory recall and other related tasks consistently better than the non-smokers," [Reported on the BBC News, 4/8/98]. Smokers may indeed smoke, in part, to improve their mood-a notion that has been quite controversial in the field. Scientists have established that people prone to depression are twice as likely to be smokers, and are less likely to succeed in quitting smoking after taking up the habit.

❑❑❑

26

Tobacco Abuse and Human Health

It is especially important to understand that harmful effects of tobacco products are dose-dependent, that they depend more on abuse than on simple use. In this light its addicting aspects acquire ominous overtones. The nicotine found in substantial amounts in tobacco products is widely considered to be a powerfully addicting drug, so much so that its addictive processes and potential have been equated with heroine, morphine and cocaine.

Since the Middle Ages tobacco leaves have been used as medicinal herbs in ointments, poultices, mouth rinses and smoke. Oral ulcers, caries and "toothache" were all treated with this wonder drug. A product is still readily available today, called "Dental Snuff" which was advertised more than a century ago as a cure for toothache, gingivitis, facial neuralgia, caries, and scurvy. Given this positive image throughout most of its history its universal popularity is not surprising. The negative health effects of tobacco use, we would remind the reader that a popular emphasis on negative aspects is a rather new phenomenon.

It's rapid absorption through the lungs of cigarette smokers is widely accepted, but its equally ready absorption through the oral mucosa under the alkaline conditions normally found in cigar, pipe and smokeless tobacco use is less publicized. Once in the blood stream, of course, nicotine acts on the central nervous and cardiovascular systems in identical fashion regardless of the method of absorption. These systemic effects have been extensively explored and reported, but the local oral effects of various tobacco habits have received little attention outside of the potential for cancer. The purpose of the present article is to summarize current concepts of such local effects, especially the nonmalignant changes in the mouth.

1. Oral Mucosal Changes from Tobacco Use

A wide variety of mucosal changes have been noted in habitual users of smoked and smokeless tobacco. These changes most likely result from the many irritants, toxins, and carcinogens found naturally in cured or burned tobacco leaves, but may also arise from the mucosal drying effects, the high intraoral temperatures, intraoral pH changes, local alteration of membrane barriers and immune responses, or altered resistance to fungal and viral infections. The following are significant mucosal changes noted in tobacco users:

Oral Cancer

Squamous cell carcinoma is quit naturally the most worrisome mucosal change encountered in smokers, but it is not intended here to dwell on this aspect of tobacco's effect. Certain premalignant lesions are mentioned below, but suffice it to say that tobacco's role in producing squamous carcinomas of the oral, pharyngeal and laryngeal mucosae is well established. At least 80% of oral cancer patients are smokers; oral cancer patients who continue smoking after treatment are much more likely to develop a new head and neck cancer; animal studies have produced oral cancers in a variety of animals via forced cigarette smoking; and case-control studies in humans have determined that a smoker is as much as 7 times more likely than a nonsmoker to develop oral malignancy.

It is important to remember that oral cancers develop by way of cocarcinogenesis, multiple factors are usually involved. Excessive alcohol intake by itself has little impact on the prevalence of oral carcinoma, but alcohol is a powerful "promoter" capable of altering mucosal cells to make them much more susceptible to malignant transformation by tobacco carcinogens. Even the alcohol in concentrated mouthwashes is thought to play a role in this process.

Tobacco chewers also develop oral carcinomas. The oral cancer risk for these users is about 4 times greater than nonusers, although the data proving this is very sparse. A very unique low-grade oral malignancy, verrucous carcinoma, is so seldom diagnosed in nonchewers that it is aptly called the "snuff dipper's cancer" his carcinoma typically develops in the exact site of chronic tobacco placement. It differs from the usual oral cancer in that it enlarges very slowly, is essentially nonmetastasizing, and consists of very mature epithelial cells, i.e. has little dysplasia. The mortality rate for verrucous carcinoma is among the lowest for any cancer of the human body, so low that it would be a blessing were it the usual type of oral cancer diagnosed in smokeless tobacco users. Unfortunately, this is not the case. Verrucous carcinoma is extremely rare (1/1,000,000 persons annually), hence, the typical smokeless tobacco cancer is the same as the typical smoker's oral cancer, i.e. invasive squamous cell carcinoma.

Leukoplakia

While the exact etiology of oral leukoplakia still eludes us, tobacco smoking is by far the most broadly accepted factor. Approximately 80% of leukoplakia patients are smokers and when large groups of adults are examined we find that smokers are much more likely to have leukoplakia than nonsmokers (23% vs. 4%). Pipe smokers and heavy cigarette smokers have greater numbers of lesions and larger lesions than other smokers, especially after many years of tobacco abuse. Sixty percent of smoke-induced leukoplakias, furthermore, disappear 6-12 months after affected patients stop smoking.

Leukoplakia in smokers is usually seen as a well-defined white plaque of the oral . The average patient is 50-60 years of age and male, while the average lesion is 1.4 cm. in diameter and has been present for 2-4 years at the time of diagnosis. The overall cancer transformation potential of leukoplakias is approximately 4%, but certain clinical forms of leukoplakia are much more prone to cancer than others. Speckled leukoplakia (erythroleukoplakia), for example has a cancer potential of at least 25%, in some studies as high as 41%. Do leukoplakias in smokers become malignant more frequently than those in nonsmokers? We presently lack the data to say with confidence, but it appears that leukoplakias in nonsmokers, or those remaining after smoking cessation, are more likely to transform than those in smokers. Hence, smokers have much more leukoplakia but their lesions seem to be less aggressive.

Nicotine Palatinus (Stomatitis)

Once a common mucosal change of the hard palate, nicotine palatinus has become something of a rarity as cigar and pipe smoking have lost popularity. Even before it became an uncommon lesion, however, it's status as a precancerous condition had been lost. While this lesion is most certainly produced by heavy and frequent contact with tobacco smoke and is just as certainly a white keratotic change, it does not transform into malignancy and is a response to the heat of tobacco smoke rather than the chemicals in the smoke. It is completely reversible within a few months of quitting the smoking habit, even when present for many decades before the habit is stopped. Today we primarily see this entity in heavy cigarette smokers (more than two packs per day) and its intensity is typically much less than lesions of a few decades ago.

Smokeless Tobacco Keratosis

Snuff pouch or smokeless tobacco keratosis is, like nicotine palatinus, no longer considered to be a true leukoplakia. It is a white keratotic plaque, as is leukoplakia, but there the similarity ends. It typically has a semitranslucent appearance rather than a flat whiteness, and it has a microscopic appearance different from the simple excessive keratin by which leukoplakia is characterized. It also is located only in areas of direct contact with snuff or chewing tobacco and is almost always completely reversible when the affected patient quits his or her habit. While there can be no doubt that it is a precancerous entity, its malignant potential is less than true leukoplakia by a magnitude of ten (4% vs. O.4% estimated lifetime risk for leukoplakia and smokeless tobacco keratosis, respectively). More wrinkled and more intensely white smokeless tobacco keratoses are considered to be "higher grade" lesions, but these clinical grades seem to have little correlation with the eventual development of cancer. Today the only smokeless tobacco keratosis feature known to have an increased risk of cancer transformation is continued presence after tobacco cessation. Those lesions remaining after two months without smokeless tobacco should be considered to be true leukoplakias, possibly developed from or hidden by the previous smokeless tobacco keratosis.

Smoker's Melanosis

Smoking is capable of stimulating oral mucosal melanocytes to produce excessive melanin, thereby creating patches of brown pigmentation on gingival or buccal mucosae in 5-22% of heavy smokers. The number and intensity of these smoker's melanoses are dose-dependent and smoking cessation seems to reverse the process completely. *In vitro* testing has demonstrated that it is the nicotine itself which activates one of the steps in melanin production. To date no case has transformed into a carcinoma or melanoma and no case has been reported in nonsmoking users of smokeless tobacco, but an occasional lesion may have leukoplakia superimposed on it. In the latter event, we use the term "melanoleukoplakia" (or "leukomelanosis") and presume that at least the keratotic aspect of the plaque will behave as any other leukoplakia without melanosis.

Submucous Fibrosis

Oral submucous fibrosis is a precancerous condition characterized by a progressive stiffening of the oral mucosa to the point wherein affected persons have difficulty opening their mouths. A 10-year malignant transformation rate of 8% has been determined in India, where tobacco chewers routinely mix betel leaf, areca nuts, and/or slaked (shell) lime with their tobacco. This entity, which frequently has leukoplakia superimposed on it, is seldom seen in the U.S. but is found in the mouths of 0.4% of India's villagers.

Leukoedema

A poorly demarcated, grayish-white, opalescent change of the buccal mucosa, bilaterally, is frequently seen in persons with darkly pigmented skin. Always a benign lesion and usually considered a simple variation of normal, it is nevertheless capable of an increased whiteness and size in smokers. Axell and Henricsson found leukoedema in 60% of Swedish adult smokers but in only 36% of nonsmokers.

Hairy Tongue

The condition of elongated filiform papillae mimicking hair on the dorsum of the tongue is frequently seen in heavy smokers Theoretically tobacco smoke prevents the epithelial cells from sloughing in a normal fashion and they accumulate into a very thickened and white (unless stained brown, black, etc.) surface which extends as long "hairs" at the tips of the papillae. We have no clinical research definitively linking hairy tongue with smoking but it frequently decreases or disappears when the habit is stopped. Smokeless tobacco use is also thought to contribute to this condition, but there is no scientific evidence to prove the association.

2. Gingival & Alveolar Bone Damage

Gingivitis

Frequent studies have concluded that tobacco smokers are more likely to demonstrate gingival redness, hemorrhage and inflammatory enlargement (edema) than nonsmokers with similar oral hygiene habits. Others, however, have found no increased gingival bleeding in smokers, possibly because of the vasoconstriction produced by nicotine. Smokeless tobacco users likewise have no more gingivitis than nonusers, but this habit is well known to produce a characteristic painless loss of gingival tissues in the area of tobacco contact. This gingival "recession" frequently includes bony destruction but is seldom associated with the inflammatory changes noted clinically in routine gingivitis. It may result from a combination of factors, including abrasion and the local tobacco-induced release of collagenase and other protein-destroying enzymes. Logic would seem to indicate that tobacco use must have some effect on the gingival and periodontal structures. Tobacco contains, after all, several cytotoxic substances which have been shown to be secreted in salivary and gingival crevicular fluids. Smokers also have markedly more calculus than nonsmokers, and heavy smokers have more calculus than light smokers. Nicotine's well-known vasoconstriction effects have been shown to be strong in periodontal tissues, even in the face of elevated blood pressures, and gingival circulation has been shown to decrease by as much as 70% during the smoking of a cigarette. Tobacco smoking, furthermore, also suppresses human immune responses, including responses to oral microbial toxins. Oral leukocytes, especially neutrophils, have a diminished ability to move, to phagocytize, or to secrete enzymes in smokers.

Periodontitis

While research has been equivocal relative to the production of gingivitis, overwhelming evidence from a large number of investigations has made it obvious that smokers have considerably more severe periodontitis than nonsmokers. After controlling for different oral hygiene habits, patient age, gender, and socioeconomic levels, smokers still have deeper pockets, more alveolar bone loss, more tooth mobility, and greater tooth loss than nonsmokers. Solomon *et al* have gone so far as to state that the periodontal status of smokers is at least 15 years more advanced than the status of nonsmokers of the same chronological age. Even in persons under 40 years of age, smokers are known to have twice the periodontitis and tooth loss that nonsmokers have.

As previously mentioned, smokeless tobacco users are also prone to a localized periodontal destruction or recession resulting from the combination of nicotine's chemical effects and the effects of physical abrasion on local gingival tissues. This association between smokeless tobacco use and gingival/periodontal recession in the site of chronic tobacco placement is a rather strong one, and recession correlates well with both the quantity of daily use and the duration of the smokeless tobacco habit. It should be emphasized, however, that routine gingivitis in the area of tobacco contact appears necessary to either initiate or promote this change, a change identified in 31-92% of young adult smokeless tobacco users.

Acute Necrotizing Ulcerative Gingivitis (Anug)

ANUG is strongly correlated with tobacco use. Ninety-eight percent of ANUG patients are smokers, and persons who smoke more than 10 cigarettes per day have a tenfold increase in ANUG prevalence compared to nonsmokers. The exact cause of ANUG is, of course, unknown, but it occurs most frequently in teenagers and young adults and may result from defective neutrophil function allowing bacterial and possibly viral (cytomegalovirus) invasion of gingival tissues. The vasoconstrictive action of nicotine and other tobacco components is thought to contribute strongly to the painful tissue necrosis and ulceration seen in this disease, but emotional stress and poor oral hygiene appear to play roles almost as important. Smoking seems to contribute as much to unsuccessful periodontal therapy as to the promotion of periodontitis. In fact, Miller has very recently concluded that no postoperative factor contributes more to the failure of periodontal soft tissue grafting than does this habit, a feature which also holds true for skin grafts outside the mouth. Apparently active smoking is required. Patients who stop smoking before periodontal surgery and refrain from smoking for several weeks thereafter have a graft success rate similar to that of nonsmokers.

Delayed Alveolar Wound Healing

Delayed wound healing in smokers is noted in other situations as well. Routine periodontal surgery and tooth extraction are also adversely affected, possibly because of the direct contact of smoke constituents on open wounds, or reduced blood flow, or impaired leukocyte function, or diminished fibrinolytic activity, or the drying effect of mouth-breathing, or the clot-disrupting effect of the negative intraoral pressure produced during tobacco smoking. Regardless of the cause, smokers have significantly less blood filling of extraction sockets and develop five times more dry sockets than do nonsmokers.

3. Tooth Decay, Abrasion, Erosion

Caries (Tooth Decay)

Nineteenth century dental surgeons presumed that tobacco smoking protected against caries and as late as 1968 Schmidt felt able to state that caries reduction was the only advantage of smoking. Most recent investigators, however, have concluded that cigarette smoking is most certainly associated with an increased caries rate but that a cause-and-effect relationship is still not proven. Smokers have a significantly greater number of carious or repaired tooth surfaces than nonsmokers, and heavy smokers are more affected than light smokers. Smokers also have higher plaque rates than nonsmokers. It has been suggested that smokers as a group have poorer oral hygiene habits and skills, make fewer visits to dentists, and have lesser overall health standards than nonsmokers. As this has not been adequately controlled for in past investigations it may be the real reason for the increased caries rate.

More than one-fifth of the content of some brands of smokeless tobacco is sugar and a few brands are even capable of serving as growth media for several bacteria implicated in the production of caries. Smokeless tobacco users, moreover, have significantly greater numbers of caries-associated bacteria at the site of quid placement when compared to nonusers. The caries that results from smokeless tobacco use is frequently cervical or root caries, but regardless of the caries type carious destruction is seen far more frequently in smokeless tobacco users who also have gingivitis. While the evidence would appear to conclusively associate smokeless tobacco use with dental caries, a cause-and-effect relationship is not conclusively established and several well-designed contemporary investigations have found little or no correlation.

Abrasion

Tooth abrasion resulting in a notching of incisal edges and cusp tips is a well-established consequence of holding a pipe in the same location while smoking. But long-term smokeless tobacco users also demonstrate excessive and frequently generalized wear of occlusal and incisal surfaces, especially in persons employed in dusty environments, such as coal mines and textile mills. This abrasion, produced by sand and other gritty materials remaining in processed tobacco, usually takes many years to develop, but the present authors have seen severe abrasion after only four years of smokeless tobacco use.

Erosion

The chemical dissolution of enamel has occasionally been reported in tobacco smokers and chewers, but only as a secondary consequence of using breath mints or as idiopathic cases. Erosion does not seem to be a serious problem and, in fact, chewers may be somewhat protected by the natural alkalinity of smokeless tobacco.

4. Alteration of Local Physiology & Microbiology

Diminished Blood Flow

Several physiological oral alterations produced by tobacco use have already been mentioned. The average smoker experiences a 60-70% reduction in oral mucosal blood flow while smoking, and laser Doppler evaluation of gingival blood flow has determined that smokeless tobacco users have, on the average, a 45% decrease compared to nonusers. Decreased flow lasts for 2-3 hours after smoking a single cigarette and is the most likely explanation for the known lack of gingival hemorrhage in smokers with otherwise obvious signs of poor oral hygiene and/or advanced periodontitis.

Poor Periodontal Attachment

The cytotoxic substances absorbed from tobacco, especially nicotine and cotinine (its major metabolite), are found in secreted saliva and crevicular fluids. Lab studies suggest that the attachment of fibroblasts to cementum is altered by nicotine and new investigations are underway to further evaluate the local effects of tobacco cytotoxins.

Increased Salivary Flow

While cigarette smoking typically causes a noticeable short-term increase in salivary flow rates, the long-term influence of tobacco use is unclear. Certainly, intense smokeless tobacco use has been shown to result in degenerative changes of more than 40% of minor salivary glands located in the site of chronic tobacco placement, and immunoglobulin A (IgA) levels are depressed in smokers, but the clinical impact of these changes appears minimal.

Bacteria

Smoking reduces the overall concentration of bacteria in the mouth, but the mix of oral microflora does not appreciably alter, nor is the rate of plaque formation enhanced, at least not when differing oral hygiene levels are controlled for. Smokeless tobacco usage does, however, appear to enhance the presence of several periodontal pathogens, namely *Veillonella, Bacteroides,* and *Fusobacterium.*

5. Candidiasis and Median Rhomboid Glossitis

The great majority (83%) of oral candidiasis patients are moderate to heavy cigarette smokers. *Candida,* a normal component of approximately 50% of the mouths of healthy, normal individuals (without dentures), can be variably affected by smoking, becoming either more or less intense in different individuals. Heavy smokers who also abuse alcohol are more likely to lack oral *Candida* than any other group. Median rhomboid glossitis, a candida-induced tongue change, is most frequently (85%) seen in smokers and is markedly improved upon smoking cessation.

Sinusitis

Tobacco smokers are much more prone to maxillary sinusitis than nonusers and this condition typically resolves or is significantly reduced when the smoking habit is discontinued. This effect is probably secondary to a tobacco-induced edema of the sinus membranes and by diminished ciliary activity of respiratory epithelial cells. The edema tends to narrow sinus openings or orifices while the weakened ciliary movement prevents mucous and bacterial transport out of the sinuses.

Altered Taste

We all have predetermined thresholds for the sensations of taste and smell. Tobacco smoking is well known to greatly increase these thresholds, thereby diminishing our ability to detect various tastes and smells. Relative to taste, a dose-related association is known for bitter sensations, and to a lesser extent for salty tastes, but there appears to be little change in the ability to detect sweet or sour substances. The diminished taste acuity becomes progressively worse with each additional year of tobacco use. Smokeless tobacco use seems to have no or little effect on taste acuity. Interestingly, persons who are able to quit smoking express a very strong preference for sweet foods, a feature which probably helps to explain the weight gain so commonly noted after smoking cessation.

6. Aphthous Ulcers and Fever Blisters

Recurring aphthous ulcers, or canker sores, are exquisitely painful and very common, affecting 10-20% of the general population and as much as 50% of health professionals. Rather than producing or inducing this ulcerative disorder, tobacco smoking seems to prevent its occurrence or diminish its effects. Aphthae frequently begin to appear or reappear in persons who stop smoking, and almost all (96%) affected individuals are nonsmokers. Canker sores are likewise less prevalent in smokeless tobacco users than in nonusers. The reasons for this "protection" are unclear, but may be related to an increased mucosal keratinization or a reduced immune attack against the bacterial antigens thought to trigger ulcer formation.

Swango and Kleinman are the only investigators to study the relationship between tobacco users and secondary herpes simplex or fever blisters. They found a slightly higher prevalence in smokeless tobacco users when compared to nonusers (1.3% vs. 0.8%).

7. Esthetics Considerations

Halitosis

Both smoked and smokeless tobacco usage produce unpleasant breath odors or "bad breath". With smoking the halitosis is produced predominantly by the retention and subsequent exhalation of inhaled smoke in the lungs. Regardless of the method of absorption, however, numerous noxious elements characteristically escape from the lungs via the blood/air interchange. Pipe and cigar tobacco contains more sulfur than cigarettes; hence users tend to have a more offensive halitosis than cigarette smokers.

Tooth and Prosthetic Stains

Tobacco stain, a brown/black extrinsic stain, is typically found on the enamel surfaces of smokers and tobacco chewers. It is especially pronounced in cervical areas and the lingual aspects of the mandibular incisors. Its intensity is more dependent on the amount of available plaque and calculus than on the amount of tobacco used, but as a general rule smokers have almost twice as much tooth staining as nonsmokers and heavy smokeless tobacco users have nine times as much as light users. Smoked tobacco staining of dentures is a special problem because it is often too heavy and too deeply embedded to be removed by denture cleansers. This is especially true for persons with poor oral hygiene, possibly because of the softening of plastics secondary to plaque-produced acids. Smoking is also known to contribute to or accentuate the staining properties of chlorhexidine solutions and gels.

Cancer and Tobacco

Genetically Modified tobacco used in developing personalized cancer vaccine

Tobacco is better known as a cause of cancer rather than a potential cure. In a recent research development in Arizona and elsewhere believe tobacco plants may hold the key to developing a personalized cancer vaccine as well as treatments for other diseases. Plant-based biotechnology already has been used to produce drugs for diseases such as cystic fibrosis and Gaucher's disease, but some of these personalized drugs can be pricey for consumers. This is outcome of the experiments which is part of a growing field of plant based biotechnology and the vaccine has been successfully used in human clinical trials and it showed an immune response in 70 percent of non-Hodgkin's lymphoma patients without harmful side effects. In Germany that initially will grow the tobacco plants to make personalized vaccines for lymphoma patients. The vaccine is produced by taking DNA from a person's cancer cells, genetically modifying the DNA strands and transferring the virus to a tobacco plant. The plant responds by producing a protein that can help a cancer patient battle the disease. Scientists estimate that such vaccines can take just six to 10 weeks to produce from the time of a patient's biopsy. That would give doctors a quick time frame to administer the vaccine rather than revert to traditional cancer treatment such as chemotherapy, which has harmful side effects.

□□□

27

Tobacco Abuse - Indian Scenario

The tobacco plant, Nicotiana, has probably been responsible for more deaths than any other herb. At present, tobacco smoking is causing over 3 million deaths a year worldwide, and if current smoking trends continue the annual mortality will exceed 10 million by around 2030. Add to this the mortality from cancers caused by oral uses and the death toll becomes still higher. Undoubtedly, tobacco is the most important avoidable cause of premature death and disease in the world.

Introduction

Humankind has had a long history of using plants and weeds for their intoxicant, medicinal or social properties. Throughout the world and through the ages, plants have been smoked, chewed, applied on, or ingested, and their use has been accompanied often by elaborate social and religious rituals. Today, the use of one of these plants, tobacco, is one of the most widely recognized human social habit and the addiction it causes is also one of the most difficult to overcome. Indians used the 'Indian weed' in many ways including chewing the leaves, but they smoked it also using a calumet or peace pipe. After the Spanish successfully introduced tobacco to Europe and the British aristocracy adopted pipe smoking as a popular passtime, its increasing cultivation formed the economic basis of the first successful English colonies in North America. Virginia tobacco created a substantial and highly profitable English market, despite smoking's ill repute. In 1604, King James I pronounced in his "Counterblast to Tobacco" that smoking was "loathsome to the eye, hatefull to the Nose, harmfull to the brain, [and] dangerous to the lungs". However, it was good for tax rolls, and by the 1630s tobacco was firmly established as Virginia's staple crop and chief source of revenue. Thomas Jefferson would be puzzled today if he were to decide on the conflicting economic and health issues of tobacco use in the context of his admonition that "the greatest service that can be rendered any country is to add a useful plant to its culture".

Tobacco Abuse

It is the nicotine that causes smokers to become addicted to tobacco, and the chemical itself is lethal in small doses. When tobacco smoke is inhaled, the nicotine passes quickly to every organ of the body. The brain and nervous system are stimulated by small doses and depressed by larger ones. Nicotine increases the heart rate and the blood pressure, and may contribute directly to the excess of thrombosis and atheroma in smokers. Nevertheless, nicotine replacement therapy is used in helping people to stop smoking, because it spares them the many other harmful contents of tobacco smoke—for example, the carcinogenic polycyclic aromatic hydrocarbons and N-nitroso compounds; irritant substances such as acrolein; benzene; formaldehyde; ammonia; acetone; acetic acid, and carbon monoxide.

Tobacco use is the most popular form of substance abuse used by teenagers worldwide. The question is, why? Tobacco use has so many harmful affects that you would think we would steer away from smoking. However, smoking has become one of the most popular drugs put out there for teenagers to use, and yes, tobacco is a drug. Teenagers must become aware of what Cigarettes are made of, and the terrible harm that smoking can do to our well being. Hopefully this web site will do just that. So please continue reading! It is only after teenagers become aware of the problems that the abuse may stop, and the prevention may continue. World Health Organization experts have called for stricter laws, improved implementation of these laws and increased taxes to help reduce tobacco abuse in the region. World

Health Organization, said: "We need stricter surveillance and monitoring and awareness among the masses to stop tobacco abuse. Tobacco control needs strong political commitment as well as the participation of civil society. There is need to monitor tobacco use, protect people from tobacco smoke, offer help to quit tobacco use, warn about the dangers of tobacco, enforce bans on tobacco advertising and promotion and raise taxes on tobacco products.

India has six million children, aged 4 to 14, and working fulltime in the bidi industry. Bihar, Gujarat and Manipur have registered more than 40 per cent use of non-cigarette tobacco products by 13 to 15-year-old boys. Tobacco will kill about 10 million people every year by 2020 and 70 per cent of these deaths will occur in the developing countries and if the current smoking pattern continues in India, tobacco will kill about a million men aged between 25 and 69 by 2025" said Judith Mackay of the World Lung Foundation, a NGO. In India alone, 5,500 youths succumb to tobacco addiction each day thus making tobacco a source of growing concern not only for the government but also for social organizations working for human welfare, according to a research carried out by Healis and World Health Organisation (WHO) across the country. Among youth, .6 per cent of India's boys, 1.1 per cent of its girls smoked, and 42.9 per cent of youth were exposed to passive smoking. Tobacco killed more than half of its regular users and was responsible for five million deaths worldwide every year. "Children start using tobacco because the environment in which they are, encourages them to do so. The global as well as Indian research findings make it quite clear youngsters are more likely to start using tobacco if they are exposed to advertisement of tobacco products, association of their role models with tobacco products and possession of objects advertising tobacco. Studies conducted by WHO established that environmental smoke and lack of awareness regarding the health hazards of tobacco were the main factors leading to increasing tendency among youths towards smoking. Usage of tobacco, especially smoking, among youths has turned into a trend which has caught up rapidly. The trend requires youngsters of this generation to take a few puffs or be branded as "uncool" by peers. Thus an assuming step towards pleasure turns into a life-long addiction. This trend has also led to a conspicuous rise in use of tobacco amongst youths.

Why We Should Not Smoke (Short Term Effects)

Tobacco smoke consists of tar, nicotine, carbon monoxide and other gases. The amount of nicotine and other substances that is absorbed through the lungs depends on how much of the smoke is inhaled. Nicotine is a stimulant and users feel that tobacco helps them relieve boredom, tiredness and also helps them reduce stress and anxiety. The effects are almost immediate but fade quickly, which encourages constant use. Some people may experience nausea and dizziness when they inhale tobacco smoke. While the user may believe that the smoking is helping their emotional stress and anxiety, they are only harming themselves in the long run. Many users of tobacco are also worried that if they do quit they will be subject to weight gain. The media has brought upon the thought that being as thin as models on magazine covers is healthy (see Eating Disorders) , but on the contrary this is very unhealthy, another reason to stop smoking! Gaining weight is not a health problem; it is the thinness and the smoking that will cause the harmful effects.

Among teenagers, smoking increases problems from asthma, and the respiratory problems associated with smoking are more pronounced in asthmatics. Asthma has been found to be more common in 11 to 16 year-olds who smoke, and the progress of asthma throughout childhood and early adulthood does not improve as greatly in smokers than in non-smokers. The most serious health hazard of tobacco for youth is addiction and its long-term effects. Although lack of research makes it difficult to say with certainty how quickly nicotine addiction develops in young tobacco users, existing evidence indicates that most smokers become addicted while still in adolescence. This addiction results in an increased risk of developing and prematurely dying from tobacco-related diseases. The more a person smokes/ the longer a person smokes, the more likely they are to suffer from sicknesses. Smoking releases epinephrine, a hormone which creates physiological stress in the smoker, rather than relaxation. The addictive quality of the drug makes the user feel he must smoke more to calm down, when in effect the smoking itself is causing the agitation. Therefore tobacco is likely to cause physical and psychological dependency in a short space of time if it is smoked regularly. Women who smoke during pregnancy are more likely to have smaller babies and run a bigger risk of losing the child. This is why it is strongly encouraged to Stop Smoking! When you stop, the chances of developing cancer and heart disease are immediately decreased. After a number of years, the risk of a person who has stopped smoking is the same as a person who has never smoked. However, it is even more strongly encouraged to not even start. If you never start you are not putting yourself in danger, and you do not have to put yourself through withdrawal when you quit.

Symptoms of Withdrawal (Why we should never start smoking to begin with)

Physical withdrawal symptoms include changes in: body temperature, heart rate, digestion, muscle tone and appetite. Psychological withdrawal symptoms include: irritability, anxiety, sleep disturbances, nervousness, headaches, fatigue, nausea and cravings for tobacco that can last days, weeks, months, years, or an entire lifetime.

Fight Against Tobacco Abuse

Whereas the prevalence of tobacco use has declined11 in the general Canadian population in the last decade, the trends among the Native population are alarming. Apart from having much higher consumption rates for smoke and smokeless tobacco, Native people also tend to start smoking earlier than non-Natives. In addition, there seems to be a certain lack of concern or awareness concerning the serious tobacco-related risks to the present and future health of many Native communities. Given the current levels of tobacco use among Native people, any effective reduction strategy will have to include at least three components, namely Protection, Prevention and Cessation.

Protection

Since Environmental Tobacco Smoke (E.T.S.) is a serious health hazard to which non-smokers as well as smokers are continuously exposed, the first step is to ensure that the community has as many smoke-free places as possible. Naturally, Aboriginal leaders have to be enlisted in enforcing the existing laws pertaining to tobacco, whether they be provincial, territorial or municipal. Each community can in addition, make and enforce regulations as it sees fit. Obviously such laws and regulations would make exceptions for traditional ceremonies using sweetgrass, tobacco or other sacred plants. In general, however, a number of measures could be taken" pertaining to:

1. Non-Smoking Policies

- All community buildings should be smoke free, especially health centers, hospitals, community centers, schools, band council offices, etc.
- Smoking on school buses should be prohibited.

- Staff working in any of the community services should be non-smokers. Smokers should be encouraged to stop and offered assistance (financial or otherwise) to do so. - Smoking in workplaces should be prohibited or restricted to a designated area.

2. Sales of Tobacco Products to Minors

- No sales of cigarettes or smokeless tobacco to minors in community stores.
- No cigarette vending machines in community.
- Vendors in violation of law or regulations should be prosecuted (suspend license or fine). - No sales of cigarettes in "kiddy packs" of 5's or 15's.

3. Lobbying Activities

- Native leaders could lobby within Native organizations and with governments for tougher legislation regarding tobacco abuse and youth.

4. Role Models

- Native leaders and elders should develop a strong stance regarding the non-traditional use of tobacco and at the same time teach youth about the traditional use of tobacco.

5. Socio-Economic Development

- Native leaders should involve the community in developing economic initiatives, encouraging youth participation in healthy cultural / sports activities and assist in recreational programming.

6. Community Involvement in

- Designing a strategy to prevent abuse of tobacco and help those who wish to quit with particular emphasis on youth and parents. By opting for a comprehensive approach in dealing with the non-traditional use of tobacco, the community can provide a basis which will encourage families to make their homes smoke free and healthier. All the research shows that parents and older siblings' attitudes and behaviour regarding the use of tobacco have a great impact on younger children because the latter will naturally imitate what they see. Parents, therefore have an important role to play and should strive to:

- Be good role models by either not smoking, or quitting, and sharing their experience with the children. If parents smoke, they should not do so indoors and not in front of the children. Discuss the issue of tobacco and explain in a simple, matter-of-fact way the effects of tobacco abuse including smokeless tobacco. Involve the children in a family strategy to have a smoke-free house. Encourage children to participate in sports, hobbies and cultural activities. Praise children's achievements to build self-esteem.
- Teach them how to say "No" and handle peer pressure effectively.
- Make it clear that everyone deserves a smoke-free environment.

Prevention

We know that tobacco use starts in adolescence and even earlier among Native children who often begin using snuff or chewing tobacco as young as 5 years of age. By age 13 to 15, they become regular smokers and are likely to remain so into their adult life. On the other hand, very few people start smoking past their 20th birthday. Therefore, prevention efforts must target children and teens so they will not start and become dependent on nicotine. To be effective, an anti-tobacco approach must take into account the many factors" which affect the initiation of tobacco use in children and youth.

Socio-economic Factors

- Poverty, unemployment, and lack of educational / sports / cultural programs.

Environment Factors

- Parents and siblings using tobacco in non-traditional ways, and perception of such behaviour as the norm.
- Peer pressure and behaviour.

Personal Factors

- Low academic achievement.
- Low self-esteem.
- Risk-taking or rebellious behaviour.
- Low participation in sports / cultural activities.

Since the tobacco industry relies on recruiting great numbers of young people into the tobacco market for its survival, the common themes found in tobacco promotion are designed to appeal to youth.

Myths and Facts About Tobacco Control

- What are the common myths in the economics of tobacco control that this report aims to dispel?
- Once a government decides that interventions are justified and chooses to do so, what should be the goals of its intervention measures?
- What are the recommended strategies for governments to achieve these goals?
- How cost-effective are the suggested tobacco control measures?
- Given the relatively low quit rates even in countries where cessation interventions are available, would it be better to implement strategies solely aimed at preventing children from smoking?
- It is mostly up to each government to decide whether to adopt the above recommendations or not. Is there a role for international agencies to assist governments in the process of choosing and implementing tobacco control policies?

What are the common myths in the economics of tobacco control that this report aims to dispel?

Below are the common myths about tobacco:

Myth 1: Tobacco is only an issue for affluent people and affluent countries.

Reality: Smoking is declining among males in most high-income countries. In contrast, it is increasing in males in most low- and middle-income countries and in women worldwide. Within individual countries, tobacco consumption and tobacco-related disease burdens are usually greatest among the poor.

Myth 2: Governments should not discourage smoking other than making its risks widely known. Otherwise, they would interfere with consumers' freedom of choice.

Reality: First, many smokers are unaware of their risks, or they simply underestimate or minimize the personal relevance of those risks, even in high-income societies where the risks are relatively widely known. Second, most smokers start when they are children or adolescents – when they have incomplete information about the risks of tobacco and its addictive nature – and by the time they try to quit, many are addicted. Third, smoking imposes costs on non-smokers. For these reasons, the choice to smoke may differ from the choice to buy other consumer goods and governments may consider interventions justified.

Myth 3: Smokers always bear the costs of their consumption choices.

Reality: Not necessarily so. They do impose certain costs on non-smokers. The evident costs include health damage, nuisance and irritation from exposure to environmental tobacco smoke. In addition, smokers may impose financial costs on others (such as bearing a portion of smokers' excess healthcare costs) . However, the scope of these costs is difficult to measure and they vary in place and time, so this report makes no attempt to quantify them In high-income countries, smokers' healthcare costs on average exceed non-smokers' in any given year. It has been argued that, because smokers tend to die earlier than nonsmokers, their lifetime health care costs may be no greater than those of nonsmokers; however, recent reviews in high-income nations conclude that smokers' lifetime healthcare costs do indeed exceed nonsmokers', despite their shorter lives. If healthcare is paid for to some extent by the public sector, smokers will thus impose their costs on others.

Myth 4: Tobacco control will result in permanent job losses for an economy..

Reality: Successful control policies will lead to only a slow decline in global tobacco use (which is projected to stay high for the next several decades). The resulting need for downsizing will be far less dramatic than many other industries have had to face. Furthermore, money not spent on tobacco will be spent on other goods, generating alternative employment. Studies for this report show that most countries would see no net job losses and that a few would see net gains if consumption fell.

Myth 5: Tobacco addiction is so strong that simply raising taxes will not reduce demand; therefore, raising taxes is not justified

Reality : Scores of studies have shown that increased taxes reduce the number of smokers and the number of smoking-related deaths.

Price increases induce some smokers to quit and prevent others from becoming regular or persistent smokers. They also reduce the number of ex-smokers returning to cigarettes and reduce consumption among continuing smokers. Children and adolescents are more responsive to changes in the price of consumer goods than adults-that is, if the price goes up, they are more likely to reduce their consumption. This intervention would therefore have a big impact on them. Similarly, people on low incomes are more price-responsive than those on high - incomes, so there is likely to be a bigger impact in developing countries where tobacco consumption is still increasing. Models developed for this report show that tax increases that would raise the real price of cigarettes by 10 percent worldwide would cause 40 million smokers alive in 1995 to quit and prevent a minimum of 10 million tobacco-related deaths.

Myth 6: Governments will lose revenues if they increase cigarette taxes, because people will buy fewer cigarettes.

Reality: Wrong. The evidence is clear: calculations show that even very substantial cigarette tax increases will still reduce consumption and increase tax revenues. This is in part because the proportionate reduction in demand does not match the proportionate size of the tax increase, since addicted consumers respond relatively slowly to price rises. Furthermore, some of the money saved by quitters will be spent on other goods which are also taxed. Historically, raising tobacco taxes, no matter how large the increase, has never once led to a decrease in cigarette tax revenues.

Myth 7: Smuggling and illicit production will undermine the effects of raised tobacco taxes.

Reality : Smuggling is a serious concern. But even in the face of smuggling, the evidence from a number of countries shows that tax increases still increase revenues and reduce cigarette consumption. Furthermore, governments can adopt effective policies to control smuggling. Such policies include prominent tax stamps and local-language warnings on cigarette packs, as well as the aggressive enforcement and consistent application of tough penalties to deter smugglers.

Myth 8: Governments should not raise cigarette taxes because such increases will have a disproportionate impact on poor consumers.

Reality: Existing tobacco taxes do consume a higher share of the poor consumers' income than of rich consumers. However, policymakers' main concern should be over the distributional impact of the entire tax and expenditure system, and less on particular taxes in isolation. Poor consumers are usually more responsive to price increases than rich consumers, so it is likely that their consumption of cigarettes will fall more sharply, and their relative financial burden may be correspondingly reduced.

Myth 9: In response to higher cigarette taxes, smokers will switch to cheaper brands or cheaper tobacco products and thus there will be no reduction in overall tobacco consumption.

Reality : This behavior, which is also known as "substitution", establishes a legitimate concern. However, not all smokers will engage in this behavior. Price increases will discourage non-smokers from taking up smoking and induce many smokers to quit or reduce consumption. Consequently, there will be reductions in overall consumption and prevalence. Only a certain portion of smokers will not be affected and some of them manage to maintain their levels of tobacco consumption through substitution. Non-price measures, nicotine replacement therapy and other cessation interventions can help curb tobacco use among this group.

Myth 10: Tax rates for cigarettes are already too high in most countries

Reality : The question of what the "right" level of tax should be is a complex one. The size of the tax depends in subtle ways on empirical facts that may not yet be available, such as the scale of the costs to nonsmokers, income levels, and also on varying societal values, such as the extent to which children should be protected. It also depends on what a society hopes to achieve through the tax, such as a specific gain in revenue or a specific reduction in disease burden. For the time being, a useful yardstick may be the tax levels adopted as part of the comprehensive tobacco control policies of a number of countries where cigarette consumption has fallen. In such countries, the tax component of the price of a pack of cigarettes is between two-thirds and four-fifths of the retail cost. Currently, in the high-income countries, taxes average about two-thirds or more of the retail price of a pack of cigarettes. In lower-income countries, taxes amount to not more than half the retail price of a pack of cigarettes, which are still very much below the level in high-income countries.

Myth 11: Measures to reduce tobacco supply are effective ways to reduce consumption

Reality : While interventions to reduce demand for tobacco are likely to succeed, measures to reduce its supply are less promising. This is because, if one supplier is shut down, an alternative supplier gains an incentive to enter the market. The extreme measure of prohibiting tobacco is unwarranted on economic grounds, unrealistic and likely to fail. Although crop substitution is often proposed as a means to reduce the tobacco supply, there is scarcely any evidence that it reduces consumption, since the incentives to farmers to grow tobacco are currently much greater than for most other crops. However, it may be a useful strategy where needed to aid the poorest tobacco farmers in transition to other livelihoods, as part of a broader diversification program. Similarly, the evidence so far suggests that trade restrictions, such as import bans, will have little impact on cigarette consumption worldwide. Instead, countries are more likely to succeed in curbing tobacco consumption by adopting measures that effectively reduce demand, and applying those measures symmetrically to imported and domestically-produced cigarettes. However, there is one supply-side measure which is key to an effective strategy for tobacco control: action against smuggling. Control of smuggling will improve governments' revenue yields from tobacco tax increases.

Myth 12: Tobacco controls will simply compound the poverty of rural economies that are heavily dependent on tobacco farming

Reality : The market for tobacco is likely to remain substantial for at least the next several decades and, while any future gradual decline in consumption will clearly cut the number of tobacco-farming jobs, those jobs will be lost over a decades or more, not overnight. Adopting sound agricultural and trade policies can help farmers in poor countries compete fairly for the world market. Governments are justified to prudently help the poorest of tobacco farmers with the adjustment costs of a gradual decrease in demand for their product. Many governments have helped with such adjustment costs for other industries.

Once a government decides that interventions are justified and chooses to do so, what should be the goals of its intervention measures?

Appropriate intervention measures should be taken to deter children and adolescents from smoking, protect nonsmokers and give adults all the information they need to make an informed choice. Governments might also consider helping those who want to quit by making it easier for them to obtain nicotine replacement therapy (NRT) and other cessation interventions.

What are the recommended strategies for governments to achieve these goals?

Where governments decide to take strong action to curb the epidemic, a multi-pronged strategy should be adopted. Its aims should be to deter children from smoking, to protect non-smokers, and to provide all smokers with information about the health effects of tobacco. The strategy, tailored to individual country needs, should include: (1) raising taxes so that they constitute at least two-thirds to four-fifths of the total retail price of cigarettes; (2) publishing and disseminating research results on the health effects of tobacco, adding prominent warning labels to cigarettes, adopting comprehensive bans on advertising and promotion, and restricting smoking in workplaces and public places; and (3) widening access to nicotine replacement and other cessation therapies.

How cost-effective are the suggested tobacco control measures?

Tobacco control is highly cost-effective as part of a basic public health package in low and middle income countries. Measured in terms of the cost per year of healthy life saved, tax increases would cost between $5 and $17 for each year of healthy life saved depending on various assumptions. This compares favorably with many health interventions commonly financed by governments, such as child immunization. Non-price measures are also cost-effective in many settings. Measures to liberalize access to nicotine replacement therapy, for example by changing the conditions for its sale, would probably also be cost-effective in most settings. However, individual countries would need to make careful assessments before deciding to provide government subsidies for NRT and other cessation interventions for poor smokers.

Given the relatively low quit rates even in countries where cessation interventions are available, would it be better to implement strategies solely aimed at preventing children from smoking?

A strategy aimed solely at deterring children from smoking is not practical and would bring no significant benefits to public health for several decades. Most of the tobacco-related deaths that are projected to occur in the next 50 years are among today's existing smokers. Governments concerned with health gains in the medium term may therefore consider adopting broader measures that help adults to quit.

It is mostly up to each government to decide whether to adopt the above recommendations or not. Is there a role for international agencies to assist governments in the process of choosing and implementing tobacco control policies?

International agencies such as the UN agencies should review their existing programs and policies, to ensure that tobacco control is being given due prominence; they should sponsor research into the causes, consequences and costs of smoking, and the cost-effectiveness of interventions at local level; and they should address tobacco control issues that cross borders, including working with the World Health Organization's new Framework Convention for Tobacco Control. Key areas for action include facilitating international agreements on smuggling control, discussions on tax harmonization to reduce the incentives for smuggling, and bans on advertising and promotion involving the global communications media.

❑❑❑

28

Legislation in India to Control Tobacco

Tobacco was initially consumed as medicine but passes of time it was consumed for pleasure and euphoria. Tobacco consumption is injurious to health; tobacco is frequently blamed for its cancer producing substance. Frequent and excess intake leads to death at most productive age. In India tobacco is consumed by several ways hence tobacco related problem is probably more complex here than else where in world. Indian government is bound to check tobacco consumption by imposing several regulations and heavy taxation on tobacco product. A brief account has been presented in time line series on law inforce by Indian Government to control consumption of tobacco.

Historically, tobacco consumption has been linked with social status and commensality. The habit of rural men, usually assembled in caste-based or social class-based groups, sharing a hookah in daily gatherings, is an example of fellowship, solidarity and the consultative process. The use of tobacco by women is often considered, by different sections of society, in ways different from that of men. Among urban women, smoking is often seen as a symbol of emancipation and modernity. Among other sections of women, tobacco chewing has a higher level of social acceptance than smoking. As traditional values slacken their stranglehold in rural societies, the sociocultural influences that encourage or discourage tobacco use are altering. These need to be studied carefully to control tobacco consumption. Tobacco introduced initially as a product to be smoked, tobacco gradually began to be used in several other forms. Pan (betel quid) chewing became a widely prevalent form of smokeless tobacco use. Although some Chinese and European systems of medicine supported the use of tobacco, Ayurveda—the Indian system of medicine—never supported the use of tobacco as medication. The ill effects of tobacco use on human health were recognized even in the sixteenth century, which led to restrictions on its use. Tobacco thrived everywhere in the world despite social (and some religious) disapproval.

The total number of premature deaths caused by tobacco during the twentieth century has been estimated at about 100 million worldwide and, if current levels of tobacco use continue during the twenty-first century, the death toll is projected to go up to one billion. The WHO, which provides these estimates, also predicts that India will have the fastest rate of rise in deaths attributable to tobacco in the first two decades of the twenty- first century. Many of these deaths will occur in the productive years of adult life, as a consequence of an addiction acquired in youth. The compelling need to save many of these lives from falling prey to tobacco use addiction and the urgent imperatives of avoiding the huge health, economic, social and environmental burdens that would be imposed by tobacco on a nation that aspires for accelerated development. India tobacco problem is more complex than probably that of any other country in the world, with a large consequential burden of tobacco related disease and death. The prevalence of tobacco use among men has been reported to be high (generally exceeding 50%) from almost all parts of India (more in rural than in urban areas). Women from most parts of India report smokeless tobacco use and the prevalence varies between 15% and 60%. Among 13–15-year-old school-going children, the current use of any tobacco product varies from 3.3% in Goa to 62.8% in Nagaland. In the late 1980s, the number of tobacco -attributable deaths inv was estimated to be 630,000 per year. On conservative estimates, the tobacco attributable deaths currently range between 800,000 and 900,000 per year. The cost of the tobacco attributable burden of just three groups of diseases— cancer, heart disease and lung disease was estimated as Rs 277.611 billion (US$ 6.5 billion) in 1999. This increased to Rs 308.33 billion (US$ 7.2 billion) in the year 2002– 2003.

There is sufficient evidence to infer a causal relationship between smoking and vascular diseases such as coronary heart disease, stroke and subclinical atherosclerosis, respiratory diseases such as chronic obstructive pulmonary disease and pneumonia, adverse reproductive effects and cancer at ten sites. Government of India has come up with suitable legislation to create awareness and prevent the uses of direct or indirect preparation of tobacco. A time line is given as under

1975: The Cigarettes (Regulation of Production, Supply and Distribution) Act 1975 was the first anti-tobacco legislation promulgated by the Government of India. This Act gave birth to the statutory warning on cigarette packages and advertisements, whereby printing of the cautionary note, "Cigarette smoking is injurious to health", was made mandatory. The Act also laid down specific instructions with respect to the font size, colour, language as well as the placement/visibility of the statutory warning on the product pack, etc.

1990: In India, pan masala and chewing tobacco are classified as food material. Smokeless tobacco users far outnumber smokers in the country. In 1990, provision under the prevention of Food Adulteration Rules (1955) were applied to smokeless tobacco products, necessitating the printing of warnings such as "chewing of tobacco is injurious to health" and "chewing of supari may be injurious to health" on the product packs and advertisements. In the same year, smoking in certain public places was banned vide a Cabinet secretariat administrative order. These "public places" were identified as hospitals, dispensaries, educational institutions, conference rooms, domestic flights, NC sleeper coaches in trains, suburban trains, AIC buses, etc. State Governments were advised against allowing sale of tobacco products in and around educational institutions and hospitals.

1995: In its 22nd report (December 1995), the Parliament's Committee on Subordinate Legislation observed that the guidelines laid down by the Cigarettes (Regulation of Production, Supply and Distribution) Act 1975 were not strictly implemented. The Committee made wide-ranging recommendations, including, strong and rotational warnings in regional languages on tobacco products; ban on direct as well as indirect advertisement of tobacco products; prohibition of smoking in public places; and initiation of measures for awareness on tobacco through healthcare centers, educational institutions and mass media. These recommendations of the Parliament's Committee resulted in the modification of the proposed comprehensive legislation on tobacco control. It may be worth noting that the Committee's concurrent recommendations for an integrated and gradual approach, after doing a study for rehabilitating the millions of farmers and labour involved, has been ignored.

2000: The advertisement codes as provided in the Cable Television Networks Rules, 1994 were amended to prohibit advertisements that directly or indirectly promote production, sale or consumption of cigarettes and tobacco products.

2001: On March 7th, the Government introduced The Cigarettes and Other Tobacco Products (Prohibition of Advertisement and Regulation of Trade and Commerce, Production, Supply and Distribution) Bill, 2001. The Bill, which was limited to cigarettes, was referred to the Parliamentary Standing Committee on HRD. The Standing Committee presented its report on December 5th and increased the scope of the Bill to cover all tobacco products nationally.

2003: The Cigarettes and Other Tobacco Products (Prohibition of Advertisement and Regulation of Trade and Commerce, Production, Supply and Distribution) Bill, 2003, was passed by both houses of Parliament and gazette (34 of 2003) on 19th May.

2004: India ratified W.H.O.'s Framework Convention on Tobacco Control (FCTC) on 5th February. On 25th February, the Government of India notified rules to implement some of the provisions of the Act of 2003 with effect from 1 May 2004. The provisions relate to ban on public smoking, advertising and sales to minors as stipulated by the Act.

2008: From 2nd October 2008, the Government of India notified rules to implement totally ban on smoking at public place under the provisions of the Act of 2003. A countrywide ban on smoking in public places came into effect from 2nd October, 2008. Correspondents say the ban is blatantly flouted and poorly enforced. But health officials say that ever since they began mounting surprise raids, many smokers are giving up smoking in public.

2009: All new tobacco products manufactured in India will have to carry pictorial warnings to discourage their use. Packets of cigarettes, bidis (small hand-rolled cigarettes common in India) and chewing tobacco will now have to display the scorpion and lung symbols. Tobacco products would have to clearly display the pictorial warnings and these must cover at least 40% of the packet's display area. Pictorial warnings would be "very visible" and would have "first-hand impact" on tobacco users. Pictures of diseased, corroded gums warning tobacco users about mouth cancer were put out in the weekend's newspapers.

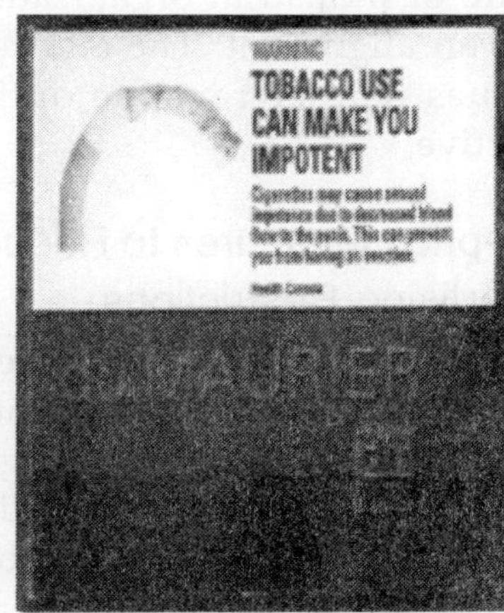

Tobacco Control Measures

To control the use of tobacco, the Central and State Governments have launched a number of schemes in recent years, with further measures under consideration.

Pricing Measures to Control Tobacco Demand

In India the motivating rationale for tobacco taxation seems to be fiscal, i.e. to maximize excise collection, rather to control tobacco use. The tobacco taxation policy also seems to be governed by concern that the people in the lowest income strata should pay less tax. The non-smoking tobacco products, commonly used among rural masses, are not taxed. Bidis, which are used by people in lower economic strata, were not taxed till the early 1990s, and currently are taxed only lightly. Non-filter cigarettes attract a much lower tax than filter cigarettes. The central government policy of keeping the prices of certain tobacco products cheaper for low-income groups ignores the potential higher occurrence of tobacco-related diseases. Discussion on base and rate structure of tobacco taxation brings into sharp focus the conflict between tax objectives: revenue maximization or tobacco consumption control. For the ease of tax administration, the narrow cigarette sector is taxed at increasingly higher rates to satisfy revenue needs, while as much as four-fifths of tobacco consumption is either lightly or not taxed. If the objective is to discourage tobacco consumption, the existing tax base and rate structure is skewed and not in alignment with production and consumption patterns. At the same time, other macro-considerations, such as employment and income generation, and the nature of industrial organization, hinder any serious effort towards such an alignment. The end result is that tobacco taxation in India pursues neither revenue optimization nor tobacco control as an objective.

Non-price Measures to Reduce Demand for Tobacco

Advertising Restrictions

The central government regulates advertising and promotion of cigarettes. The state governments regulate, at most, advertising of other tobacco products. Advertising of cigarettes is currently banned on television and commercial radio stations, and in government owned premises. Cinema, press and outdoor advertising is permitted, as is sponsorship of sports and other events and cultural activities. The use of hoardings and billboards is, however, restricted in certain areas and some sports grounds. Some state governments have started to follow the policy of the central government. For instance, in 1997, the Government of Delhi State imposed a ban on tobacco advertising. During 1997-98, some other states followed in imposing such a ban, notably Himachal Pradesh and Goa.

The central government has considered imposing a total ban on advertising of cigarettes under a new Tobacco Products (Regulation of Production, Supply and Distribution) Bill. The code would have prohibited the use of personal testimonials by those well known or those specifically working with children. Advertisements were also to be banned in any media aimed primarily at those under 18. However, the code had to be withdrawn in December, 1998, because of lack of cooperation from India's domestic manufacturers. In particular, the code came into conflict with the promotion of tobacco products at sport venues and the use of prominent celebrities for tobacco promotion. The Tobacco Institute of India is preparing a new code for possible adoption by the tobacco manufacturers, modelled on similar agreements between government and the tobacco industry in European countries, such as the United Kingdom and the Netherlands. The new code includes standardized health warnings, the publication of tar and nicotine levels, and the introduction of age restrictions for tobacco purchase.

Health Warnings

A single health warning ("Cigarette smoking is injurious to health") is mandatory on packets and any advertisements.

Other Restrictions

Smoking is currently banned in government offices, public transport (apart from air-conditioned railway coaches) and internal flights, cinema halls, theatres and government-run stadiums. The Central Ministry of Health and Family Welfare set up an expert committee consisting of 21 professionals from various disciplines to examine various facets of the tobacco industry including economics of tobacco use. Following the submission of the report of this committee, if it is accepted by the government, additional tobacco control measures could be introduced, inter alia better enforcement of the control measures already in force and some new measures.

Tobacco production is an important source of income for India's farmers. While alternative crops are available in terms of agronomic suitability, in general a move away from tobacco production would result in reduced income and food security for a considerable number of farmers. Manufacturing, particularly of bidis, is also a source of employment and hence of income for a large number of people. Thus any attempts to control the use of tobacco would need to take into account the economic impact on these sectors.

The New "Greener" Cigarette Packages

"The green packaging will go good with your black lungs and yellow teeth. The idea is that environmentalists will feel better about killing themselves. I guess the way people who smoke "light" cigarettes do. The fact is, the package is biodegradable and so is your body.""The green packaging will go good with your black lungs and yellow teeth. The idea is that environmentalists will feel better about killing themselves. I guess the way people who smoke "light" cigarettes do. The fact is, the package is biodegradable and so is your body."

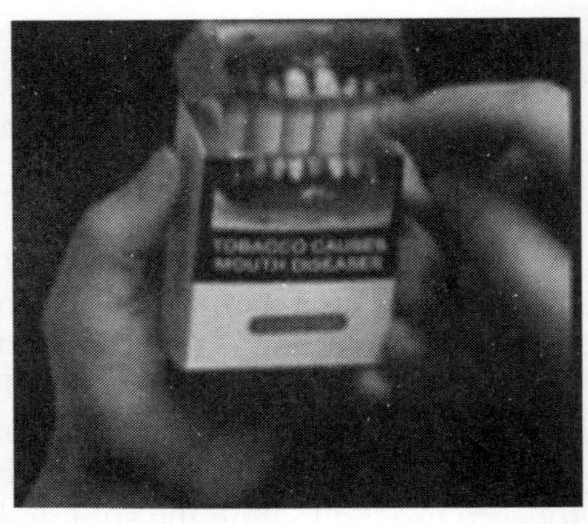

People smoke and will continue to do so, whatever the consequences. The 'corporate social responsibility', however, can't be 'fattened up' in reference to a greener packaging. Social responsibility requires a healthy product. When will the tobacco industry finally understand that some research might lead to a 'greener' product out of their so hypocritically used statement: 'corporate social responsibility'? It might safe them a future similar as to the 'old-time' automakers in Detroit by ignoring innovation for too long.

A Coffin-Shaped Pack of Cigarettes- Symbol of Warning

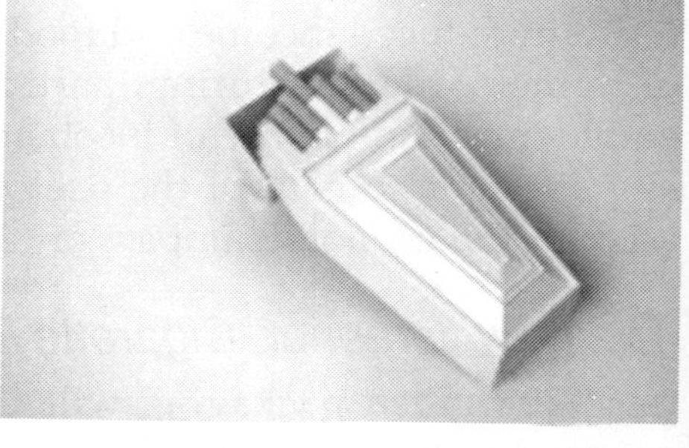

In April 2009 for a "campaign to stop smoking", Ukrainian designer Reynolds, designed a concept for a 'social responsible' packaging of tobacco products, resulting in a coffin-shaped pack of cigarettes. The idea is not so new and unique, but still quite impressive, however no tobacco company will use the design in reality, as it is meant to be a visual stimulus to quit smoking.

World No -Tobacco Day (2009 31st May)

Most people are unaware of the extent of the harm that tobacco causes even if they have some idea that it is a health risk. Tobacco is the leading preventable cause of death. More than five million people die from the effects of tobacco every year – more than from HIV/ AIDS, malaria and tuberculosis combined. It is the only legal consumer product that kills when used exactly as the manufacturer intends. Up to half of all smokers will die from a tobacco-related disease. Second-hand smoke harms everyone who is exposed to it.

Tobacco companies use packaging and other advertising techniques to make tobacco appealing, while distracting consumers from the harsh reality of how tobacco destroys health.Tobacco companies spend tens of millions of dollars every year turning new users into addicts and keeping current users from quitting. Through advertising and promotional campaigns, including the use of carefully crafted package designs, the tobacco industry continues to divert attention from the deadly effects of its products. It is a proven fact that warnings on packaging are an inexpensive and powerful way to show the truth about tobacco consumption. Warnings that include images of the harm that tobacco causes are particularly effective at communicating risk and motivating behavioural changes, such as quitting or reducing tobacco consumption. Picture warnings convey a clear and immediate message, even to people who cannot read. They reduce the overall attractiveness of tobacco packages — an important function for a product whose new users are typically young and image- and brand-conscious.

More and more countries are fighting back by requiring that tobacco packages graphically show the dangers of tobacco, as called for in the WHO Framework Convention on Tobacco Control. They use the MPOWER technical assistance package developed by WHO to help meet their commitments under this international treaty. The regional launch of the World No Tobacco Day will be held in Phnom Penh, Cambodia on 29 May 2009. For more information on the activities and events in Cambodia and other countries across the Region visit WPRO-2009.

□□□

PART-VI
Others

29

Mechanizing of Tobacco Cultivation and Production

Tobacco is the most widely grown nonfood crop in the world; it is produced in approximately 100 countries and on every continent. Tobacco is consumed around the world as cigarettes, cigars, chewing or smoking tobaccos and snuff. The tobacco has been cultivated using traditional methods, labor-intensive, small acreage, and time consuming, it is difficult to increase production and can not meet the increasing demands of the market until the mechanization of tobacco production will be accelerated with better pace.

The manual labour required to produce and market tobacco varies greatly around the world, depending, primarily on the level of mechanization used for transplanting, harvesting and market preparation. Manual labour involves risks of musculoskeletal problems from activities such as transplanting seedlings, application of suckercides, harvesting, separation of the cured tobacco into grades and lifting of tobacco bales. Training in proper lifting methods and provision of ergonomically designed tools can helpa and prevent these problems. Knife injuries may occur during cutting, and tetanus may arise in open wounds. Sharp, well-designed knives and training in

their use can reduce the number of injuries. Field labourers, usually those not accustomed to working in tobacco fields, sometimes become nauseous and/or dizzy soon after direct contact with green tobacco during harvesting, perhaps because nicotine or other substances are absorbed through the skin. In the United States, the condition is called "green tobacco sickness" and affects a small percentage of workers. Symptoms occur most often when sensitive individuals are harvesting wet tobacco and their clothing and/or exposed skin is in almost continuous contact with green tobacco. The condition is temporary and not known to be serious, but causes some discomfort for several hours after exposure. Suggestions for sensitive workers to minimize exposure during harvesting or other tasks requiring prolonged contact with green tobacco include not starting work until the leaves have dried or wearing lightweight rain gear and waterproof gloves when the leaves are wet; wearing long trousers, long-sleeve shirts and possibly gloves as precautions when working in dry tobacco; and leaving the field and washing immediately if symptoms occur. Skin diseases may occur in workers handling tobacco leaf in warehouses or barns. Sometimes workers in these storage areas, especially new workers, may develop conjunctivitis and laryngitis. Other preventive measures include good washing and other sanitary facilities, provision of first aid and medical care, and proper training.

The existence of barriers in some realms of agricultural production precludes the development of wage labor have contributed to the rapid mechanization and structural change among tobacco producers. Many of the "obstacles" to consolidation were overcome not by technological change, but by weakened political support for the tobacco program. These results suggest that in addition to economic and technological considerations, we need to assess more carefully the political foundations of the capitalist development of agriculture, in many instances these obstacles are based primarily on political factors, how the tobacco program until recently has proved to be an obstacle to consolidation and structural change in tobacco production. The tobacco program has conditioned the extent of technological development and structural change in tobacco production.

In order to reduce production costs, and effectively solve problems for tobacco farmers, tobacco growers can afford to allow, affordable machinery, agricultural machinery an active role in coordinating the use of farm machinery purchase. Mechanization can reduce these risks, but carries risks of injury from the machinery used, including transportation accidents. Well-designed tractors with safety cabs, properly guarded machinery and adequate training can reduce the number of injuries. Tobacco requires heavy labour for all production events as mentioned above. Some of the major automation is discussed here.

A. Mechanization in Tobacco Production and Processing

To produce of tobacco crops for its quality leaves requires lots of manpower at each and every step of crop husbandry right from nursery bed preparation to harvesting. Harvesting curing and processing is labour and time taking technical things which has also been discussed in brief. Possible and economin vible machnization and automation is discussed with product of reputed firms and companies engague in the production and marketing worldwide since long time. To make it easier to uinderstand the text is given as per the convinent to the users.

1. Tobacco Transplanters

Tobacco are mainly cultivated by means of transplanting, in general transplanting is done due to narrow window of seasons for transplanting and demands of huge amount of human labours, several attempts has been successfully made to fabricate tobacco transplanter, some of the mechanical and automatic transplanter are discussed in brief.

1.1 Mechanical Transplanter

The F Max Evolution Transplanting Machine is suitable for transplanting a wide variety of vegetables and tobacco (from tomatoes to cabbages, peppers). Each planting unit is equipped with an 8 cartridge rotary carousel which eases the operators task of loading. The F Max is adaptable for thin plants and for long plants and can work on soft soils as

well as hard and stony soils. The machine is available in modular versions and it can plant up to 4,000 plants/hour/unit. This machine is manufactured by Ferrari Costruzioni Meccaniche.

The Ferrari transplanter FX can work with small sods of every shape (cylindrical, conical, pyramidal) normal or pre-compressed, both in plastic or polystyrene trays. The machine was born to solve the most difficult transplanting problems with lettuce, however there is a wide variety of crop which can be transplanted with it (tobacco, cabbage, cauliflower, chicory). The only limit is represented by the plants dimension which cannot be higher than 15cm (5.9in). The FX element is equipped with a parallelogram system swinging on the ground, so it can follow perfectly also the most uneven and covered with stones soil. Each unit can transplant about 4,000 plants per hour. The machine is available in 2 versions: hauled and self-propelled. At request you can have the hauled version prepared to work on raised beds (with rear wheels adjustable in depth). Depth of work can be adjusted up till 8 cm and it is possible to work following the quincunx (diamond) planting system.

1.2 Automatic Transplanter

This machine, also known as the "FUTURA", has been developed to have high-technology characteristics. It has been studied for the automatic transplant of vegetables and tobacco in modules placed in trays. It needs exclusively one operator who puts the single planting element inside the charging slide of the tray. This device can house up to 2 plant-holder panels, according to their dimensions. Each element of the machine is provided with a system which automatically pulls out; from the longer side of the tray all plant-holder panels, according to their dimensions. Each element of the machine is provided with a system which automatically pulls out from the longer side of the tray, the all plants of the same row in beatings of 2 and from 7 to 11 plants each stroke. These plants will be received and taken from other moveable grippers who will supply the feed unit. With working capacity up to 7,500 plants/hour/row. Pressure of earthing up on the rear wheels is adjustable from 0 to 130 kg. High suitability for the tray

in commerce. "Full Plant" device for the control and elimination of the missing plant from the tray to the transplant with a system of selecting control before the download in the container. Multiple readings up to 11 plants in succession with memorization, discrimination and replacement with a mechanical device for recovery of the empty container. High tolerance for different grounds. High tolerance to plant height (up to 20cm - 7.87in. Left/ Right oleo dynamic balancing and parallelism to the ground to transplant. Available to work on flat beds as much as on raised beds and also on those already traced and fertilized.

2. Tobacco Weeder

Weeding in tobacco is laborious and time consuming job, having effect on yield and quality of its produces. Tobacco, being very delicate at early stage of growth and development appropriate weed free situation is very much desired. Use of weedicides and herbicides may leads to poor quality production hence, manual weeding is preferred. To overcome this problem several hand and tractor drawn weeder has been developed. Some of tobacco traits such as leaf spacing, leaf angle and leaf size may have a significant impact on the suitability of the different cultivars for tractor drawn weeding implement. The torsion weeder is easy to mount on existing inter-row cultivators, and consists of two flexible tines tilted towards each other, which undercut weeds close to the crop row with a vibrating and dragging movement. The intensity of the process is adjusted by changing the working depth and position of the tines vis-à-vis the ground surface. Because of its simple design, this mechanical weeder is relatively inexpensive. All this, as well as the good test results in agriculture, and the tobacco industry, make the torsion weeder a very interesting tool for improving in-row weed control. A demonstrative test involving various crops also showed excellent results.

A single cultivation with the torsion weeder reduced the proliferation of weeds by 85%. The torsion weeder has two flexible tines on each side of the crop row. The tines are slightly tilted so that they drag along the crop row. The driving speed pushes the tines outwards to the sides, so that they do not damage the crop. The vibrating tines, which go approximately two centimetres deep, break up the soil and uproot the weeds.

3. Tobacco Harvester

Traditional hand-harvesting of tobacco requires a considerable amount of labour during a five to six week period. The labour cost associated with the production of tobacco. In recent years, to overcome this problem several automatic harvester has been formulated and harvesting operation has been partly mechanized with the introduction of machine harvesters. Tobacco growers with small acreages do not consider growing tobacco to be economically viable. Mechanization of various production practices tend to decrease labour costs and make the physical work easier. Varietal differences were observed in most traits studied. Cultivars with wider leaf spacing are less prone to weight loss when harvested with machine than those with narrow leaf spacing. Narrow leaf spacing had a high number of leaf hangers after the 5th pass of the harvester. These same cultivars showed a greater weight loss after the 5th pass when compared to other cultivars. Grade index comparison suggests that there is no differential response to quality traits between machine and hand harvested trials. Among the cultivars grown India are considerable differences in topping height, maturity, number of leaves, leaf spacing leaf angle and leaf size. Some of these traits such as leaf spacing, leaf angle and leaf size may have a significant impact on the suitability of the different cultivars for machine harvesting. Very less study on the response of cultivars to machine harvesting has been conducted in the past. This was mainly due to a limited number of growers using machine harvesting. Since the number of growers adapting machine harvesting has been increasing in recent years, an understanding of the response of different cultivars to machine harvesting is needed.

4. Tobacco Combine

Automatic tobacco harvesters are high clearance self-propelled machines, designed to mechanically remove ripe tobacco leaves from the stalks. They are conceived to be as simple as possible to use, designed for high flexibility to let you adjust travel and header speed, the number of leaves to be pulled and the number of times you will pass through your crop. In the field the "soft touch" defoliators permit to remove the leaves from the plants without damage, transporting them, still wet, on a special conveyor belt, to a trailer or box where blowers automatically distribute them. Powered by a reliable John Deere 100 HP engine the 3-wheel hydrostatic drive incorporates the shortest possible turning radius. A field-tested and proven innovation that saves you considerable harvesting time. To gather tips, the exclusive cutters bar removes leaves in a downward motion similar to handpicking. Leaves are harvested cleanly at the stock, without tipping.

5. Tobacco Tying Machine

The tobacco tying machine played a significant role in the mechanization of the tobacco industry, and Bill played a small but important part in its development. Up until its invention tobacco was tied onto sticks by very skilled (and very fast) worker. The tying machine was based on a regular sewing machine - complete with needle, looper and thread. The main difference is that the needle was very large (about 8 in long and 1/4 in diameter), the thread actually heavy duty rayon twine, and the looper very large.

6. Mechanized System for Handling and Curing Tobacco

The system consisted of a tractor–towed, trailer mechanism that engaged and hoisted loads of approximately 360 burley plants of approximately 1 Mg mass. Ten slotted steel rails, 3.05 m long, holding 36 notched plants were placed onto parallel wooden beams suspended at a height of 2.13 m by wooden posts set in the ground. Burley tobacco was cured in this configuration covered by polyethylene. Mechanically harvested burley tobacco plants placed onto steel slotted receivers were retrieved from a field, transported to a field curing structure, and placed onto the structure for air curing by a single worker. Time–and–motion experiments showed that the system could retrieve tobacco from the field and place it onto a curing structure adjacent to the field at the rate of 0.1 to 0.18 ha/h. Replicated experiments also showed that the system operated with negligible leaf loss due to handling. Finally, experimental results showed that leaf grade index decreased with time that filled tobacco rails were left lying on the ground after being harvested and prior to being retrieved. This study further indicated that the estimated cost of the proposed harvesting system compares favorably with systems that require several manual laborers.

7. Mechanized Tobacco Curing and Housing

Lack of hired labor for harvesting and housing of harvested tobacco leaves leads to developing a mechanized housing system and a one-tier field curing structure that would accommodate the mechanized housing system and provide an economical alternative to the conventional curing barn. The mechanical housing system for the field curing structure requires only 39% of the labor required to load, transport, and house burley tobacco in the conventional barn. The mechanical housing system for the field curing structure requires two workers compared to three or four for conventional barns. Using Net Present Value as a means of comparing capital investment decisions showed the field curing structure with mechanized housing system to be a better investment than the conventional and tiered field curing structure systems for land areas of and above 0.7 ha (1.6 acre) and 6 ha (15 acre), respectively.

B. Mechanization and Automation in Finished Tobacco Product Production

Postharvest management of various kind of tobacco and their possible and most feasible mechanization and automation has been described in a short as per the jurdiction of this presentation. After intial processing (curing in some cases) raw tobacco leaf has to undergo a sereasre processing, at this point of time where in nicotine, sugars and product temperature and moisture play a decesecive role in finished product quality, physical testure and aroma.

1. Tobacco Leaf Processing Plant

The curing of tobacco is a slow process carefully monitored to produce tobacco leaves with specific colors, wilts and degrees of dryness. Four different methods of curing—air curing, fire curing, flue (heat) curing, and sun curing—impart distinctive flavors to the tobacco leaves. Since freshly cured tobacco has a bitter taste, most cigarette tobacco is dried, cooled, somewhat rehydrated, and then stored for two or three years. The leaves ferment as they age, becoming milder and darker. In this automated machine, pressed tobacco is being moved on computerized rollers at a processing plant in North Carolina. Primary processing line with capacity from 3000kg/hr to 12000kg/hr to over 180 users in the worldwide. Small unit with capacity of 300kg/hr to 1500kg/hr. The whole producing procedure can be divided into following sections: lamina processing section, cut-lamina processing section, stem processing section, cut-tobacco blending section, Burley tobacco processing section.

1a. Lamina Processing Section

This section includes the procedure of bale opening, loosening and conditioning, casing, storing, etc. Flexible carton opening system is used for automatically strap cutting, cover opening and lining board removing of carton and the unpacked carton can be recycled and used. After opening, the lamina bale can be conveyed to traditional slicing and conditioning processing or conveyed to on-line vacuum conditioning system and conditioning cylinder. On-line continuous vacuum conditioning system can decrease degradation, increase temperature and moisture, remove greenish odor and kill insect eggs while ensure continuous and automatic production. Conditioned and cased lamina will be conveyed into pre-blending silo, to blend different origins and grades lamina sufficiently. Foreign matter sorter developed to adopts optical recognition and digital image processing technology, can detect and remove foreign matters efficiently.

1b. Cut-Lamina Processing Section

This section includes the procedure of lamina cutting, constant flow rate control, cut-lamina expansion and drying, etc. Cutter adopts AC servo, frequency regulation and Field bus technology, which can improve the cutting quality and convenient for operation. There are two processing technology for cut-lamina expansion and drying, i.e. traditional expansion + drying cylinder with thermal plate exchanger and new developed fuel oil (gas) pipeline cut-lamina drying plant. It achieve closed-loop control over moisture content via electronic weighing conveyor and moisture meter, which can reduce over drying cut-lamina at the beginning and the end of drying process, and improve the quality of cut-lamina. The fuel oil (gas) pipeline cut-lamina drying system can achieve high expansion, decrease the cut-lamina consumption and reduce tar content. The optimized design can ensure reliable and safe operation.

1c. Stem Processing Section

This section includes the procedure of increasing temperature and moisture, storing, cutting, expansion drying, etc. Stem washing trough plant use the recycle warm water to process stems on-line, which can make stems absorb moisture sufficiently and remove sand, stone, scrap and stem skins. Cut-stem can achieve high expansion and good filling power after expansion and drying. There are two processing technology for cut-stem expansion and drying, i.e. cut-stem expansion tunnel + drying cylinder with plate thermal exchanger and Injection type cut-stem conditioning plant + fluidized bed cut-stem drier.

1d. Cut-tobacco Blending Section

All kinds of cut-lamina, cut-stem, expanded cut-tobacco and reconstitute tobacco will be blended in proportion. The blended cut-tobacco will go through metering and flavoring then conveyed into cut-tobacco silo. Cut-tobacco bin storing system developed by KSEC can combine automatic logistics with primary processing. It can replace bulking silos and suitable for multi-blend, small volume and more flexible.

1e. Burley Tobacco Processing Section

It set multi casing application before burley toasting and accelerate the casing liquid absorbing. Burley tobacco toaster has function of keeping humidity and drying. It has wide adjustable range of drying temperature and precise moisture control and high automation.

2. Automatic Tobacco Moisture Measurement for Processing

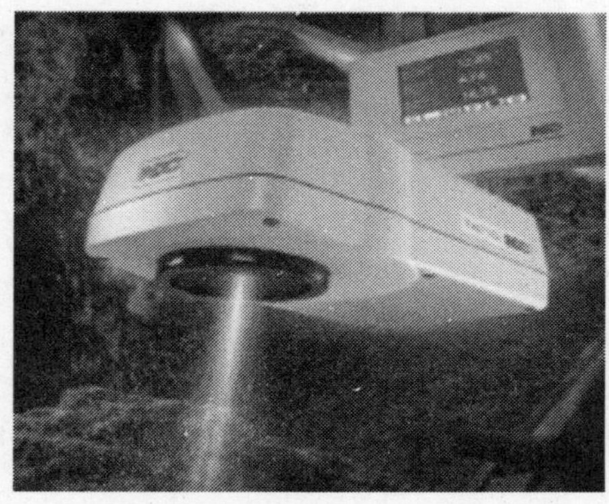

Understanding the process is key to on-line measurement performance. Gauges and measurement solutions to meet the precise product and performance requirements of the tobacco industry has been engineered. Generic moisture meter cannot provide a measurement solution for a product such as tobacco, with its inherent natural variation. The TM710e range of instruments, the company's 8th generation of tobacco analyzers, offers the industry the only no-compromise tobacco moisture measurement from very low moistures right up to 60%. With optional additional measurement parameters of nicotine, sugars and product temperature. It is the definitive tobacco gauge, with full process integration capability. And the range of supervisory and operator-level interfaces ensures that critical process information is provided instantly at point of need. For at-line measurement, the Infra Lab provides fast, accurate single or multi-component measurements.

□□□

30

Tobacco Alkaloids

Alkaloids are typically abundant in roots, bark, leaves and fruits, but they may also occur in every part of the plant. Despite advanced chemical structure elucidation, the biosynthesis of secondary compounds is only incompletely characterized, due to the long pathways which often involve multiple co-regulated enzymes which are still largely unknown. Dozens of genes involved in alkaloid pathways have been cloned, eight of them being directly involved in tropane and tobacco alkaloid metabolism.

Plants produce a wide variety of chemical compounds, which can be divided into two groups primary and secondary metabolites. Primary metabolism is considered as a set of processes involved in the fundamental maintenance of life and growth, whereas secondary metabolism consists of systems having important functions for plants in survival and competing in the environment, e.g. in protection against ultraviolet light, as well as in various defiance related reactions. Thus, plant secondary metabolites characteristically accumulate at certain developmental stages or in specific organs. Secondary metabolites often constitute less than 1% of the total dry weight. Plant secondary compounds are divided into five categories: polyketides, isoprenoids (e.g. terpenoids), alkaloids, phenylpropanoids and flavonoids. Certain secondary compounds are abundant in several plant species, such as many phenolic compounds. However, in the case of alkaloids, the production is highly specific to particular families or even species. Often secondary compounds are stored inside the plant cells in vacuoles, which account for the volume of the plant cell to a varying extent depending on the physiological state of the cell. Secondary metabolites are often unique to the plant and are structurally rather complex, low molecular weight compounds. More than 100 000 different secondary metabolite structures have been discovered from plants. The best studied secondary metabolites are alkaloids, with a total number of 21,120 found in monocots, dicots and gymnosperms. Over 12 000 alkaloids have been structurally characterized. Alkaloids accumulate about 20% of all plant species and have attracted much interest owing to their strong physiological properties enabling their use as pharmaceuticals or pesticides. Due to the versatile nature of compounds, alkaloids can be defined only rather loosely as nitrogen-containing organic substances of natural origin with a greater or lesser degree of basic characters. Alkaloids are typically classified according to the ring system that constitutes the main part of their structure. However, some structures fit in more than one category and thus a rigid hierarchical classification is difficult to define. Of the approximately 15 classes, the pharmacologically most interesting alkaloids belong to the groups of seven only and are present in table 29.1.

Table 29.1 Alkaloids groups commonly present in plant kingdom

Alkaloids groups	Examples
amino alkaloids	ephedrine, mescaline
pyridine and piperidine alkaloids	coniine, tobacco alkaloids
tropane alkaloids	hyoscyamine, scopolamine, cocaine
isoquinoline alkaloids	berberine, morphine, codeine
indole alkaloids	physostigmine, vinblastine, vincristine
quinoline alkaloids	quinine, camptothecin
steroidal alkaloids	veratramine, jervine, solasodine

Tobacco Alkaloids

Nicotine is the most abundant alkaloid in cultivated tobacco (*Nicotiana tabacum* L.), typically constituting more than 90 % of the total alkaloid pool. In addition to Nicotiana (Solanaceae) plants, nicotine has also been found from genera of various other families: Asclepias (Asclepiadaceae); Atropa, Duboisia, Lycopersicon, Withania, Cestrum (Solanaceae); Eclipta, Zinnia (Compositae); Lycopodium (Lycopodiaceae); Macuna (Leguminosae); Sedum and Sempervivum (Crassulaceae). In Nicotiana species, the predominant alkaloid in the leaves, depending on the species, is nicotine (most *Nicotiana* spp.), nornicotine (*N. tomentosiformis, N. otophora*) or rarely anabasine (N. glauca); although in roots nicotine dominates in almost all species. Other common Various minor tobacco alkaloids and their N-methyl derivatives, e.g. myosmine, N-methylmyosmine, cotinine, nicotyrine, nornicotyrine, nicotine N-oxide, 2,3-bipyridyl and metanicotine are thought to arise by bacterial action or oxidation during processing of commercial tobacco, rather than by biosynthetic action in the living tobacco plant . The biosynthesis of nicotine together with several other tobacco alkaloids takes place in plant roots, where the corresponding biosynthetic enzymes are found to be highly expressed. Alkaloids are further transported through the xylem stream to the leaves, where they gets accumulate in the vacuoles. In leaves alkaloids play an important role in several defense related reactions, such as protecting the plant against herbivores and vertebrate predators, in detoxification processes, as well as in regulating growth. Alkaloids can also represent an allocation of nitrogen pool that otherwise might be utilized in primary metabolism.

Biosynthesis of Nicotine

The genetic regulation of alkaloid biosynthesis in *N. tabacum* is known to be primarily under the control of at least two independent genetic loci, A and B. It was postulated that the A and B loci are coordinate regulators of a subset of genes encoding components of polyamine and alkaloid biosynthesis, thus the A and B loci have been referred to as NIC1 and NIC2, respectively. However it was shown that the A and B loci do not coordinate specifically the alkaloid biosynthesis, but rather regulate a diverse set of gene families of which only a small part (four out of a total 33 gene families) represents the alkaloid biosynthetic genes, whereas the majority of A-B regulated genes are involved in a variety of plant stress responses. Putrescine, a key intermediate in tobacco alkaloid biosynthesis can be synthesized directly from ornithine in a reaction catalyzed by ornithine decarboxylase (ODC, EC 4.1.1.17), or formed indirectly from arginine in a reaction sequence initiated by arginine decarboxylase (ADC, EC

4.1.1.19). Putrescine metabolism is directed to higher polyamines in one branch and to alkaloids in another branch, in a reaction catalyzed by putrescine-N- methyltransferase (PMT, EC 2.1.1.53). PMT is an important enzyme in the overall regulation of the alkaloid pathway driving the flow of nitrogen away from polyamine biosynthesis to alkaloid biosynthesis. N- Methylputrescine, the product of the PMT-catalyzed reaction, is further converted to N-methylamino butanal by a diamine oxidase DAO, or N- methylputrescine oxidase MPO (EC 1.4.3.6). It was first described from tobacco roots. Recently, the gene encoding MPO has been cloned by Heim in 2007. Interestingly, MPO has shown to catalyse the oxidation of cadaverine; thus it also accepts other substrates from the pathway. Spontaneous cyclisation of N-methylamino butanal leads to formation of the N-methylpyrrolinium ion, which further donates the pyrrolidine ring structure to nicotine. Nicotine is formed in a condensation reaction of N-methylpyrrolinium and a metabolite of nicotinic acid, putatively 3, 6-dihydronicotinic acid, the latter providing the pyridine ring supplied from the NAD biosynthesis pathway it suggested that the condensation reaction between nicotinic acid and N-methylpyrrolinium is catalysed by an NAPDH-dependent reductase called A622. It was shown that A622 is jasmonate-inducible and that the expression pattern of A622 is highly similar to that of PMT. However, the conclusive finding concerning the nature of nicotine synthasey and the final step in nicotine formation has not yet been obtained. Several genes encoding enzymes involved in the upper part of the biosynthetic pathway of nicotine have been cloned, including ADC, ODC, PMT and QPRT (quinolinate phosphoribosyltransferase).

Metabolism of other Tobacco Alkaloids

Nicotine is catabolised to nornicotine in a demethylation reaction. This reaction was suggested to involve cytochrome P450 and was further elucidated recently as CYP82E4. Nornicotine is converted to myosmine via a presumably irreversible reaction. The primary metabolite of nicotine in body tissues is cotinine; approximately 70 to 80% of nicotine is metabolised to cotinine and about 4% to nicotine N-oxide. Anabasine is synthesized from nicotinic acid and 1-piperideine deriving from lysine. Concerning the biosynthetic origin of anatabine, it was suggested that anatabine is formed by a dimerization reaction of a metabolite of nicotinic acid, 3,6-dihydronicotinic acid. Although superficially similar to nicotine, anatabine and anabasine differ from nicotine in three crucial aspects: the pyrrolidine ring is replaced by a piperidine ring, there is total absence of any N-methylated derivatives, and they are commonly present in both (S)- and (R)-forms. Anatalline was first isolated by Kisaki and co-workers (1968) from roots of N. tabacum and it was found in high concentrations in methyl jasmonate-elicited BY-2 cell cultures of tobacco in the recent studies (I). Anatalline was shown to accumulate in two isomeric forms (II). The biosynthetic origin and metabolism of anatalline is still unknown.

Table 29.2 Alkaloids and related substances in tobacco leaves

Identified substances	Structural formula	Quantity (%)
Nicotine	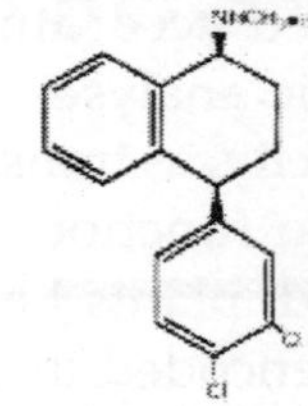	0.027-1.1
Narnicotine	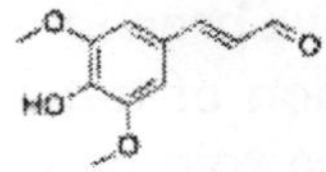	0.005-1.160
Anabasine	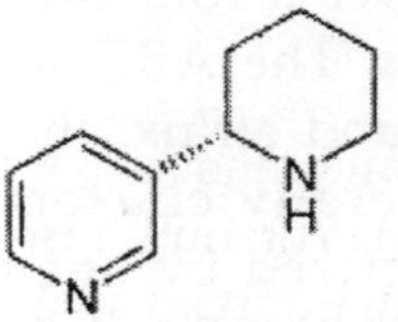	0.007-0.020
Myosmine 2,3 dipyridyl	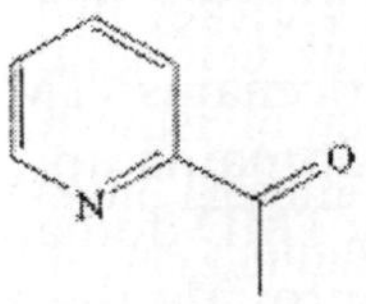	0.001-0.007
Nicotine- N oxide		0.002-0.141
3- pyridylmethyl ketone	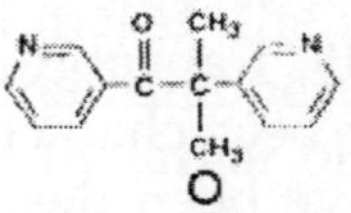	0.001-0.002
Nicotinic acid	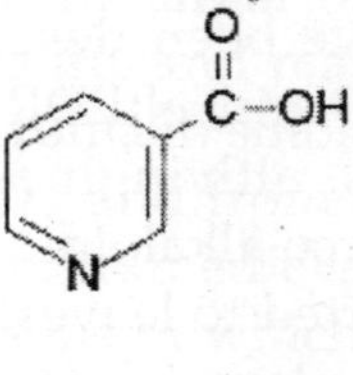	0.005-0.074
Others (isolated but not identified)		0.162-0.440

Transport of Secondary Metabolism

Transport of various chemical compounds across biological membranes is anessential physiological process in all living organisms. Membrane transport is involved in the uptake of nutrients, in the efflux of toxic compounds, subcellular localization of metabolites, as well as in communication between the cell and its environment. Plant cells are capable of producing a wide variety of secondary compounds, which in cytosol may be toxic to plant cells, and thus it is crucial that these compounds can be efficiently removed from the cytosol. They may be transported into the apoplastic space for further modifications, or into the vacuole, the common storage site of secondary compounds. The transporters are classified according to the protein family and transporter type. Phylogenetic analyses have proven to be useful for prediction of substrate specificity, as transport proteins typically cluster phylogenetically according to function. Among all the multicellular organisms hitherto sequenced, plants have the largest representation of ABC transporter proteins encoded in their genome. Altogether 131 members of ABC proteins have been found in Arabidopsis .The reason for such a large representation of transporters is still unclear, but it has been suggested to have a role in adaptive functions for survival and environmental stress. The ABC super family contains transport systems of both uptake and efflux; the members of each of the two transporter groups generally cluster together. Structurally, ABC transporters are characterized by one or two cytosolically oriented nucleotide binding-folds (NBFs) linked to multiple hydrophobic transmembrane-spanning domains (TMD). The substrate crosses the membrane via the TMD domains and in some cases the substrate specificity is determined by TMD domains. NBFs are oriented towards the cytoplasmic side and provide the site for ATP hydrolysis. ABC transporters are classified into four main subfamilies: MDR (multidrug resistance), MRP (multidrug resistance associated protein), AOH (ABC one homolog) and PDR (pleiotropic drug resistance). Plant alkaloids are often transported by ABC transporters in microbes and herbivores, but only a few of them have been characterized from plants. Altogether four PDR transporters have been described in *N. tabacum*, namely NtPDR1 , NtPDR2 , NtPDR3 and NtPDR4 . All of them are categorized as stress responsive genes, although none have been identified as nicotine transporters. Tobacco alkaloids, which are synthesized in root tissues, are further transported to leaves, where they have a defensive role for the survival of the plant. The transporter protein responsible for this transport process has not yet been identified, but based on the studies performed with tobacco hornworm it was suggested that a P-glycoprotein mediated (MDR-like) transport is involved. In different species, the same compounds may be transported by different transporter systems. Berberine uptake in *Coptis japonica* (Thunb.) was shown to involve MDR-type ABC transporter, although in *Arabidopsis thaliana* (L.), which does not produce endogenous berberine, its transport is effected by another type of transporter called MATE (multi antimicrobial extrusion/ multidrug and toxin extrusion). Transporters in plasma membrane might function in the secretion of endogenous metabolites having role in defence reactions in plants, as was suggested by Jasinski and co-workers (2001). A PDR5 homolog NpABC1 from N. plumbaginifolia (L.) was localized in the plasma membrane and was shown to be strongly expressed after addition of sclareolide or the antifungal diterpene clareol. The ABC transporter PDR5 of the yeast Saccharomyces cerevisiae confers resistance to several unrelated drugs, and substrates of the transport process also include plant-derived compounds, such as taxol, indole alkaloids, and falconoid. It was also shown that PDR5 is capable of accepting tropane alkaloids as substrates (V). Further, the effect was demonstrated with tobacco cells, in which PDR5-carrying cells showed decreased cell death after administration of hyoscyamine. The plant PDR family is still poorly investigated and remains an interesting subject concerning the secretion of plant secondary metabolites and applications in biotechnological production systems using plant cells, which are often characterized by poor secretion of the produced secondary metabolites.

❏❏❏

Appendices

Appendix I : Acronyms

ACIAR	-	Australian Centre for International Agricultural Research
AIT	-	Asian Institute of Technology
BOES	-	Bureau of Economics & Statistics
CADA	-	Command Area Development Authority
CBIP	-	Central Board of Irrigation and Power
CG Centres	-	Consultative Group Centres
CGIAR	-	Consultative Group on International Agricultural Research
CIAT	-	International Centre for Tropical Agriculture
CIFA	-	Central Institute of Fisheries and Aquaculture
CIFRI	-	Central Inland Fisheries Research Institute

CIMMYT	-	Centro Internacional de Mejoramiento de Maiz y Trigo (The International Maize and Wheat Improvement Center)
CRIDA	-	Central Research Institute for Dryland Agriculture
CRRI	-	Central Rice Research Institute
CSIR	-	Council of Scientific and Industrial Research
CSIRO	-	Commonwealth Scientific and Industrial Research Organization, Australia
CSU	-	Colorado State University
CSWCRTI	-	Central Soil & Water Conservation Research & Training Institute, Dehradun
CTRI	-	Central Tobacco Research Institute Rajahmundry
CWC	-	Central Water Commission
DSE	-	Delhi School of Economics
DST	-	Department of Science and Technology, Govt. of India
DWR	-	Directorate of Wheat Research
DWMR	-	Directorate of Water Management Research, Patna
ER	-	Eastern Region
FAO	-	Food and Agriculture Organization of the United Nations
GATT	-	General Agreement on Trade and Tariffs
GFCC	-	Ganga Flood Control Commission
IARI	-	Indian Agricultural Research Institute
IASRI	-	Indian Agricultural Statistics Research Institute
IBPGR	-	International Board for Plant Genetic Resources
ICARDA	-	International Center for Agricultural Research in the Dry Areas
ICFRE	-	Indian Council of Forestry Research and Education

ICRAF	-	International Council for Research on Agro-forestry
ICRISAT	-	International Crops Research Institute for the Semi-Arid Tropics
IDMI	-	International Data Management Institute
IFAD	-	International Fund for Agricultural Development
IFPRI	-	International Food Policy Research Institute, Washington
IIHR	-	Indian Institute of Horticultural Research, Bangalore
IIP&DM	-	Indian Institute of Pests and Disease Management
IIPR	-	Indian Institute of Pulses Research
IIVR	-	Indian Institute of Vegetable Research
IIT	-	Indian Institute of Technology
IITM	-	Indian Institute of Tropical Meteorology
ILRI	-	International Livestock Research Institute, Nairobi, Kenya
IMD	-	Indian Meteorological Department, Pune
INCID	-	Indian National Committee on Irrigation and Drainage
IPGRI	-	International Plant Genetic Resources Institute
IRRI	-	International Rice Research Institute, Los Banos, Philippines
IVRI	-	Indian Vetnary Research Institute
IWMI	-	International Water Management Institute
MOA	-	Union Ministry of Agriculture
MORD	-	Union Ministry of Rural Development
MOWR	-	Union Ministry of Water Resources
MRC	-	Mekong River Commission
NAARM	-	National Academy of Agricultural Research and Management

NATP	-	National Agricultural Technology Project of ICAR
NBFGR	-	National Bureau of Fish Genetic Resources
NBPGR	-	National Bureau of Plant Genetic Resources
NBSS&LUP	-	National Bureau of Soil Survey and Land Use Planning
NCAP	-	National Centre for Agricultural Economics and Policy Research, New Delhi
NEH	-	ICAR Complex for North-east Hill Region
NGOs	-	Non-Government Organizations
NIH	-	National Institute of Hydrology, Roorkee
NHB	-	National Horticultural Board
NRCW	-	National Research Centre for Women
NRSA	-	National Remote Sensing Agency, Hyderabad
PDCSR	-	Project Directorate for Cropping System Research
RAU	-	Rajendra Agricultural University
SAD	-	State Agricultural Department
SAUs	-	State Agricultural Universities
SIWI	-	Stockholm International Water Institute
UA	-	University of Arizona, Tueson, USA
UC	-	University of California, Davis Campus, USA
UEA	-	University of East Anglia
UR	-	University of Reading
USAID	-	United States Agency for International Development
USDA	-	United States Department of Agriculture
WALMI	-	Water and Land Management Institute
WFC	-	World Financial Centre
WMO	-	World Meteorological Organisation
WRI	-	World Resources Institute, Washington
WTC	-	Water Technology Centre
WTCER	-	Water Technology Centre for Eastern Region
ZT	-	Zero Tillage

Appendix ii : Important Organization/Institutes Engaged in Tobacco Related Research

Central Agricultural University

1.Central Agricultural University, Imphal (Manipur) 795 004

Central Universities

1.Aligarh Muslim University, Aligarh (Uttar Pradesh) 202 002

2.Banaras Hindu University, Varanasi (Uttar Pradesh) 221 005

3.Visva Bharati Sriniketan (West Bengal) 731 236

Agricultural Universities

1.Acharya N G Ranga Agricultural University, Rajendranagar, Hyderabad (Andhra Pradesh) 500 030

2.Anand Agricultural University, Anand (Gujarat) 388 110

3.Assam Agricultural University, Jorhat (Assam) 785 013

4.Bidhan Chandra Krishi Viswavidyalaya, Mohanpur, Nadia (West Bengal) 741 252

5.Birsa Agricultural University, Ranchi (Jharkhand) 834 006

6.Chandra Shekhar Azad University of Agriculture and Technology, Kanpur (Uttar Pradesh) 208 002

7.Chaudhary Charan Singh Haryana Agricultural University, Hisar (Haryana) 125 004

8.Dr Panjabrao Deshmukh Krishi Vidyapeeth, Akola (Maharashtra) 444 104

9.Govind Ballabh Pant University of Agriculture and Technology, Pantnagar, (Uttaranchal) 263 145

10. Ch Sharwan Kumar Krishi Vishwavidyalaya, Palampur (Himachal Pradesh) 176 062

11. Indira Gandhi Krishi Vishwavidyalaya, Raipur (Chhattisgarh) 492 012

12. Jawaharlal Nehru Krishi Vishwa Vidyalaya, Jabalpur, (Madhya Pradesh) 482 004

13. Junagadh Agricultural University, Junagadh (Gujarat) 362 001
14. Kerala Agricultural University, Vellanikkara, Distt Trichur (Kerala) 680 656
15. Dr Balasaheb Sawant Konkan Krishi Vidyapeeth, Dapoli, (Maharashtra) 415 712
16. Mahatma Phule Krishi Vidyapeeth, Rahuri (Maharashtra) 413 722
17. Marathwada Agricultural University, Parbhani (Maharashtra) 431 402
18. Maharana Pratap University of Agriculture and Technology, Udaipur, (Rajasthan) 313 001
19. Narendra Deva University of Agriculture and Technology, Faizabad, (Uttar Pradesh) 224 229.
20. Navsari Agricultural University, Navsari (Gujarat) 396 450
21. Orissa University of Agriculture and Technology, Bhubaneshwar, (Orissa) 751 003
22. Punjab Agricultural University, Ludhiana (Punjab) 141 004
23. Rajasthan Agriculture University, Bikaner (Rajasthan) 334 002
24. Rajendra Agricultural University, Samastipur, Pusa (Bihar) 848 125
25. Sardar Krushi Nagar Dantiwada Agricultural University, Dantiwada (Gujarat) 385 506
26. Sardar Ballabh Bhai Patel University of Agriculture and Technology, Modipuram, Meerut (Uttar Pradesh) 250 110
27. Sher-e-Kashmir University of Agricultural Sciences and Technology, Srinagar (Jammu and Kashmir) 191 12
28. Sher-e-Kashmir University of Agricultural Sciences and Technology, 45-B, Gandhinagar, Jammu (Jammu and Kashmir) 180012
29. Tamil Nadu Agricultural University, Coimbatore (Tamil Nadu) 641 003
30. University of Agricultural Sciences, GKVK Campus, Bangalore (Karnataka) 560 065

31. University of Agricultural Sciences, Dharwad (Karnataka) 580005

32. Uttar Banga Krishi Vishwavidyalaya, Pundibari , Cooch Behar (West Bengal) 736 165

Deemed-to-be Universities

1. Indian Agricultural Research Institute Pusa, New Delhi 110 012

2. Allahabad Agricultural Institute, Allahabad (Uttar Pradesh) 211 007

Important Indian Council of Agricultural Research (ICAR) Institution

National Institutes

1. Indian Agricultural Research Institute Pusa, New Delhi 110 012

Central Agricultural Research Institutes

1. Central Agricultural Research Institute, Port Blair, Andaman and Nicobar Group of Islands (Andamans and Nicobar Island) 744 101

2. Central Institute of Post-Harvest Engineering and Technology, Ludhiana (Punjab) 141004

3. Central Research Institute for Dryland Agriculture, Santoshnagar, P O Saidabad Hyderabad (Andhra-Pradesh) 500 059

4. Central Tobacco Research Institute, Rajahmundry (Andhra Pradesh) 533 105

5. ICAR Research Complex for Goa, Ela, Old Goa (Goa) 403 402

6. ICAR Research Complex for Eastern Region, Walmi Complex, Phulwari Sharif, Patna (Bihar) 801 505

7. ICAR Research Complex for North Eastern Hills Region, Umroi Road, Barapani (Meghalaya) 793 103

8. Indian Agricultural Statistics Research Institute, Library Avenue, Pusa Campus, New Delhi 110 012

Other

1. National Academy of Agricultural Research and Management, Rajendranagar, Hyderabad (Andhra Pradesh) 500 030

National Bureaux

1. National Bureau of Plant Genetic Resources, FCI Building, Pusa New Delhi 110 012

2. National Bureau of Soil Survey and Land Use Planning, Shankar Nagar, Amravati Road, Nagpur (Maharashtra) 440 010

General

1. National Centre for Agricultural Economics and Policy Research, Library Avenue, Pusa, New Delhi 110012

2. National Research Centre for Women in Agriculture, 1199, Jagamara, Bhubaneswar (Orissa) 751 031.

Appendix iii : Tobacco Related and other Important Websites

Tobacco related websites

1. www.tobaccoinstitute.com/
2. www.tobacco.org/
3. www.tobacco-control.org/
4. Tobacco Directory www.calug.com/tobacco/
5. www.tobaccofreekids.org/
6. Tobacco Cigars - http://www.tobaccocigars.com/
7. Cuban Cigars - http://www.lacasadelcubano.com/
8. Cigarette Papers - http://www.dansrollies.co.uk
9. Best Cigars - http://www.cigarchoiceguide.com
10. Cigar Blog | Wine Blog - http://www.pythonbytes.com
11. Cigar Smoking - http://www.esperanzacigars.com
12. www.TobaccoWeek.org
13. www.dh.gov.uk/en
14. www.drugabuse.gov/
15. www.umtrn.sph.umich.edu/TobaccoWebsites.php
16. www.keywen.com/en/TOBACCO
17. www.baseballandtobacco.com/
18. www.tcfor.catcp.org/
19. www.ag.ca.gov/tobacco/resources.php
20. www.tobaccofreekids.org/

Other Important Websites

1. Agriculture Factbook http://www.usda.gov/factbook/
2. Agricola http://agricola.nal. usda. gov/
3. CABI http://www.cabLorg/
4. Consultative Group on International Agriculture Research www.cgiar.org

5.A GRIS http://www.agris.com

6.Agricultural Network Information Centre (AgNIC) http://www agnic.org/

7.A gNIC Partnership on Agricultural Biotechnology http:// agnic.umd.edu

8.D epartment of Agriculture and Co-operation http: //agricoop.nic. in

9.Food and Agriculture Organization of1he United Nations http: //www.fao.org!

10. Indian Council of Agricultural Research http://www.icar.org.in

11. American Society of Agronomy http://www.agronomy.org

12. Crop Science Society of America http://www.crops.org/

13. Soil Science Society of America http://www.soils.orgl

14. Agricultural Statistics http://www.usda.gov/nass/pubs/ agstats.htm

Appindix iv: Tobacco Type and Producing States in India

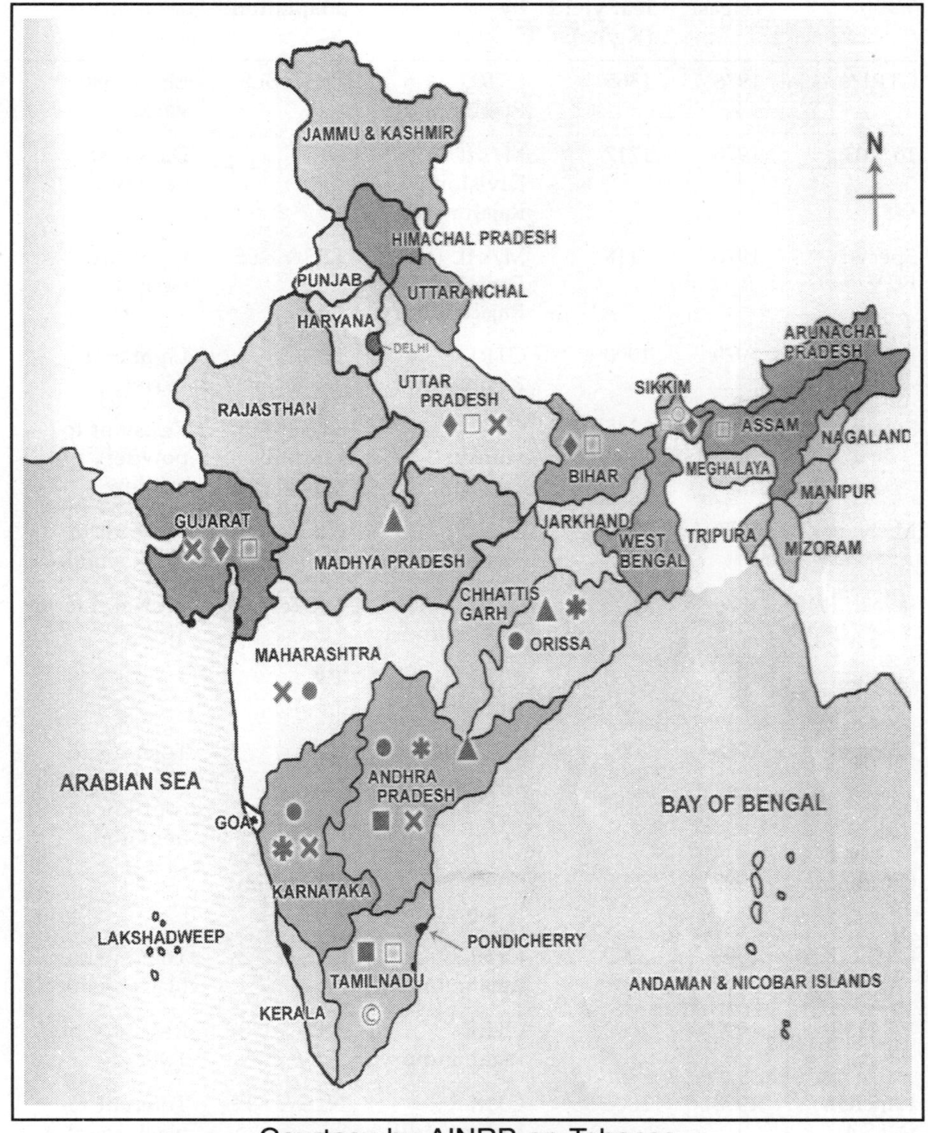

Courtesy by AINRP on Tobacco

Appendix v: FCV Tobacco Varieties Released through All Indian Coordinated Research Project /All India Network Research Project

Variety	Year of release	Cured leaf yield (Kg/ha)	Developed by	Area of adaptation	Remarks
CTRI Special	1976	1365	CTRI, Rajahmundry	TBS & SLS	Light cast variety
16/103	1976	1717	M/s IL TD Division, Rajahmundry	NLS	Dark cast variety
Special FCV	1976	1118	M/s IL TD Division, Rajahmundry	TBS & KLS	Light cast variety
Jayasri	1979	1990	CTRI, Rajahmundry	TBS	Light cast variety
Swarna	11984	11450	CTRI RS, Hunsur	KLS (high rainfall zones)	Resistant to powdery mildew
Mc Nair 12	1986	1880	CTRI, Rajahmundry	NLS	Tolerant to black shank
Jayasri (MR)	1986	1503	CTRI, Rajahmundry	TBS & SLS	Resistant to to TMV
Hema	1987	1560	CTRI RS, Guntur	TBS	
Bhavya	1988	2000	I CTRI RS, Hunsur	KLS	Tolerant to black shank & root-knot nematode
Gauthami	1992	2000	CTRI, Rajahmundry	TBS & SLS	
CM12 (KA)	1993	2000	CTRI, Rajahmundry	NLS	Tolerant to black shank
VT 1158	1993	2000	CTRI, Rajahmundry	NBS & CBS	Resistant to TMV
Kanchan (NLS 4)	1998	2000	CTRI RS, Jeelugumilli & CTRI RS, Hunsur	NLS & KLS	Tolerant to black shank & root-knot nematode

Contd...

Thrupthi	1998	1800	RARS, Shimoga	KLS	Tolerance to drought
Rathna	2001	2000	CTRI RS, Hunsur	Medium and low rainfall regions of KLS	Field resistant to black shank
Kanthi (Cy 79)	2006	1600-2000	CTRI RS, Kandukur	SLS & SBS	Withstands moisture stress to certain extent
Hemadri (11-1624)	2006	2500	CTRI RS, Guntur		TBS
Siri (Cy 135)	2006	2900	CTRI, Rajahmundry		TBS

Source: Adopted form Anonymous (2006a)

Appendixvi : Bidi Tobacco Varieties Released Through All Indian Coordinated Research Project /All India Network Research Project

Variety	Year of release	Cured leaf yield (Kg/ha)	Developed by	Area of adaptation	Remarks
GT4	1976	2600 (rainfed) 2841 (normal)	BTRS, Anand	Gujarat in Karnataka, Kolhapur and	Tolerant to drought
NPN 190	1979	1964	ARS, Nipani	Sangli districts in Maharashtra Karnataka,	
Anand 119	1984	2625	BTRS, Anand	Maharashtra and A.P. Karnataka,	for Gujarat and a National variety in 1984
Anand 2	1984	2555	BTRS, Anand	Maharashtra and A.P. in Karnataka, Kolhapur and	for Gujarat and a National variety in 1984
Spoorthy (PL 5)	1984	2000	ARS, Nipani	Sangli districts in Maharashtra	
GT5	1985	3300	BTRS, Anand	Gujarat	Shy suckering, high nicotine and smooth smoke
GT7	1993	2535	BTRS, Anand	tobacco growing areas of Gujarat	Drought tolerant
GTH1	1995	3644	BTRS, Anand	tobacco growing areas of Gujarat	root-knot, high nicotine
Bhavyasree	1997	2000	ARS, Nipani	Karnataka	Tolerant to frog-eye spot
GT9	2001	3077	BTRS, Anand	Gujarat	Mosaic resistant
NBD 43	2005	1742	ARS, Nipani	Karnataka	

Source: Adopted form Anonymous (2006a)

Appendix vii : Natu Tobacco Varieties Released Through All Indian Coordinated Research Project /All India Network Research Project

Variety	Year of release	Cured leaf yield (Kg/ha)	Developed by	Area of adaptation	Remarks
Prabhat	1977	1500	CTRI, RS, Guntur	Natu tobacco areas of Andhra Pradesh	Resistant to TMV, cigarette Natu
Vishwanath	1986	2429	CTRI, RS, Guntur	Natu tobacco growing belt of Guntur	Suitable for cigarette natu
Natu Spl.	1992	1600	CTRI, RS, Guntur	Natu tobacco growing belt of AP	
Bhairavi (NG- 73)	2004	2600	CTRI, RS, Guntur	Suitable to Cigarette Natu areas of AP.	

Source: Adopted form Anonymous (2006a)

Appendix viii : Chewing & *Rustica* tobacco Varieties Released Through All Indian Coordinated Research Project /All India Network Research Project

Variety	Year of release	Cured leaf yield (Kg/ha)	Developed by	Area of adaptation
Gandak Bahar (HP 6-20)	1976	2280	CTRI RS, Pusa	chewing tobacco areas of north Bihar
Sona (HP-63-3)	1977	3178	CTRI RS, Pusa	chewing tobacco areas of north Bihar
Vairam (HV 67-18)	1977	2800	CTRI RS, Vedasandur	Pit- cured chewing tobacco areas of Tamil Nadu
DD 437	1977	1865	CTRI RS, Dinhata	Motihari tobacco growing areas of West Bengal
Sonar Motihari (HD 65-40)	1977	1690	CTRI RS, Dinhata	Motihari tobacco growing areas of northern West Bengal
GC 1	1981	2693	BTRS, Anand	*Rustica* / Calcutti tobacco growing areas of Gujarat. Suitable for *Lal* and *Kala Chopadia*
Thangam	1980	3226	CTRI RS, Vedasandur	For smoke-cured growing areas of Tamil Nadu
Bhagyalakshmi	1980	3532	CTRI RS, Vedasandur	Sun cured tobacco areas of Tamil Nadu
Maragatham	1981	3013	CTRI RS, Vedasandur	For smoke cured tobacco areas of Tamil Nadu
Prabha	1981	2200	CTRI RS, Pusa	Hookah and chewing tobacco areas of north Bihar
GT6	1986	2712	BTRS, Anand	Middle Gujarat, Suitable for*Lal* and *Kala Chopadia*

Contd...

Pusa Tobacco-76	1990	2600	CTRI RS, Pusa	districts of north Bihar
Meenakshi	1992	4000	CTRIRS Vedasandur	Sun curing chewing tobacco tracts of Dindigul and Madurai districts in Tamil Nadu
Vaishali Special	1993	2778	CTRI RS, Pusa	'Bandi' type chewing tobacco growing tract of North Bihar. Tolerant to black shank
GCT 2	1994	3512	BTRS, Anand	*Rustica* tobacco growing areas of Gujarat
GT 8	1998	4469	BTRS,	Gujarat, Anand, Suitable for Kala chopadia
GCT3	2001	5986 (North Gujarat) 3285 (Middle Gujarat)	BTRS, Anand of Gujarat	*Rustica* tobacco growing areas. High yielding and suitable for *Lal Chopadia*
Abirami	2006	4000	CTRI RS, Vedasandur	For Southern, Central and Western Zones of Tamil Nadu except coastal belt, Rich in solanesol

Source: Adopted form Anonymous (2006a)

Appendix ix : Cheroot Tobacco Varieties Released Through All Indian Coordinated Research Project /All India Network Research Project

Variety	Year of release	Cured leaf yield (Kg/ha)	Developed by	Area of adaptation	Remarks
Sendrapatty Special	1986	2100	CTRI, RS, Vedasandur	Broad leaf Oosikappal country cheroot tobacco growing areas of Tamil Nadu	

Source: Adopted form Anonymous (2006a)

Appendix x : Cigar Wrapper Tobacco Varieties Released Through All Indian Coordinated Research Project /All India Network Research Project

Variety	Year of release	Cured leaf yield(Kg/ha)	Developed by	Area of adaptation	Remarks
Cigar wrapper S5	1977	1476	CTAI AS, Dinhata	Cigar wrapper tobacco growing areas of Northern West Bengal	

Source: Adopted form Anonymous (2006a)

Appendix xi : Recommended Manurial Scheduling for Various Types of Tobacco

States	Tobacco type	Fertilizer Scheduled (ha)
Andhra Pradesh	FCV- Black soils	30 N, 60 P_20_5 and 80 K_20 kg
	FCV- Red soils	50 N, 60 P_20_5 and 80 K_20 kg
	Burley	100 N, 60 P_20_5 and 80 K_20 kg
	Natu	200 N, 50 P_20_5 and 50 K_20 kg Nitrogen in 4 equal splits at sowing, 30,45 and 60 DAS
Karnataka	FCV- Black soils	30 N and 40 P_20_5 kg, Nitrogen as basal dose by plant row plough furrow method.
	FCV- Red soils	40 N, 80 P_20_5 and 80 K_20 kg, Fertilizers applied by double band placement or dollop method.
	Bidi	120 N, 60 P_20_5 and 40 K_20 kg, One-fourth of N through organic sources and the rest through fertilizers.
Gujarat	Bidi	200 N and 50 P_20_5 kg, Organic sources such as castor cake, FYM, green manure etc provide 50 kg N ha-1 and the reminder to be given through inorganic sources.
Tamil Nadu	Cigar and cheroot	25 t of FYM along with 100 kg N (2/3 as ammonium sulphate and 1/3 as ground nut cake), half before planting and the remaining half 6-7 weeks after planting.
Bihar	Chewing	50 t ha-1 of FYM and 120 kg N , half as mustard cake and the remaining half as ammonium sulphate in two splits at planting and 40 days after plating.
West Bengal	Chewing and Hookah	25 t ha of FYM and 150 N and 150 P_20_5 kg. Nitrogen in two splits at planting 50 days after plating
	Wrapper	20 t ha of FYM and 125 N and 150 P_20_5 and 200 K_2O kg. Nitrogen in two splits at planting 50 days after plating

Appendix xii. Cropping Pattern According to Land Utilization

Crops	2002-03 Area (000′ ha)	% share to total cropped area	2003-04 Area (000′ ha)	% share to total cropped area
I.	**133,009**	**56.27**	**141,495**	**80.07**
Food crop				
1. Total food grains	**116,105**	**65.70**	**124,843**	**70.64**
(A) Cereals & millets	**97,765**	**53.62**	**100,328**	**56.77**
Rice	42,034	23.79	42,634	24.13
Jowar	9,341	5.29	9,360	5.30
Bajra	7,932	4.49	10,624	6.01
Maize	6,761	3.83	7,405	4.19
Ragi/ marua	1,518	0.86	1,766	1.00
Wheat	25,203	14.26	26,587	15.04
Barley	703	0.40	739	0.42
Other cereals & millets	1,273	0.72	1,213	0.69
(B) Pulses	**21,340**	**12.08**	**24,515**	**13.87**
Gram	5,919	3.35	7,139	4.04
Tur (Arhar)	3,362	1.90	3,550	2.01
Other pulses	12,060	6.82	13,826	7.82
2. Sugar	**4,825**	**2.73**	**4,434**	**2.50**
3. Condiments & spices	**3,056**	**1.73**	**3,101**	**1.75**
4. Total fruits & vegetables	**8,957**	**5.07**	**9,046**	**5.12**
(i) Total fruits	**3,728**	**2.11**	**3,751**	**2.12**
(A) Fresh fruits	**3,063**	**1.73**	**3,066**	**1.74**
(B) Dry fruits	**664**	**0.38**	**683**	**0.39**
(ii) Vegetables	**5,230**	**2.96**	**5,295**	**3.00**
5. Other food crops	**64**	**0.04**	70	**0.04**
II. Non-food crops	**43,711**	**24.73**	**49,149**	**27.81**
1. Oil seeds	**2,393**	13.54	**26,286**	**14.87**
Ground nut	6,013	3.40	6,151	3.48
Castor seed	696	0.39	724	0.41
Sesamum (Til or Gingelly)	1,556	0.88	2,003	1.13
Rapeseed & Mustard	4,318	2.44	5,445	3.08
Linseed	450	**0.25**	47	0.27

Contd...

Linseed	450	**0.25**	47	0.27
Coconut	1,906	1.08	1,924	1.09
Niger seed	415	0.23	432	0.24
Safflower	369	0.21	368	0.21
Soybean	6,083	3.44	6,553	3.71
Sunflower	1,638	0.93	1,986	1.12
Others	490	**0.28**	224	0.13
2. Fibres	**8,836**	5.00	**8,713**	**4.93**
3. Dyes & tanning material	**43**	0.02	**46**	**0.03**
4. Drug, narcotics & plantation crops	**2,187**	1.24	**1,326**	**1.32**
Tobacco	393	0.22	431	0.24
Tea	505	0.29	595	0.34
Coffee	315	0.18	316	0.18
Rubber	543	0.31	546	0.31
Others	407	0.23	408	0.23
5. Fodder crops	7,999	4.53	10,803	6.11
6. Green manure crops	58	0.03	58	0.03
7. Other non-food crops	655	0.37	918	0.52
Total area sown under all crops (I+II)	**176,719**	**100.00**	**190,644**	**100.00**

Source : Agricultural Statistics at a Glance, 2006, Directorate of Economics and Statistics, Ministry of Agriculture, Govt. of India (Website: http://www.dacnet.nic.in/eands).

Appendix xiii. State-wise Consumption of Plant Nutrients per Unit of Gross Cropped Area (Provisional)

Zone/State	2004-2005				2005-2006			
	N	P_2O_5	K_2O	Total	N	P_2O_5	K_2O	Total
North	**104.5**	**31.0**	**6.0**	**141.5**	**108.5**	**34.6**	**7.2**	**150.3**
Haryana	123.6	37.7	3.1	164.4	132.7	39.5	4.5	176.7
Himachal Pradesh	32.3	9.0	7.4	48.6	31.9	10.2	8.3	50.4
Jammu & Kashmir	45.8	19.6	2.9	68.3	56.1	23.5	4.2	83.7
Punjab	151.6	40.0	5.4	197.0	158.2	46.6	8.0	212.7
Uttar Pradesh	92.7	28.7	7.0	128.4	94.6	31.9	7.8	134.4
Uttrakhand	65.0	16.3	5.8	87.1	67.6	18.1	6.6	92.3
Chandigarh	3.3	-	-	3.3	-	-	-	-
Delhi	12.3	0.5	-	12.7	10.2	-	-	10.2
South	**74.8**	**36.3**	**28.1**	**139.2**	**91.4**	**44.7**	**32.5**	**168.5**
Andhra Pradesh	93.5	43.6	23.6	160.8	123.1	55.8	27.5	206.4
Karnataka	57.3	31.8	23.9	112.9	65.9	38.0	29.3	133.2
Kerala	29.9	14.1	24.2	68.2	27.9	15.0	25.6	68.5
Tamil Nadu	90.8	39.8	48.5	179.1	105.2	48.2	53.4	206.8
Pondicherry	618.4	272.7	245.1	1136.2	602.4	294.6	253.8	1150.8
A & N Islands	6.0	3.8	1.6	11.3	8.0	3.1	2.0	13.1
Lakshadweep	-	-	-	-	-	-	-	-
East	**48.6**	**17.3**	**12.6**	**78.6**	**51.3**	**20.1**	**14.4**	**85.8**
Arunachal Pradesh	1.7	0.8	0.4	2.9	1.7	0.8	0.4	2.9
Assam	19.7	12.9	9.3	41.9	23.6	14.1	12.3	50.0
Bihar	78.5	9.0	5.3	92.8	87.3	16.7	12.5	116.6
Jharkhand	34.1	19.2	2.0	55.3	38.6	18.4	3.2	60.2
Manipur	78.6	11.1	4.4	94.0	55.7	6.4	3.4	65.4
Meghalaya	10.5	7.0	0.7	18.1	11.0	6.3	0.8	18.1
Mizoram	6.2	5.2	3.1	14.5	9.9	8.5	4.9	23.3
Nagaland	0.8	0.6	0.2	1.6	0.9	0.6	0.2	1.7
Orissa	25.9	9.0	-6.2	41.1	28.2	10.5	7.0	45.7
Sikkim	3.6	1.5	-	5.1	1.7	0.9	0.2	2.9
Tripura	28.0	7.7	4.6	40.4	29.8	8.7	7.0	45.6
West Bengal	65.0	35.0	30.0	130.0	63.0	36.9	27.9	127.7
West	**38.4**	**18.8**	**5.7**	**62.9**	**40.9**	**19.5**	**7.0**	**67.4**
Gujarat	66.7	26.2	8.5	101.4	73.8	29.0	10.3	113.2
Madhya Pradesh	31.2	19.9	2.8	53.9	28.3	16.3	3.0	47.5
Chhattisgarh	40.4	17.0	6.1	63.5	40.7	18.3	6.6	65.6
Maharashtra	43.0	23.7	11.6	78.4	47.8	26.0	14.8	88.7
Rajasthan	25.0	9.4	0.8	35.1	28.6	11.2	0.9	40.7
Goa	15.1	8.8	10.2	34.1	14.7	8.3	9.7	32.7
Daman & Diu	3.3	-	-	3.3	-	-	-	-
Dadra & Nagar Haveli	22.3	11.7	1.3	35.3	23.0	12.7	2.0	37.7
All-India	**61.4**	**24.3**	**10.8**	**96.5**	**66.7**	**27.3**	**12.7**	**106.7**

Note : Consumption of plant nutrients per hectare have been worked out on the basis of gross cropped area available for the year, 2003-04.

Source : Fertilizer Statistics, 2005-06, the Fertiliser Association of India, New Delhi

Appendix xiv. Agro-ecological Regions

Eco-system	Region	States represented	Eco-region	Soil type
Arid	1. Western Himalaya	Jammu & Kashmir, HP	Cold arid	Shallow Skeletal
	2. Western plain, Kutch and part of Kathiawar Peninsula	Gujarat, Rajasthan, Haryana, Punjab	Hot arid	Desert & Saline
	3. Deccan plateau	AP, Karnataka	Hot arid	Red & Black
Semi-arid	4. Northern plain and Central Highlands including Aravallis	Gujarat, Rajasthan, UP, MP, Haryana, Punjab	Hot Semi-arid	Alluvium derived
	5. Central (Malwa) Highlands, Gujarat plains & Kathiawar Peninsula	Gujarat, MP	Hot Semi-arid	Medium & deep black
	6. Deccan Plateau	Karnataka, AP, Maharashtra, MP	Hot Semi-arid	Shallow and medium (with inclusion of deep) black
	7. Deccan (Telangana) Plateau and Eastern Ghats	AP	Hot Semi-arid	Red & Black
	8. Eastern Ghats, TN uplands and Deccan (Karnataka) Plateau	Karnataka, TN, Kerala	Hot Semi-arid	Red loamy
Sub-humid	9. Northern plain	Bihar, UP, Punjab	Hot sub humid (dry)	Alluvium derived
	10. Central Highlands (Malwa, Bundelkhand & Eastern Satpura)	MP, Maharashtra	Hot sub humid	Black & Red
	11. Eastern Plateau (Chhattisgarh)	MP	Hot sub humid	Red & Yellow
	12. Eastern (Chhota Nagpur) Plateau and Eastern Ghats	Orissa, W. Bengal, Bihar, MP, Maharashtra	Hot sub humid	Red & lateritic

Contd...

	13. Eastern plain	UP, Bihar	Hot sub humid (moist)	Alluvium-derived
	14. Western Himalayas	J&K, HP, UP	Warm sub humid (to humid with inclusion of per humid)	Brown forest and **podzolic**
Humid -per humid	15. Bengal and Assam plains	West Bengal, Assam	Hot sub humid (moist) to humid (inclusion of per humid)	Alluvium-derived
	16. Eastern Himalayas	Arunachal Pradesh, Sikkim, W. Bengal	Warm per humid	Brown and Red hill
	17. North Eastern Hills (**Purvanchal**)	**Tripura, Mizoram, Meghalaya**	Warm per humid	Red and **lateritic**
Coastal	18. Eastern Coastal plain	TN, **Pondicherry**, AP, **Orissa,W.** Bengal	Hot sub humid to semi-arid	Coastal Alluvium-derived
	19. Western Ghats & Coastal plain	Kerala, Goa, Daman & Diu, Maharashtra, Gujarat, Kerala	Hot humid per humid	Red, **lateritic** and Alluvium-derived
Island	20. Island of Andaman-**Nicobar** Islands and **Lakshadweep**	Andaman-**Nicobar** and **Lakshadweep**	Hot humid per humid	Red loamy and sandy

Source : Fertiliser Statistics, 2005-06, the Fertiliser Association of India, New Delhi.

Appendix xv. Agro-climatic Regions / Zones in India

S.No.	Agro-climatic regions/zones	States represented
I	Western Himalayan region	Himachal Pradesh, Jammu & Kashmir, Uttarakhand
II	Eastern Himalayan region	Arunachal Pradesh, Assam, Manipur, Meghalaya, Mizoram, Nagaland, Sikkim, Tripura, West Bengal
III	Lower Gangetic plain region	West Bengal
IV	Middle Gangetic plain region	Uttar Pradesh, Bihar
V	Upper Gangetic plain region	Uttar Pradesh
VI	Trans Gangetic plain region	Chandigarh, Delhi, Haryana, Punjab, Rajasthan
VII	Eastern plateau and hills region	Chhattisgarh, Jharkhand, Madhya Pradesh, Maharashtra, Orissa, West Bengal
VIII	Central plateau and hills region	Madhya Pradesh, Rajasthan, Uttar Pradesh
IX	Western plateau and hills region	Madhya Pradesh, Maharashtra
X	Southern plateau and hills region	Andhra Pradesh, Karnataka, Tamil Nadu
XI	East coast plains and hills region	Andhra Pradesh, Orissa, Pondicherry, Tamil Nadu
XII	West coast plains and ghat region	Goa, Karnataka, Kerala, Maharashtra, Tamil Nadu
XIII	Gujarat plains and hills region	Gujarat, Dadra & Nagar Haveli, Daman & Diu
XIV	Western dry region	Rajasthan
XV	Island region	Andman & Nicobar Islands, Lakshdweep

Appendix xvi. Land use Classification in India (in million ha)

Classification	1950-51	1960-61	1970-71	1980-81	1990-91	1995-96	1999-00	2000-01#	2001-02#	2002-03#	2003-04#
I. Geographic area	328.73	328.73	328.73	328.73	328.73	328.73	328.73	328.73	328.73	328.73	328.73
II. Reporting area for land utilization statistics (1 to 5)	284.32	298.46	303.76	304.15	304.86	304.88	305.62	305.0	305.01	305.24	305.84
1. Forests	40.48 (14.2)	54.05 (18.1)	63.91 (21.00	67.47 (22.2)	67.80 (22.2)	68.82 (22.6)	69.12 (22.6)	69.62 (22.8)	69.51 (22.8)	69.64 (22.8)	69.70 (22.8)
2. Not available for cultivation (a+b)	47.52	50.75	44.64	39.62	40.48	41.37	41.50	41.55	31.78	42.08	42.22
(a) Non agricultural uses	9.36 (3.3)	14.84 (5.0)	16.48 (5.4)	19.66 (6.4)	21.09 (6.9)	22.36 (7.3)	23.59 (7.7)	23.81 (7.8)	24.07 (7.9)	24.28 (8.0)	24.49 (8.0)
(b) Barren and uncultivable land	38.16 (13.4)	35.91 (12.0)	28.16 (9.3)	19.96 (6.6)	19.39 (6.4)	19.01 (6.2)	17.51 (5.7)	17.74 (5.8)	17.71 (5.8)	17.80 (5.8	17.73 (5.8)
3. Other uncultivated land excluding fallow land (a+b+c)	49.45	37.64	35.06	32.31	30.22	28.64	32.25	27.71	27.37	27.41	27.00
(a) Permanent pastures and other grazing land	6.68 (2.3)	13.97 (4.7)	13.26 (4.4)	11.97 (3.9)	11.40 (3.7)	11.06 (3.6)	10.80 (3.6)	10.83 (3.6)	10.59 (3.5)	10.51 (3.4)	10.45 (3.4)
(b) Miscellaneous tree crops and groves	19.83 (7.0)	4.46 (1.5)	4.30 91.4)	3.60 (1.2)	3.82 (1.3)	3.48 (1.2)	3.71 (1.2)	3.32 (1.1)	3.37 (1.1)	3.36 (1.1)	3.37 (1.1)
(c) Cultivable waste land	22.94 (8.1)	19.21 (6.4)	17.50 (5.8)	16.74 (5.5)	15.00 (4.9)	14.10 (4.6)	17.74 (4.5)	13.56 (4.4)	13.41 (4.4)	13.54 (4.4)	13.18 (4.3)
4. Fallow land (a+b)	28.12	22.82	19.88	24.75	23.36	23.85	25.03	25.04	24.94	33.46	26.04
(a) Fallow land other than current fallow	17.44 (6.1)	11.18 (3.8)	8.76 (2.9)	9.92 (3.3)	9.66 (3.2)	10.02 (3.3)	10.43 (3.4)	10.19 (3.3)	10.30 (3.4)	11.76 (3.9)	11.24 (3.7)
(b) Current fallow	10.68 (3.8)	11.64 (3.9)	11.12 (3.5)	14.83 (4.9)	13.70 (4.5)	13.83 (4.5)	15.78 (5.2)	14.84 (4.9)	14.64 (4.8)	21.70 (7.1)	14.80 (46.1)

Contd...

5. Net area sown	118.75 (41.8)	133.20 (44.6)	140.27 (46.3)	140.00 (46.0)	143.00 (46.9)	142.20 (46.7)	140.96 (46.1)	141.16 (46.3)	141.42 (46.4)	132.66 (43.5)	140.88 (46.1)
6. Area sown more than once	13.14	19.57	25.52	32.63	42.74	45.27	48.48	44.55	48.83	44.06	49.76
7. Gross cropped area (5+6)	131.89	152.77	165.79	172.63	185.74	187.47	189.44	185.71	190.24	176.72	190.64
8. Cropping intensity*	111.1	114.7	118.2	123.3	129.9	131.8	134.4	131.6	134.5	132.2	135.3
III. Net irrigated area	20.85	24.66	31.10	38.72	47.78	53.40	57.11	54.84	55.87	53.15	55.10
IV. Gross irrigated area	22.56	27.98	38.19	49.78	62.47	71.35	78.81	75.87	77.94	72.97	76.82

Note: 1. # Provisional 2. * Cropping intensity is percentage of the gross cropped area to the net area sown 3. Figures in parentheses indicate percentage to reported area. 4. The decline in net area sown in 2002-03 reflects the impact of the severe drought during 2002.

Source: Agricultural Statistics at a Glance, 2006, Directorate of Economics and Statistics, Ministry of Agriculture, Govt. of India (Website: http://www.dacnet.nic.in/eands).

Appendix xvii. All-India Area Under Food Grains-cultivated and Irrigated-with Land Man Ratio

Year	Gross area under food grains (000' ha)		% Share of area under food grains to Gross		Land man ratio$
	Cultivated	Irrigated	Gross cultivated area	Gross irrigated area	
1950-51	97,321	18,317	73.8	81.2	-
1955-56	110,560	20,626	75.1	80.4	0.34
1960-61	115,581	22,065	75.7	78.9	0.32
1965-66	114,103	24,032	73.5	77.8	0.29
1970-71	124,316	30,117	75.0	78.9	0.28
1975-76	128,181	34,093	74.8	78.6	0.25
1980-81	126,667	37,851	73.4	76.0	0.24
1985-86	128,023	40,407	71.7	74.4	0.22
1990-91	127,835	44,866	68.8	71.0	0.20
1995-96	121,015	49,543	64.6	69.4	0.18
2000-01*	121,048	53,721	65.2	70.8	0.17
2001-02*	122,780	53,808	64.5	69.0	0.16
2002-03*	113,860	50,163	64.4	68.7	0.16
2003-04*	123,447	52,635	64.8	68.5	N.A.
2004-05*	120,078	N.A.	N.A.	N.A.	N.A.

Note: N.A.: Not Available

*: Provisional, $: Land man ratio = Arable land & land under permanent crops / population

Source : Fertiliser Statistics, 2005-06, the Fertiliser Association of India, New Delhi.

Appendix xviii. India's Position in World Agriculture

Item	India	World	India's Share %	Rank	Next to
1. Area(million ha)**					
Total area	329	13428	2.4	Seventh	Russian Federation, Canada, U.S.A., China, Brazil, Australia
Land area	297	13067	2.3	Seventh	Russian Federation, China, Canada, U.S.A., Brazil, Australia
Arable land	162 F	1404	11.5	Second	U.S.A.
Irrigated area	55 F	277	20.6	First	
2. Population (million)					
Total	1050	6225	16.9	Second	China
Agriculture	755	3234	23.3	Second	China
3. Economically active population (million)					
Total	460	3037	15.1	Second	China
Agriculture	270	1333	20.3	Second	China
4. Crop production (million t)					
Total Cereals	232	2075	11.2	Third	China, USA
Wheat	65	556	11.7	Second	China
Rice (Paddy)	132	589	22.4	Second	China
Coarse grains	35	930	3.8	Forth	U.S.A., China, Brazil
Total pulses	12	57	21.1	First	
Groundnut	8*	36	22.2	Second	China
Rapeseed & Mustard	4	36	11.1	Third	China, Canada
5. Fruits & Vegetables (million t)					
Vegetables & Melons	82F	842	9.7	Second	China
Fruits excluding Melons	46	480	9.6	Second	China
Potatoes	23	311	7.4	Third	China, Russian Federation
Onion (Dry)	5F	53	9.4	Second	China
6. Commercial crops (million t)					
Sugarcane	290	1333	21.8	Second	Brazil
Tea	0.89	3.21	27.7	First	
Coffee (green)	0.28	7.20	3.9	Sixth	Brazil, Vietnam, Indonesia, Colombia, Mexico
Jute & allied fibres	1.98	3.23	61.3	First	
Cotton(lint)	2.10	19.53	18.8	Third	China, USA
Tobacco leaves	0.60	6.2	9.7	Third	China, Brazil

Note : FAO Estimates, * Unofficial Figures, ** Figures relate to 2002
Source : Agricultural Statistics at a Glance, 2006, Directorate of Economics and Statistics, Ministry of Agriculture, Govt. of India (Website: http://www.dacnet.nic.in/eands).

□□□

Glossary

A horizon : The surface horizon of a mineral soil having maximum biological activity or eluviation (removal of materials dissolved or suspended in water) or both.

'A' Line means the male sterile parent used in crossing to evolve / produce hybrid seed. It is the male sterile line.

Abscidc acid (ABA) : An important growth-inhibitor of plants, initially called dormin, has wide range of physiological effects including promotion of senescence and abscission, causing dormancy in buds and seeds, closure of stomata, retardation and inhibition of growth known as *'antigibberrelin'* since it inhibits gibberrilin-stimulated growth.

Abscission layer : A zone of cells in the petiole or other plant structure whose cells separate and thereby bring about leaf fall, fruit drop etc.

Abscission : Detachment of fruit, leaf or other parts from a plant.

Absolute advantage : The ability to produce a greater physical output with a given set of resources than a similar set of resources can produce elsewhere.

Absolute growth rate (AGR) : The rate of increase in size of a growing plant (or part of it) in a given time under specific condition.

Absorption : The process by which a substance is taken into or includes within another substance, like intake of water by soil or intake of nutrients by plants.

Acclimatization : An adjustment in the morphology or physiology of an organism in response to a change in the environment.

Accumulation : The uptake of substances against a concentration gradient and hence an energy-requiring process.

Achene : A dry indehiscent, one-seeded fruit in which the ovary wall remains free from the seed-coat.

Acid equivalent *:* refers to the portion of a formulation (herbicide) which can be theoretically converted into acid. It is written as a.e.

Acid soil : A soil with a pH reaction of less than 7.0 (usually less than 6.6). An acid soil has a preponderance of hydrogen ions over hydroxyl ions, and blue litmus paper turns red in contact with moist acid soil.

Acid-sulphate soil is the one which intrinsically postulates a pH value lower than 3.5 and simultaneously represents a peaty saline soil. The accumulation of soluble salts as well as free-acid substances chiefly comprising ferric sulphate and aluminum sulphate form these soils. To illustrate, *Kari* soil in Kerala represents a monumental acid sulphate soil. This soil is highly acidic, deep black in colour, contains high organic matter, is poorly drained and above all remains under water during the monsoon season.

Active absorption *:* refers to the movement of ions and water into the plant root due to metabolic processes of the root, frequently against an activity gradient.

Active ingredient is the portion of a formulation (herbicide) which is directly responsible for the pesticidal (herbicidal) influence. It is written as a.i.

Acute toxicity is generally used with respect to herbicides and refers to the intensive and rapid killing of plants by herbicide.

Adaptation is the ability of an organism to alter the organs quantitatively as well as qualitatively to meet the requirement of life. The ability to bring the inner forces into adjustment with the exterior forces is also called adaptation.

Adenosine triphosphate (ATP) : ADP with an additional phosphate group attached by a high-energy bond, its decomposition into ADP and phosphate results in the release of energy needed for cellular metabolism.

Administrative regulation : Rules and regulations established by administrative decision.

Adsorption : The attraction of ions or compounds to the surface of solid. Solid colloids adsorb large amounts of ions and water.

Adventitious bud is formed at an uncommon location on a plant or it develops at an unusual time during plant growth and development.

Adventitious roots are the roots which appear at an unusual location or are produced at an unusual time during plant growth, like the nodal roots in grasses.

Adventitious : Organs arising in unusual positions, as buds from roots.

Aeration (soil) : The process by which air and other gases in the soil are renewed. The rate of soil aeration depends largely on the size and number of soil pores and on the amount of water clogging to permit rapid aeration, is said to be well aerated, while a poorly aerated soil either has few large pores or has most of its pores blocked by water.

Aerobic (condition) refers to the presence of more oxygen as a part of the environment or growth only in the presence of molecular oxygen (referred to as aerobic organisms) or the occurrence of a process in the presence of molecular oxygen (aerobic decomposition).

Age of tobacco seedling : Seedling of tobacco become fit to be transplanted when it attain the age of 7-9 weeks in case of *N. tabacum* and 5-6 week for N. *rustica*.

Agent middlemen Negotiate the transfer of goods from seller to buyer without themselves taking title to those goods.

Aggregate demand curve : A curve showing the negative relationship between the price level and the quantities of real goods and services demanded during a given period of time.

Aggregate supplies curve : A curve showing the positive relationship between the quantities of real goods and services supplied by businesses and the price level during a given period of time.

Agribusiness : Involves the manufacture and distribution of farm supplies; production operations on the farm; and the storage, processing, and distribution of farm commodities and items made for them.

Agricultural bargaining : A group of producers organized to gain greater power in the market for its members.

Agricultural economics : An applied social science dealing with how mankind uses technical knowledge and scarce productive resources to produce food and fiber and to distribute them to society for consumption over time.

Agricultural fundamentalism : The school of thought holding that there is something special and unique about the farm way of life.

Agricultural options : A type of financial instrument that gives the holder the right to buy or sell futures contracts.

Agricultural policy : Public policies developed to achieve specific objectives desired for agriculture.

Agrology : The study of applied phases of soil science and soil management.

Aleurone (layer) : is a type of protein generally present as a thin layer of the endosperm in maize and the small grained seeds.

Alfisols : Soils with grey to brown surface horizons, medium to high supply of bases, and B horizons of alluvial clay accumulation. These soils form mostly under forest or savanna vegetation in climates with slight to pronounced seasonal moisture deficit.

Allelopathy is any direct or indirect harmful effect that one plant has on another mutually on each other through the production of harmful compounds that escape into the environment.

Alternative uses of tobacco : Tobacco is one of the best sources of secondary metabolites; viz., nicotine, solanesol, organic acid nicotine sulphate, protein and oil as well as other bio compound like enzymes and vitamins. These entire products are derived from leaf or leaf waste.

Amino acids are nitrogen containing organic acids that couple together to form proteins. Each acid molecule has one or more amino group (NH2) and at least one carboxyl group (-COOH). Some amino acids contain sulphur.

Amphi-photoperiodic plants : Plants which flower either under shorter or longer photoperiods, but not in the intermediate range, for example, *Media elegans* and *Setaria verticillata.*

Amylopectin is the type of starch in which glucose molecules are connected so as to form a branched chain.

Amylose : Amylose is the type of starch in which glucose molecules are connected so as to form a straight chain.

Anabasine is present in trace amounts in tobacco smoke, and can be used as an indicator of a person's exposure to tobacco smoke. Anabasine is a nicotinic acetylcholine receptor agonist. In high doses, it produces a depolarizing block of nerve transmission, which can cause symptoms similar to those of nicotine poisoning and, ultimately, death by asystole. In larger amounts it is thought to be teratogenic in swine. The intravenous LD_{50} of anabasine ranges from 11 mg/kg to 16 mg/kg in mice, depending on the enantiome

Anabasine : Anabasine is a pyridine alkaloid found in the Tree Tobacco (*Nicotiana glauca*) plant, a close relative of the common tobacco plant (*Nicotiana tabacum*). It is similar to nicotine. Its principal (historical) industrial use is as an insecticide.

Annidation is the complementary effects of resources by exploiting the environmental supplies in differing ways by the components of a community

Anthesis consists of the process of dehiscence of anthers and spells out the period of the distribution of pollen.

Anti-auxin is a chemical or substance which inhibits the effect of auxin in the stimulation of cell enlargement.

Antitranspirants : Substances that retard or inhibit transpiration, e.g. PMA.

Apical dominance : Inhibiting effect of a tenninal bud upon the development of lateral buds.

Apical meristem : Tissue at the tip of a root or shoot where active cell division occurs.

Apomixis : The development of an embryo from cell other than a fertilized egg. The setting of seed without fertilization as nuclear embryo in citrus.

Aquatic weed denotes a foreign aquatic (water) plant making an inroad to spend at least a part of its life cycle in water.

Arable land : land currently being cropped or capable of being cropped without requiring additional costs of development.

Arbitrage : The simultaneous purchase and sale of a commodity in two different markets to take advantage of differences in the prices of that commodity in the markets.

Arid climate represents a highly dry climate characterized by an average annual rainfall of 250 millimeters or even less. The evapotranspiration far exceeds the rainfall in these areas.

Assets : Iems of money value owned by a business, including land, buildings, tractors, combines, and so on.

Auxins : A group of growth-regulators that may stimulate cell growth, root development and other growth process including seed germination.

Available nutrient refers to that portion of any element or compound in the soil that can be readily absorbed and assimilated by the growing plants. It is different from an exchangeable nutrient.

Average fixed cost (AFC) : Fixed cost per unit of output, determined by dividing total fixed cost by output at each level of output.

Average physical product (APP) : Units of output produced per unit of input for each level of variable input use.

Average total cost (ATC) : Total cost per unit of output, determined by dividing total cost by output at each level of output.

Average value product (AVP) : The value of output per unit of input for each level of variable input use.

Average variable cost (AVC) : Variable cost per unit of output, determined by dividing total variable cost by output at each level of output.

Back cross : In plant breeding a cross of hybrid with one of its parents and the purpose is to transfer a specific gene from an undesirable variety to another commercially desirable one lacking in that particular character.

Balance of trade : The value of merchandise exports minus the value of merchandise imports.

Band treatment is the application of herbicide directly either in the crop row or on the specified strip.

Bank reserves : The amount that a bank has on deposit with the Federal Reserve Bank or held as cash in its vault.

Barn : A farm building used for storage of tobacco leaves or other farm produces & implements.

Barn cured : Tobacco leaves cured with in a bran, a structure for controlled caring.

Basis : The difference between a futures price and the cash price of a commodity at a particular time and place.

Biological resources : Natural resources, such as forests and fisheries, that produce a harvestable yield from the resource stock.

Bio-type means a group of plants possessing an identical genetic constitution which may be homozygous or heterozygous.

Biroype : A population within a species that has a distinct genetic variation.

Blind cultivation is the cultivation done before the emergence of the plants from the soil.

'B'line is the fertile counter parent of the' A' line. It is also called the maintainer line.

Boron toxicity : Symptoms are the appearance of a yellow discolouration of the leaf tips that spreads along the margins. Large brown elliptical spots appear along the leaf margins. Affected parts turn brówn and wither. Boron toxicity occurs on coastal soil, arid region soils, soils irrigated with high boron water and geothermal areas.

Breeder seed : It is the seed directly controlled by the originating or sponsoring plant institution or individual, and the source of the production of seed for the certified classes. It is genetically purest seed stock.

Broom rape Orobanche : Tobacco crop is very often infested by Orobanche (*Orobanche cernua*) or Broom rape a phanerogamic, total root parasite. It is reddish brown or yellowish brown unbranched herb growing to a height 60 cm with leaves scale like ovate to lanceolate, 6-20 mm long, acute, sessile and devoid of chlorophyll. The stem is pubescent.

Bud is a very young or unelongated stem.

Budget line : Shows all possible combinations of goods that a consumer can buy with a given money income.

Bulk density is the ratio of the weight of oven dried soil to its bulk maximum volume. It is also called the volume weight.

Buoyancy : The upward force excreted on the volume of fluid by virtue of the density difference between the volume of fluid and that of surrounding fluid.

Capital : Productive resources (goods) that are available, as a result of past human decisions, to produce other want-satisfying goods.

Capitalism : A type of economic organization in which private individuals or groups own and manage all resources.

Capitalized value : The present value of a resource obtained by discounting the value of its future net income stream.

Carotene is the yellow colouring material found in a number of fruits and vegetables, also in leaves and green plant parts. It is the precursor of Vitamin A.

Caryopsis is the one seeded fruit of the plants of the grass family.

Centre of origin : An area where the given plant species shows the greatest genetic diversity.

Cereal means a grass which is primarily grown for its seeds used as feed or food. *Chlorophyll* is the green pigment contained in the chloroplast and consists of large and complex molecules which can absorb and capture radiant energy.

Certified seed : It is the progeny of breeder, foundation or registered seed, which is handled to maintain satisfactory genetic purity and identity in a manner acceptable to certifying agency.

Ceteris paribus : Holding some variables constant, while letting specific variables change.

Change in demand : A shift in the entire demand schedule.

Change in quantity demanded : A movement along a given demand curve in response to a price change.

Change in quantity supplied : A movement along a given supply curve in response to a price change.

Change in supply : A shift in the entire supply schedule.

Chemical de-suckering : De-suckering can be done manually or by using suckercide, a chemical to avoid time consuming laborious and more expensive process. Systemic suckercide, which restrict sucker growth without killing. Maleic hydrazide (MH) is the only true systemic because it is absorbed by leaves and translocated through the plant to small sucker buds. Rate of Maleic hydrazide (MH) application is 100 ppm.

Chemical rabbing : Sterilization is done by applying fungicide, pesticides and herbicides. Drench the seed bed with 0.2% Blitox or Bordeaux mixture before or after sowing of seed bed to protect germinating seed against soil borne fungi. Application of 40 g of 4% endosulfan dust per 10m^2 bed or drench with chlorphuriphos 20%EC @ 2 ml / liter of water 24 hours before seed sowing.

Choice : Deciding on one action over another or one good in preference to another, according to certain criteria.

Chronic toxicity refers to the effect of an herbicide. The toxicity occurs at slow rate over a period of time due to herbicides applied to a plant.

Cigarette Beetle (*Lasioderma sericomi*) : Commonly known as the cigarette weevil, is a serious pest of stored tobacco. Cigarette beetle infests the cured leaf in storage. The larvae of the beetles feed on tobacco leaves making large galleries and when infestation is heavy, turn the tobacco into powder.

Clone is a plant produced vegetatively by rooting a certain part of a single plant. The plant, thus produced is identical to the parent plant in all the characteristics.

Coleoptile is indicative of the first leaf immediately above the cotyledon. It encloses the tip of the stem and other leaves, that is, the sheath.

Collective farm : Typically, a farm in the USSR, owned by the government, but operated by a number of families, which share in the farm-derived revenue.

Common market : A customs union that also provides for the free mobility of factors of production between the member nations.

Common property resources : Resources that are owned by the government in the name of the public.

Comparative advantage : A situation in which an individual, region or nation is relatively superior at producing some goods because of its lower opportunity costs, and gains by trading for goods that another is relatively more proficient at producing.

Complements Using two goods in combination because the presence of one good enhances the benefit obtained from the other.

Conglomerate merger : The merging of two or more firms that operate in unrelated industries.

Conjunctive use of water represents the planned use of surface and sub-surface water to augment canal supply, combat water logging and also to facilitate irrigation with poor quality ground water. The conjunctive use of surface and ground waters has been practiced in India to a limited extent.

Conservation tillage means any tillage sequence which reduces the loss of soil and water relative to conventional tillage including minimum tillage, mulch tillage, no-tillage, plough-planting, sod-planting, sub-surface tillage and *wheat* track planting.

Conservation : Preserving or extending the productive life of a resource.

Constant returns : A constant input-output ratio.

Consumer Price Index : The average price of a market basket of goods and services in index form to permit comparisons of the prices of goods and services over a period of years.

Consumption component : The utility or satisfaction an individual derives from an education.

Consumptive use of water is the total amount of water used by the vegetation of a given area in transpiration, building of plant tissues and that which evaporates from the adjacent soil, snow or intercepted precipitation on the area, in a specified time. It is generally expressed in inches or centimeters with respect to time. Evapotranspiration has been often used synonymously with CUW, although strictly speaking CUW also includes the water used for building the plant tissues. Quantitatively, there is only a slight difference and hence both are used interchangeably.

Contact herbicide is an herbicide which causes the death of a plant or its part only after coming in contact with the plant or the plant parts covered by the applied herbicide.

Contract production : Contractual agreements between producers and their input suppliers or product marketing firms.

Cooperative : An association of member-owners operating a business that provides services at cost to its patrons.

Cost-benefit ratio *(CBR)* equals the per rupee investment which is derived by dividing the gross returns by the total cost of cultivation.

Cotyledon means the first leaves arising from the embryo. Their number varies from one in monocotyledonous plants to two or more in dicotyledonous plants.

Critical period (stage) denotes the period during which a crop is affected severely due to soil moisture or nutrient(s) stress and the loss cannot be made good by adequate supply in other periods or stages.

Critical stages of tobacco for irrigation : Grand growth initiation stage i.e. 25-30 days after planting and development stage i.e. 55-60 days after planting (after topping and piercing) were found to be most critical stages for irrigation.

Crop coefficient : Ratio between crop evapotranspiration to pan evaporation when crop is raised in large field under optimum growing conditions.

Crop geometry : Crop geometry defines the pattern of distribution of plants over the ground or the shape of the area available to the individual plant.

Crop water-use efficiency : The ratio of crop yield to the amount of water depleted by the crop plants through evapotranspiration.

Cropping system management : Cropping system management is basically the design of spatial and temporal combinations of crops in an area.

Cropping system : Cropping system is the cropping pattern used on a farm and their interactions with farm resources, other farm enterprises and available technology that determine their makeup.

Cross fertilization is the fertilization of an ovule or a number of ovules of one flower with the pollen from another flower, or it may be defined as the fertilization of the flower of one plant by the pollen from another plant.

Cross-pollinate : To apply pollen of one flower to the stigma of another, commonly refers as pollinating the flowers of one plant by pollen from another plant.

Cultivar : It is equivalent to 'variety' and is defined as an assemblage of cultivated plants which is clearly distinguishable by any character (morphological, physiological, cytological, chemical or others) and when reproduced (sexually or asexually) retains its distinguishing characters. The term is derived from cultivated variety.

Curing of tobacco : It is process of drying decomposition of chlorophyll until the green colour disappears hydrolysis of starch in to sugar and respiration of fermentation of sugar harvested leaves.

Cusec : The quantity of water flowing at the rate of 1 cubic foot per second. One cubic foot of water weights 62.4 lbs or 6.24 gallons.

Cytokinins : A term used for substances which promote cell-division and exert other growth regulatory functions, e.g. zeatin (naturally occurring) and 6-benzyl amino-purine (synthetic).

Cytoplasmic male sterility : A type of male sterility conditioned by the cytoplasm rather than by nuclear genes and transmitted only through the female parent.

Damping off (*Phytophthora palmivora*) : Damping off is a common disease occurring in tobacco nurseries, the infection may spread rapidly, and damp off the entire bed of seedlings within three or four days. It causes extensive damage in the tobacco tracts.

Day cusec : The quantity of water in cusecs flowing for 24 hours.

Delta (water) : The total depth of water required by a crop.

Deoxyribose nucleic acid (DNA) : The nucleic acid of the chromosomes, the main carrier of genetic information.

Deoxyribose : A 5-carbon sugar which is present in deoxyribose nucleic acid.

Depth of irrigation : Water depth required to bring soil water content of root zone to field capacity.

De-suckering : The process of removal these suckers resultant of topping is known as de suckering. To reap maximum benefit of topping, suckering should be avoided and suckers are removed periodically. The process of removal these suckers commence after 6-7 days of topping operations and it will continue for 2-3 times for effective result.

Dihybrid : Hybrid that is result of crossing of homozygous parents that differ with respect to 2 loci.

Diploid : A plant with 2 sets of chromosomes.

Disposable personal income The after-tax income of individuals available for spending on goods and services.

Distribution efficiency : Ratio of water made directly available to the crop and that released at the inlet of a block of fields.

Diversion canal : A canal to divert water from one point to another. In irrigation practice, it extends from the point of diversion at the main canal to the beginning of the distribution system.

Double cross : A hybrid between 2 single crosses involving 4 different inbred lines.

DP 401 : First improved variety of chewing tobacco developed through selection by CTRI, RS, Pusa during 1961. The typical genotypic characters are its long broad, sessile and elliptical leaves. It gives a total cured leaf yield of 1670 kg and first grade leaf yield of 1050 kg per ha.

Drainage requirement : The amount of water that drain must convey in a given time for satisfactory crop growth.

Drainage water : Also called gravitational water, free water and excess water. The water that the soil is unable to hold against the force of gravity, Also water from surface, ground or storm water flowing into a drain.

Drainage : The removal of excess surface or ground water from land by means of surface or subsurface drains.

Duty (water) : It is the area irrigated by 1 cusec discharge of water during the crop period; it is equal to twice the base divided by delta.

Economic profit : The amount by which a firm's revenue exceeds its total (explicit plus implicit) costs. Also referred to as *pure profit*.

Ecotype : Plant type or strain within a species, resulting from exposure to a particular environment.

Effective irrigation : A controlled and uniform application of water to crop land in required amount at required time, with minimum cost to produce optimum yields without waste of water and adverse effect on soil.

Entrepreneurship : Organizing resources to produce and market goods and services.

Equilibrium price : The price in a market at which quantity supplied and quantity demanded are equal.

Equilibrium : A condition in which opposing forces within a system just offset one another.

European Community (EC-101) : A group of countries which have reduced or abolished tariffs among themselves and established a common and uniform tariff to outsiders. The EC includes 10 countries : West Germany, France, Italy, Belgium, the Netherlands, Luxembourg, Great Britain, Denmark, Ireland, and Greece.

European Free Trade Association (EFTA) : A group of countries which have reduced or abolished tariffs among themselves, but have not established a common tariff to outsiders. The EFTA includes : Sweden, Norway, Switzerland, Austria, Portugal, Iceland, and Finland.

Evolution : The development of a race, species or other group. Exotic (plant) : A newly introduced plant not native to a place. Ft : The first filial generation of a cross between 2 individuals.

Exchange : Function includes all the activities throughout the market channels as buyers and sellers make their transactions.

Expansion path : A line connecting the least cost combination points along the production surface. Also, a line connecting the most profitable combination points for each of a number of production possibilities curves.

Expansionary fiscal policy : Increasing government expenditures or reducing taxes to cause economic activity to expand and increase the nation's GNP.

Expansionary monetary policy : Government action to increase the money supply and lower the interest rate so as to expand economic activity and increase the nation's GNP.

Expenditure method : Measuring GNP by totaling the value of all final goods produced in the economy during an accounting period.

Explicit cost : An expenditure made for the use of a resource.

Exports : Products and services sold to foreign countries.

External growth : Expanding the size of a firm by adding another firm through its purchase or other means of merger.

Externalities : Costs or benefits that are external to the decision maker and are imposed on others.

Facilitating functions : Those activities throughout the marketing system that improve its operational and pricing efficiency.

Facilitative organizations : Institutions that provide information or facilities for marketing commodities.

Factor markets : Markets in which resources are bought and sold. One loop in a circular flow diagram that measures the earnings of all resources used in producing a nation's goods and services.

Factor-factor relationship : The relationship between two factors or inputs used *in* production.

Factor-product relationship : The functional relationship between a factor of production and its product.

Family farm : A farm in which the family provides most of the labor required in its operation.

Farm (1974 census of agriculture) : Any establishment which, during the census year, had or normally would have had agricultural product sales of $1000 or more.

Fertilizer grade refers to the guaranteed minimum analysis expressed in percentage of the major plant nutrient elements contained by a fertilizer material or by a mixed fertilizer generally in terms of N.

Fertilizer requirement represents the quantity of certain plant nutrient elements required, over and above the amount supplied by the soil to augment plant growth to a predetermined optimum level.

Field irrigation efficiency is the ratio expressed in percentage of irrigation water available for use of a crop to that delivered to a field. It excludes ditch losses of water.

Final product value : The market value of a final product.

Final product : The finished product ready for sale to the ultimate user or consumer.

Firm : A decision-making business entity that uses resources hired from households to produce goods and services for sale to households or other consuming units.

Fiscal policy : The use of government policy to achieve specific economic goals by manipulating expenditures or the tax rate.

Fixed costs : Those costs incurred for resources which do not change as output is increased or decreased.

Flow resources : Resources whose available quantities are constantly being replenished. For example, using the wind's power or the heat energy from the sun, does not reduce the supply of these resources to others.

Form utility : The utility or satisfaction generated by the processing function in marketing as it delivers products in the desired forms.

Free market : Amarket system in which buyers and sellers decide and act at their own initiative and in their own economic interests.

Free trade area : Group of nations that abolishes trade restrictions among themselves without also imposing common tariffs on other nations.

Full employment : A rate of unemployment no greater than that due to the normal functioning of the labor market.

Fund or stock resources : Resources whose quantities are fixed in their natural state. Coal, oil, and iron ore are examples of these nonrenewable resources.

Futures contracts : Standardized contracts for future delivery of commodities that are traded on organized exchanges.

Gamete : A sex cell which, after union with another develops into a new plant.

Gap filling of tobacco seedling : Gap filling operation is done to maintain optimum plant population by replacing dead plants those are unable to bear transplanting shocked. Gap filling should be done after 8-10 days of transplanting.

Genetics : The science of heredity including the study of its chemical foundation, its development, expression and its bearing on variation, selection, adaptation, evolution, breeding and the activities of man.

Genotype indicates the hereditary constitution of an individual (plant/animal) which dictates the characteristics of that individual. Its expression is influenced by the environmental conditions. The characteristics may be the plant type, shape as well as arrangement of leaves, colour of the flower and seed in plants, and skin colour, body make up, colour of eye balls in animals.

Genus : A group of closely related species. In a scientific name the genus is the first of the 2 names given for an organism.

Germination means the resumption of growth by the embryo and the development of the seedlings from the seed placed in conditions necessary for the resumption of growth.

Germplasm is the hereditary material or the genetic traits which can be incorporated into new cultivars.

Glabrous means having a smooth surface *i.e.,* without hairs. *Glumes* are the pairs of the bracts located at the base of a spikelet in the members of the Gramineae or grass family.

GNP Price Deflator Index : An index used to remove the effects of inflation or deflation to determine changes in real GNP over time.

Green tobacco sickness (GTS) : Workers engaged in tobacco cultivation suffers from occupational illness known as green tobacco sickness. Green tobacco sickness (GTS) is a result of dermal exposure to dissolved nicotine from wet tobacco leaves. Worker engaged in various processes gets cuts and abrasions on their palms and the skin around their nails get peeled off.

Gross national product (GNP) : The total market value of all final goods and services produced in an economy during a given period of time.

Growth curve : A graph in which the size of some plant characteristics (height, weight, leaf area etc.) is plotted against time or age. The rates of growth of any single cell or mass of cells are sigmoid (S-shaped); slow at first, increase with time to a maximum and then all off at about the same rate to zero. The period of maximum growth is called' grand period' of growth.

Growth-regulator (plant) : Organic substances which in minute amounts may participate in the control of growth processes. In plant 4 types have been recognized to date : auxins, gibberellins, kinins and inhibitors.

Growth-retardant : A chemical (such as CCC) that selectively interferes with normal hormonal promotion of growth but without appreciable toxic effects.

Guard cropping : Means growing main crop in the centre surrounded by hardy and thorny crop, such as safflower around pea, mesta around sugarcane and sorghum around maize.

Habitat refers to the sum total of the conditions of environment, apart from the organism in consideration or to the sum total of all the external conditions effecting the development, special response and growth of plants.

Haploid : Having 1 set of chromosomes, as in gametes.

Hard pan is a hard and impermeable layer formed in the soil profile due to the accumulation of clay or salts like calcium carbonate which move down into the profile in a dissolved form in rain or irrigation water.

Hectare : A metric measure of land area (one hectare equals 2.4710 acres).

Hedging :establishing opposite positions in the futures and cash markets as a protection against adverse price changes.

Heterogeneous : Units within a group have dissimilar characteristics.

Heterozygous : Not true-breeding for a specific hereditary character. Homogeneity test : A test to measure variation within a seed lot.

Homogeneous : All units within a group are alike in their characteristics.

Homozygous : Having identical genes at the same locus in each member of a pair of homologous chromosomes.

Hookah tobacco (*Nicotiana rustica*) : Bihar is one of the major producers of hookah tobacco and grown mainly in Purnia Katihar and Saharsa districts. Around 10 per cent of total area of tobacco is under this category. The average yield is very poor and is being only 900-1100 kg/ha.

Horizontal merger : A combination of two or more firms operating in the same industry.

Hormone : An organic substance produced naturally in higher plants, controlling growth or other physiological functions at a site remote from its place of production and active in minute amounts.

Household responsibility system (HRS) : China's restructured system for production and marketing decision making by individual households rather than at the commune (production team) level.

Household : The basic economic entity in society that consumes the final goods and services produced by the system.

Human capital : The educational investment that improves the knowledge and productivity of people.

Hybrid vigour : The enhanced vigor of progeny resulting from the crossing of 2 different inbred plants.

Hybrid : The fist generation progeny or offspring from a cross of different varieties or strains or inbred lines.

Hydathode : A pore generally at the tip of a vein, capable of exuding liquid water.

Hydration : The association of one or more molecules of water with an ion, molecule or micelle.

Hydroponics : The growing of plants in water solutions of essential nutrients.

Hydrotropism : The growth movement in response to unequal distribution of water, as when roots bend towards moist soil.

Hypocotyl : The part of the embryo or seedling below the cotyledon node and above the root or the transition region connecting the stem and root.

Ideotype : Refers to plant type in which morphological and physiological characteristics are ideally suited to achieve high production potential and yield reliability.

Imperfect competition : Any market structure in which firms do not exhibit the characteristics of perfect competition.

Imperfect substitutes : Resources whose marginal rate of substitution changes as their proportions are changed, ceteris paribus.

Implicit cost : A cost for which there is no cash outlay at the time a resource is being used, or for which no cash payment is required.

Import quota : A maximum limit on the quantity of a commodity that can be imported.

Imports Goods and services purchased from foreign countries.

Improved seed : The genetically and physically pure seed of an improved crop variety. Its different categories are nucleus seed, breeder's stock seed, foundation seed, registered seed and certified seed.

Improved varieties of chewing tobacco for Bihar : Since inception of Hookah and Chewing Tobacco Research Station in 1950 seven high yielding better quality varieties of chewing tobacco (*Nicotiana tabacum*) viz. DP 401 (1961), Gandak Bahar (1976), Sona (1977), Prabha (1981), PT-76 (1990) Vaishali Special (1993) and Lichichavi (2001) has developed and released by this Research Station. In case of hookah tobacco (*Nicotiana rustica*) three superior cultivars of viz. R-12, NP-220 and NP-22 have also been isolated and recommended for general cultivation in Bihar and adjoining areas.

Inbred : The progeny of either a single cross-pollination plant obtained by selfing or two closely related plants obtained by inbreeding.

Income effect : The change in a consumer's real income that occurs as the price of good changes.

Income elasticity : The responsiveness of quantity purchased to a one percent change in income, ceteris paribus.

Income method : Measuring GNP by totaling all labor and other resource earnings throughout the economy during an accounting period.

Independent variable : A variable whose changes cause the quantity of another variable to be changed.

Indifference curve : Shows all the combinations of goods that yield an individual the same amount of satisfaction or utility.

Indifference map : The utility surface shown by a family of indifference curves.

Induced mutations : The mutations artificially produced with the help of mutagens.

Inheritance : The reception or acquisitions of characters or qualities by transmission from parent to offspring.

Insecticide is a chemical material applied as a spray or dust to kill harmful insects.

Insurance hedging (or hedging) : Establishing an equal but opposite position in the futures market from the one taken in the cash market.

Intensive cropping : Intensive cropping is the process of growing a number of crops on the same piece of land during the given period of time. In other words, when the area is limited and the number of crops to be grown is increased within a definite period of time, this cropping method is termed as intensive cropping.

Intercropping : This is a process of growing subsidiary crops between two widely spaced rows of main crop. The main object of this type of cropping is to utilize the space left between two rows of main crop and to produce more grain per unit area. Parallel cropping, companion cropping, multistoried cropping and synergistic cropping are major kind of intercropping system.

Interdependent demand curves : A market in which each firm's demand curve is influenced by the pricing decisions of other firms in that market.

Interest rate : The price of borrowed money.

Interest : A charge made for borrowed money. The rate at which we discount future economic goods.

Internal growth : Expanding a firm's capital and facilities as the market for that firm's product grows.

International trade : Tansactions with foreign nations in which goods and services are bought or sold.

Internode's is the portion of the stem which is located between two nodes

Irrigation efficiency represents the ratio, expressed in percentage, of water stored in the root zone depth of the soil to the water delivered in the field from the farm supply source.

Irrigation requirement of tobacco : In, general 2-4 irrigations at 30 days interval starting from grand growth initiation stage were found to be sufficient. At pre flowering stage, little deep irrigation may be sufficient to obtain optimum yields of high quality.

Irrigation requirement refers to the amount of water exclusive of precipitation and the ground water contribution required for crop production. It represents the net irrigation requirement plus other economically unavoidable water losses. It is usually expressed as the depth of water for a given time.

Isolation distance is the recommended distance to be maintained between the seed crop and the likely source of contamination.

Kolkhozes : Collective farms in the USSR.

Labor : The physical effort of humans that is combined with other resources to produce valued goods and services.

Land : All the productive attributes of the earth's surface, including space and the natural environment.

Law of comparative advantage : The principle that each person, area, or nation will gain economically if each specializes in producing those goods and services where they have the greatest relative advantage, then trades with other similarly specialized individuals, areas, or nations.

Law of diminishing marginal utility : As an individual consumes additional units of a specific good, holding everything else constant, and the amount of satisfaction from each additional unit of those good decreases.

Law of diminishing returns : As successive amounts of a variable input are combined with a fixed input, the total product will increase, reach a maximum, and eventually decline.

Least cost combination : A combination of two or more resources in such a way that the resource cost of producing a given level of output is a minimum.

Length-of-run : A planning concept relating to the proportion of a firm's inputs that may be fixed or varied in a given time period.

Liabilities : The total claims of creditors against the value of assets owned by a business.

Lichchivi : New chewing tobacco variety Lichchivi has been released in 2001for cultivation in all chewing tobacco growing areas of Bihar and U.P. Its specific feature is its thick bodied cured leaf. Its total average yield is 28 q/ ha with 15 q/ha first grade leaf outturn.

Local variety : A mixture of different types and is well adapted to local environments. It is endemic to an area with its origin dating back to several hundred years.

Long position : Purchases of futures contracts not offset by sales.

Long run : A time period long enough that all factors of production can be varied.

Macroeconomic equilibrium : A situation where aggregate supply equals aggregate demand and there are no forces in the economy acting either to increase or decrease the nation's real GNP.

Macroeconomics The area of economics that deals with output, employment, incomes, or other activities in the aggregate.

Management : The process of controlling or directing a situation. Also one of the four factors of production, with responsibility for decision making.

Marginal cost (MC) : The change in total cost when output is changed by one unit.

Marginal factor cost (MFC) : The amount added to total cost when one more unit of the variable input is used in production.

Marginal physical product (MPP) : The amount added to total physical product when another unit of the variable input is used in production.

Marginal rate of product substitution (MRPS) : The rate at which one product substitutes for another, measured along the production possibilities curve.

Marginal rate of substitution (MRS) : The rate at which one good (or resource) can be substituted for another at the margin without changing satisfaction (or output).

Marginal revenue : The amount added to total revenue when an additional unit of output is produced and sold.

Marginal value product (MVP) : The amount added to total value product when another unit of the variable input is used.

Market demand curve : A horizontal summation of individual demand curves for a product.

Market equilibrium : That point where the market demand curve intersects the market supply curve. The market clearing price where quantity demanded equals quantity supplied.

Market orders and agreements : Government authorized programs that producers can use to determine the terms and conditions under which a commodity can be marketed.

Market power : A firm's ability to influence product prices either on the buying or selling side of the market.

Market supply curve :A horizontal summation of individual supply curves for a product.

Market : Consists of buyers and sellers with facilities to communicate with each other.

Marketing agreement : A voluntary agreement among handlers of a commodity with respect to the quality, quantity, and price of that good.

Marketing efficiency : A comparison of the value of output to the value of inputs used in the marketing process.

Marketing margin : The difference between the price that consumers pay for the final product and the price received by producers for the raw product.

Marketing order : A price agreement that, once issued, is binding on all handlers of the commodity covered by that order.

Marketing : Business activities that direct the flow of goods and services from producer to consumers.

Merchant middlemen : Purchasers of goods from wholesalers for later sale to buyers.

Merger : Combining two or more formerly independent firms under a single management entity.

Metric ton : A metric measure of weight (one metric ton equals 2204.622 pounds).

Microeconomics : The area of economics that deals with individual decision units people, firms, or markets-within the economy.

Middleman : Functions performed in moving goods and services from producer to consumer.

Mixed cropping : Mixed cropping is the process of growing two or more crops together in the same piece of land. This system of cropping is generally practiced in areas where climatic hazards such as flood, drought, frost etc., are frequent and common. *e.g.*, wheat + gram or wheat+bar1ey or wheat+ mustard or they may mature at different times, *e.g.*, arhar + jowar + mung and til or groundnut+ bajra etc. Mixed cropping may be classified into the mixed crops, companion crops, guard crops and augmenting crops based on their method of sowing.

Monetary policy : Manipulation of the money supply and the interest rate to achieve specific economic goals.

Monopoly : Single seller in an industry.

Monopsony : Single buyer in an industry.

Mutation : Sudden, heritable variation in an animal resulting from an abrupt change in the genotype nature.

National income accounts : Dollar-term measures of the nation's economic performance.

National income : Net national income minus indirect taxes, after accounting for depreciation; it is thus labor, management, and proprietary earnings plus net interest, net rent, and corporate profits.

Natural resource : A factor of production provided by nature. Examples are land, water, minerals, wind, tides, and so on.

Natural selection : An important feature of certain theories of evolution, according to which agents other than man determine which members of a population, will survive.

Net national product : The net value of all goods and services produced in the economy after deducting depreciation from GNP, to account for resources used up during the accounting period.

Net worth : The excess of assets over liabilities representing the owner's residual claim to assets.

Nick : In hybrid seed production, the parents are said to nick when they produce higher yield as a result of synchronized flowering.

Nicotine : $C_{10}H_{14}N_2$. Alkaloid commonly found in tobacco leaves and stem, commercially used as insecticide as nicotine sulphate.

Nicotinic Acid : $C_5H_5N_2$, obtained by oxidation of nicotine

Nonrecourse loan : A loan in which the lender, such as the CCC, has no recourse beyond the physical commodity in payment for that loan.

Normal profit : A situation in which each of the firm's resources is earning a return just equal to its opportunity cost.

Nornicotine : Nornicotine is a secondary tobacco alkaloid that is produced by the N-demethylation of nicotine. Nornicotine production and accumulation in tobacco are undesirable because nornicotine serves as the precursor in the synthesis of the well characterized carcinogen N'-nitrosonornicotine during the curing and processing of tobacco. Although nornicotine is typically a minor alkaloid in tobacco plants, in many tobacco populations a high percentage of individuals can be found that convert a substantial proportion of the nicotine to nornicotine during leaf senescence and curing.

Nucleic acid : A long molecule formed by lining many nucleotide molecules. The 2 kinds occurring in cells are DNA and RNA.

Nucleus seed : The original seed produced for the first time by the plant breeder. It is 100% pure in all genetically and physical quantities and used as a parent for the multiplication of breeder's stock seed.

Nursery bed sterilizations : Sterilization of nursery beds take care of weeds, seeds of other undesired crops, seed borne diseases and pests, thus it provide protection to young tiny and tender tobacco seedlings. This can be done by rabbing and chemical treatments.

Off-type : Plants or seeds deviating significantly from the characteristics of a variety as described by the breeder in any observable respect.

Oligopolistic market : A market characterized by the small number of firms producing for that market.

Oligopoly : A firm in an oligopolistic market.

Open market operations : The purchase and sale in the open market of government securities by the Open Market Committee.

Opportunity cost : The value of other opportunities given up in order to produce or consume any good.

Options : Contracts that give the holder the right to buy or sell a property.

Out cross : The mating of a hybrid with a third parent, also an off type plant resulting from pollen of a different sort contaminating a seed field.

Papery tobacco leaves : Early planting results in production of papery leaves accompanied with on severe leaf spot infestation, while delay in planting results in production of small, leathery and brittle leaves.

Parity : That price which gives a unit of agricultural commodity the same purchasing power as it had in a specified base period.

Peat soil is an organic soil containing more than 50 per cent organic matter. The word peat means an unconsolidated soil material consisting largely of undecomposed or only slightly decomposed organic matter accumulated under conditions of excess moisture. Highly decomposed organic material is referred to as 'muck'.

Perfect complements : Resources that must be used in a given ratio in order to produce a product.

Perfect substitutes : Resources that substitute for one another in production at a constant rate.

Permanent wilting point is the moisture content expressed in percentage of oven dried soil at which nearly all plants wilt and do not recover in a humid chamber unless water is added from an external source. This is the lower limit of the available soil moisture.

Permeability denotes the property of a porous medium to transmit the fluids. It is a broad term and can be further specified as hydraulic conductivity and intrinsic permeability. Hydraulic conductivity is the proportionality factor 'K' in Darcy's law and represents the rate of flow of water in a porous medium under a unit hydraulic gradient. It changes with the quality and quantity of water, but the values of the intrinsic : permeability, the property of the porous medium, remains constant for any fluid.

Personal income : The total income of people before the payment of income taxes.

pH is the negative logarithm of hydrogen ion activity.

Phenology : The study of the timing of recurring biological events, the causes of their timing with regard to biotic and abiotic factors and the interrelation among the phases of the same or different species.

Phenotype represents an organism bearing a physical or external resemblance as contrasted with its genetic composition. That is a group of organisms manifesting identical outward or physical traits.

Photoperiodism means the response of plants to a photoperiod or the length of exposure to light.

Physical function : The processing, storage, and transportation of commodities in the marketing system.

Physical supply : The total available quantity of a good or resource.

Place utility : The utility or satisfaction derived from the transportation function in marketing that delivers products in the place desired by consumers.

Plant breeding : The applied branch of botany dealing with the crop improvement and production of new improved crop varieties far-better than original in all aspects.

Plant population : Plant population defines the number of plants per unit area or the size of the area Available to the individual plants.

Plough pan represents the sub-surface soil layer having a higher bulk density and lower total porosity than the soil layers above as well as below this layer, as a result of the pressure applied by normal ploughing or other tillage operations.

Positive economics : A scientifically testable conclusion of what *is or what can be.*

Potentially arable land : land that can be brought into production after undergoing sufficient physical development.

Poverty level : Official designation of an income below that which is needed to provide the kind of living our society considers to be a basic right.

Present value : The present value of money or an economic good to be received sometime in the future.

Price ceiling : A government decreed maximum price for a commodity or service.

Price elasticity of demand : The percentage change in quantity demanded due to a one percent change in price, ceteris omnibus.

Price elasticity of supply : The percentage change in quantity supplied due to a one percent change in price, ceteris omnibus.

Price floor :A government decreed minimum price for a commodity.

Price : The rupees value per unit of a resource, good or service as determined in the market or by other means.

Puckering : Maturity of hookah tobacco is indicated by formation of yellowish brown spots commonly known as puckring

Pusa toabcoo-76 : New high yielding variety of chewing tobacco was released in 1990 s for general cultivation in Bihar. It gives 2633 kg total cured leaf and 1728 kg first grade leaf yield per ha. The main characteristic feature of this variety is semi open, long and broad light green leaves with pointed tips.

Quality of water : Tobacco quality is deteriorated due presence of chlorine in any input. Water containing up 100 ppm is permissible for irrigation. Water available for irrigation in most tobacco growing districts of North Bihar is of good quality with very low chloride content i.e. below 30 ppm.

Rabbing : Rabbing is process of burning the trashes, weeds or any other available organic refuse on the surface of nursery beds to sterilize the tobacco nursery beds. For rabbing 15-20 cm thick layer is uniformly spread over the soil surface and then it 'is burnt. Its also improve physical properties of soils.

Rhizome : Underground stem, usually horizontal and capable of producing new shoots and roots at the nodes.

Roguing : Removal of weeds, off-type and diseased plants from a standing field crops is known as roguing.

Root knot (*Meloidogyne javanica*) : This is a common disease of tobacco characterized by knotting of roots, stunting and chlorosis of the plant and reduction in crop yield, as is common for similar root-knot diseases on other crops. It is caused by the root-knot nematode *Meloidogyne incognita* and by M. Chitwood in some places.

Rural development : All the developmental activities that result in making rural America a better place in which to live and work.

Sand leaves : Inferior quality bottom leaves are termed as sand leaves, they are very trashy and thus it is uneconomical to use them hence 2 or 3 leaves from bottom must be removed once when the plants are 30 cm tall and again when plants are 60 cm tall to maintain quality of tobacco leaves .

Scarcity : A basic economic condition in which our wants exceed the resources available to satisfy those wants.

Seed certification : A means to maintain and make available to the public, sources of high quality seeds and propagating materials of superior varieties so grown and distributed as to insure genetic identity. This is done by means of inspections of fields and seeds and by regulations for checking on the production, harvesting and cleaning of each lot of seed.

Seed control : Seed law enforcement which applies to all seeds sold commercially. Correct labeling with respect to the kind of seed, germination, purity and other quality factors are ensured.

Seed priming : Soaking of seeds in drying in water or any other chemical to improve germination is known as priming. In case of tobacco freshly harvested seed of tobacco may include a high proportion of seeds, which can only be germinated in the light .The light requirements can be removed by priming the seeds in dark (soaking and drying).

Seed rate of tobacco : A seed rate of 4 g should be sufficient for 10m^2 nursery bed. Prior to sowing the seed should be mixed with fine sand and are sown uniformly by an experienced person and then leveled in order to cover the optimum depth of seeding at 0.25 cm. After seeding and covering compact the seed beds by planking. Higher seed rate should be avoided as it results in over crowding of seedlings which in turn lowers the quality of seedlings and induces diseases like damping off.

Seed : Sexually or vegetative propagated planting materials which are used for seeding and planting and as such should be free from (pests and diseases) any infection and should give a good crop stand by good seeding.

Seeds treatment : tobacco seeds should treated before sowing with 2.5 % formalin solution or 0.25% solution of Dithen Z-78 or Dithen M -45.

Selection : In a population the preservation of certain individuals that have desirable characteristics.

Selective controls : Federal Reserve System controls over interest rates, margins, and installment credit in order to achieve federal policy objectives.

Self-pollination : Transfer of pollen from the stigma of the same flower or another flower on the same plant.

Sequential cropping with tobacco : Growing at least two crops in sequence with tobacco on the same field in a year is commonly known as Sequential cropping.

Short position : Sales of futures contracts not offset by purchases.

Short run : A time period where one or more of the factors of production are fixed.

Short-day plants : The plants that normally develop flower with photoperiod (day-light period) of 8-10 hours, i.e. less than the critical day-length.

Silk (corn) : The stigma and style of the female corn flower through which the pollen-tube grows to reach the embryo sac.

Sink capacity : The capacity of the spikelets or reproductive or economic plant part like earhead, cob, tuber etc. to accept photosyntheses.

Sink : The term usually applied to refer to the plant parts that are involved in utilization and storage of photosyntheses, for example grains, fruits etc.

Slip (propagation) : A shoot or leaf cutting to be rooted for vegetative propagation.

Social costs : The sum total of internal and external costs of an economic activity.

Socialism : A type of economic organization in which the government owns all resources and directs all economic activity.

Soil organic matter represents the organic fraction of the soil that includes plant and animal residues at various stages of decomposition, cells and tissues of the soil organisms, and the substances synthesized by the soil population. Commonly, it is determined as the amount of the organic material contained in a soil sample passed through a 2 millimeter sieve.

Soil productivity refers to the capacity of a soil for producing a specified plant or sequence of plants under a specified system of management. Productivity emphasizes the capacity of a soil to produce crops and should be expressed as yield(s).

Solarization : Extremely high light intensities which have inhibitory effect on photosynthesis.

Sowing time of tobacco : The optimum sowing time in North Bihar condition is 2nd or 3rd week of August for chewing tobacco and September last to October 2nd week is best time for hookah tobacco.

Spacing of tobacco planting : The transplanting of tobacco should be done in line and spacing varies from variety to variety time and place. Under Bihar condition the optimum spacing for a genotype like pusa tobacco-76 was found to be 105 cm X 90 cm and for genotypes like vaishali special it was 90 cm X 75 cm.

Spangles : Maturity of bidi tobacco is judged by pronounced development of brown spots called spangles.

Speculation : The purchase or sale of title to goods or financial obligations in the expectation of favorable price movements.

Speculative middlemen : Own and hold commodities in the expectation of a price change in their favor.

Spillover effect : Benefits or costs that are external to the decision maker and "spill over" to others in the economy.

Stagflation : A situation in which a high rate of inflation and high unemployment occur simultaneously.

State farm : Large farms owned and operated by the government (sovkoz in Soviet Russia).

Substitutes Two different goods (or resources) between which a choice is made to satisfy human wants (or to produce a product).

Substitution effect : The substitution of one good for another as their relative prices change.

Summer deep ploughing : summer deep ploughing (twice) in the month of april-may, found highly beneficial in minimizing weeds and *Orobanche menace*, reducing insect pests and diseases problems and improving water and nutrient conserving capacity of the soil.

Supply curve : The amount of a good or service a producer is willing to offer for sale at different prices, holding everything else constant.

Systems of cropping : System of cropping is the way in which different crops are grown. Sometimes a number of crops are grown together or they are grown separately at short intervals in the same field, *e.g.*, mixed cropping and Intensive cropping.

Tariff : A tax on imported goods.

Technical efficiency : The physical input-output relationships with which marketing functions are performed.

Thermoperiodism refers to the response of plants to rhythmic fluctuation in temperature. The diurnal temperature regimes are called thermoperiodicity. *Three-way hybrid* is a hybrid evolved by crossing a single cross hybrid with an inbred line.

Tillage means the mechanical manipulation of soil mechanically to provide and promote good tilth for crop production. It includes primary tillage which involves lifting, twisting and turning of soil to incorporate crop residues and animal waste, but the secondary tillage involves harrowing to kill weeds and to break the clods.

Tiller is an erect or semi-erect secondary stem which emerges from a basal auxiliary or adventitious bud or an erect shoot that grows from the crown of grass.

Tilth refers to the physical condition of the soil in relation to plant growth and thus includes all the physical conditions which influence the crop growth and development.

Time preference : Our desire for economic goods now rather than in the future.

Time utility : The utility or satisfaction derived from the storage function in marketing that delivers a product at the desired time.

Tobacco Caterpillar (*Spodoptera litura* Fab) : It is serious pest of tobacco in the nursery when the seedlings are small; the caterpillars act as cutworms. They are capable of defoliating the whole nursery in a few days time.

Tobacco fruits : Fruits are know as capsule which are varies in size and shape the nature of apex are blunt or conical the shape may be elliptical ovoid or conical.

Tobacco leaf curls (*Ruga tabaci*) : This virus disease is widespread in India and occurs in severe forms. It is characterized by the downward curling of young leaves, leaf margins turn downwards and come together at the bottom, exposing the middle upper surface of the leaf blade. The inflorescence is greatly condensed. The disease usually appears in the field about 4-6 weeks after transplanting.

Tobacco mosaic virus (*Marmor tabaci*) : First recognized virus disease of plants, it is most common on tobacco crops in India, appearing in every tobacco tract of the country. The first-symptoms are light discolouration along the veins of the youngest leaves. Soon the leaves develop a characteristic light and dark green pattern, the dark green areas usually associated with the veins. The virus is easily transmitted in the field by mechanical means and by wind and water. Various cultural operations cause wounds on the plant, which become the portals of entry of the virus.

Tobacco seeds : Seeds shapes are spherical or broadly elliptical with variation in degree of ridging on the seed coat. The colour of seed varies from light brown to dark brown.

Tobacco soils : Well drained, neutral to slightly alkaline soil with good nutrient supplying capacity is considered ideal for chewing and hookah tobacco cultivation. The optimum pH ranges from 5 to 6.5. Quality of the leaves is affected by soil salinity. The crop is sensitive to water logging and demands well-aerated and drained soils.

Tobacco transplanting : Optimum time of planting for existing/ ruling varieties of chewing tobacco is from 25th September to 10th October.

Tobacco : Tobacco is herbaceous annual, having 71 type of tobacco in two species (20 types in *rustica* and 51 types in *tabacum). Nicotiana rustica* is some time behaves like perennial. The plant of *N. rustica* is more robust than *N. tabacum.* Plant has great plasticity and its height range in between 90 cm to 240 cm in *tabacum* and 50 cm to 140 cm in case of *rustica.*

Top cross : A cross between an inbread line and an open pollinated strain of any crop , it is very common in maize breeding programme.

Topping : Topping is the process of removal of flower heads either or alone with some upper leaves from plants. Topping increase the leaf size, thickness and dry weight of leaves the main objectives of topping are to improve the size, body texture and quality of leaves of tobacco.

Total cost (TC) The sum total of all the firm's fixed and variable costs at each level of output.

Total fixed cost (TFC) : The implicit cost of fixed resources used to produce a good or service.

Total physical product (TPP) : The total quantity of a product that is produced at each level of variable input use.

Total revenue (TR) : A firm's total value of output, obtained by multiplying the units of product produced by the price of that product.

Total value product (TVP) The total value of output produced at each level of variable input use.

Total variable cost (TVC) : Total of the costs of all the variable resources used in production.

Transfer costs : The costs arising from moving commodities from one market to another.

Turgid : Describing a cell or tissue which is firm and plump because of the internal pressure resulting from the osmotic uptake of water.

Type (plant) : A group of varieties of similar characters or an admixture of seed of indistinguishable varieties but having some characteristics in common like grain type sorghum, yellow maize etc.

Utility : The satisfaction an individual gets from consuming goods and services.

Value added : The contribution to final product value by each stage in the production process.

Value judgment : A conclusion that is based upon the values that one holds.

Value : The worth, in the market or to a person, of a good or service.

Values : Group or personally held principles regarding qualities which are worthwhile or desirable.

Variable costs : Those costs which increase or decrease as varying amounts of the resources are used.

Variance : Mean square deviations of variates from their mean. The square of the standard deviation.

Variety : A subdivision or a group of plants within a species which is characterized by growth, plant fruit, seed or other characters by which it can be differentiated from other seeds of the same kind.

Vegetative stage : Growth stage of a crop plant prior to flowering.

Vernalization : Low temperature treatment of seeds and seedlings, which has the effect of artificially 'aging' the plant and reducing the time between sowing and flowering.

Vertical coordination : The linkage of successive stages in producing or marketing a product in a single decision entity.

Vertical integration : the linkage of firms in different stages of production or marketing under the ownership of a single firm.

Vertical merger : Combining two or more firms in different stages of production or marketing into one firm.

Viability of seeds : It is represented by germination percentage which expresses the number of seedlings which can be produced by a given number of seeds times 100.

Viability : The capability of a plant structure (seed, cutting etc.) to show living properties like germination and growth.

Vitamins : Substances required in minute amount for the normal metabolism of plants and animals.

Water killing : Relatively high water transpiration rates in evergreens during a period when absorption of water can proceed only at a relatively slow rate may lead to a type of cold injury called winter killing.

Water logging denotes the state of soil saturation with free water and exclusion of the air from the soil. The water accumulation on the soil surface may occur due to inadequate surface as well as sub-surface drainage.

Water requirement of tobacco : The total water requirements (ETm) for maximum yield vary with climate and length of growing period from 400 to 600mm.

Weed represents an alien plant introducing itself and growing at a place (field *agronomically*) where it is unwanted. It encompasses all kinds of unwanted plants *viz.* trees, shrubs, broad and narrow leaved plants, sedges., aquatic as well as parasitic plants. In a wheat field, barley is a weed.

Wet year means the year in which the rainfall exceeds twice the normal.

Wettable powder is a solid formulation which when mixed with water forms a suspension of finely divided solid particles dispersed in the liquid.

Wilt : Loss of freshness and drooping of foliage of a plant due to inadequate supply of moisture, excessive transpiration or by a disease which interferes with the utilization of water by the plant.

Wilting point : It is soil-moisture condition at which the release of water .to the plant roots is just barely too small to counterbalance the transpiration losses. On an oven-dry basis it is the moisture content of soil at which plants wilt and fail to recover their turgidity when placed in a dark humid atmosphere. It is also called wilting coefficient or permanent wilting point.

Xanthophylls : A yellow or orange carotinoid pigment associated with chlorophyll in chloroplasts, also present in certain chromoplasts.

Xerophytes : A plant with structural and physiological features which permit it to grow in a dry habitat or under water-stress conditions.

Zinc deficiency : Symptoms appear 2-4 weeks after sowing as blanching of the midrib of the emerging leaf, especially at base. Brown spots appear on the older leaves, which later on coaheses give brown color. Tillering and growth are depressed. In severe deficiency the plant dies. Zinc deficiency is associated with calcareous, alkali. Peat, and volcanic soils and soils that remains wet or waterlogged most of the year. Incidence is more severe where high rates of nitrogen and phosphorus are applied.

Zygote : A fertilized egg formed after the fusion of male and female gametes.

□□□

Suggested Readings

Abedian Iraj, van der Merwe Rowena Wilkins, Nick and Jha, Prabhat (1998). *The Economics of Tobacco Control. Towards an optimal policy mix*, Applied Fiscal Research Centre, University of Cape Town.

Agrawal R.L. 1996. *Seed Technology*. Oxford and IBH Publishing Co., New Delhi.

Allen R.G., Pereira, L.S., Raes, D., Smith, M. 1998. Crop Evapotranspiration. Guidelines for Computing Crop Water Requirements. FAO Irrig. Drain. Pap. 56, FAO, Rome, 300p.

Alltucker Ken. 2008. A New weapon to fight cancer- tobacco plants. *The Arizona Republic*. Aug. 22, 2008.

Amarnath S. 1993. Pusa tobacco -76 A new high yielding chewing tobacco variety for Bihar. *Tob Dev. J.*, **2**:7-9.

Amarnath S. 1990b. Combining ability analysis for stomatal activities in chewing tobacco (*N. tobacum* L). *Tob. Res.*, **16**: 127-28.

Amarnath S. and K.D. Singh. 1994a. "Vaishali special" - A new chewing tobacco variety to replace 'Bondi local' in Bihar. *Tob. Dev.* J., **2:** 4-7.

Amarnath S. and K. D. Singh. 1994b. Genetic improvement in chewing tobacco (*N. tobacum* L) for North Bihar. *Tob. Res.*, **20**: 21-5.

Amarnath S. 1987. Genetic variability in chewing tobacco (*N. tobacum* L). *Madras Agric.* J., **74:** 421-25.

Amarnath S. 1990 a. Genetic divergence in chewing tobacco. *Tob. Res.*, **16**: 25-28.

Amarnath S. and R. Sinha. 1985. Varietal performance of chewing tobacco of varying levels of cropping under rainfed condition of North Bihar. *Tob. Res.*, 11: 1-5.

Amarnath S. 1988a.Performance of recently evolved promising lines of chewing tobacco for North Bihar. *Indian Tob. J.*, **20** : 13-15.

Amarnath S. 1988b. Problems and progress in breeding of chewing tobacco. *Indian Tob. J.*, **20:** 8-10.

Amarnath S. and R. Sinha. 1982. Importance of leaf parameters in chewing tobacco varieties. *Tob. Res.*, **8** : 34-8.

Amarnath S. and N.S. Murty . 1988. Path coefficient analysis in chewing tobacco. *Indian J. Genetics.*, **48** : 393-96.

Amarnath S., K. D. Singh and A. K. Pandey. 2004. Response of advanced breeding line of chewing tobacco to varying fertility levels under rainfed and irrigated condition. *Tob. Res.*, 30 : 22-26.

Amarnath S., K.D. Singh and A. K. Pandey. 1993. Changing scenario of tobacco production in Bihar. *Tob. Dev. J.*, **1** : 3-6.

Amarnath S., N. S. Murthy and K. D. Singh. 1995. Heterosis in chewing tobacco. *Tob. Res.*, **21** : 93-6.

Amarnath S., R. Sinha and S.N. Srivastava. 1988. Induced variation in chewing tobacco. *RAU. Res. J.*, **6:** (1&2): 38-45.

Andino V., E. Hernandez and V. Garcia 1992. The critical period of competition of weed on Virginia tobacco "Speight G.28" variety. *Cultivos Agroindustriales* (Cuba) 65-75.

Anonymous 2006 c. *Agricultural Statistics at a Glance.* Directorate of Economics and Statistics, Ministry of Agriculture, Govt. of India.

Anonymous 2006 d. *Fertiliser Statistics.* 2005-06, The Fertiliser Association of India, New Delhi.

Anonymous . 2004b. *Annual Report*, ICAR- Research Complex for Eastern Region, Patna 2003-04.

Anonymous . 2005a. *Annual Report*, All India Network Research Project Central Tobacco Research Institute, Rajahmundry.

Anonymous . 2005c. *Annual Report*, ICAR- Research Complex for Eastern Region, Patna 2004-05.

Anonymous . 2006a. Annual Report, All India Network Research Project Central Tobacco Research Institute, Rajahmundry.

Anonymous. 1980. Disease of flue cured tobacco. Leaf let No.3. Central Tobacco Research Institute, Rajahmundry.

Anonymous. 1960. *Indian tobacco – A monograph* Indian Central Tobacco Committee in of Food and Agriculture Govt of India madras.

Anonymous. 2007a. *The Tobacco News* January February.

Anonymous. 1953. *Plant Diseases.* The yearbook of Agriculture US Deptt. of Agriculture Washington DC.

Anonymous. 2001. *Annual Report,* Central Tobacco Research Institute, Rajahmundry.

Anonymous. 2002a. *Annual Report,* Central Tobacco Research Institute, Rajahmundry.

Anonymous. 2002b. *Annual Report,* ICAR- Research Complex for Eastern Region, Patna 2001-02.

Anonymous. 2003a. *Annual Report,* Central Tobacco Research Institute, Rajahmundry.

Anonymous. 2003b. *Annual Report* , ICAR- Research Complex for Eastern Region, Patna 2002-03.

Anonymous. 2004a. *Annual Report,* Central Tobacco Research Institute, Rajahmundry.

Anonymous. 2005b. *Annual Report,* Central Tobacco Research Institute, Rajahmundry.

Anonymous. 2006b. *Annual Report,* ICAR- Research Complex for Eastern Region, Patna 2005-06.

Anonymous. 2007b. *Annual Report,* ICAR- Research Complex for Eastern Region, Patna 2006-2007.

Arcury T. A., S.A. Quandt , D.I. Garcia, J.S. Preisser, D. Norton, and A. P. Rao. 2002. Clinic based case control comparison of green tobacco sickness among minority farm workers: Clues for prevention. *Southern Med j.,***95**: 1008-11.

Ashraf M. 1994. Breeding for salinity tolerance in plants. Critical Review of Plant Science 13: 17-42.

Ayars J.E., Phene, C.J., Hutmacher, R.B., Davis, K.R., Schoneman, R.A., Vail, S.S., Mead, R.M., 1999. Subsurface drip irrigation of row crops: a review of 15 years research at the Water Management Research Laboratory. Agric. Water Manag. 42: 1-27.

Bai D. Reedler R. and J. E. Brandle .1995. Identification of two RAPD markers tightly linked with the *Nicotiana debneyi* genes for resistance to black root rot of tobacco. *Theor. Appl. Genet.*, **91**: 1184–1189.

Ballard, T., J. Ehlers, E. Freund, M. Auslander. V. Brandt and W. Halperin. 1995. Green tobacco sickness: occupational nicotine poisoning in tobacco workers, *Arch Environ Health.*, **50**:384-389.

Balsubramaniyan, P. and S. P. Palaniappan. 2003. *Principle and practice of Agronomy.* Agro bios (India) publishing House Jodhpur.

Barkley, Paul W. 1977. Economics: The Way We Choose. New York: Harcourt Brace Jovanovich, 652 pp.

Barry, Peter, John A Hopkin and C. B. Baker. 1978. Financial Management in Agriculture, 3rd ed. Danville, 111.: The Interstate Printers and Publishers, 1983, 491 pp. Barlowe, Raleigh. Land Resource Economics, 3rd ed. Englewood Cliffs, N.J.: Prentice-Hall, 616 pp.

Benowitz N.L. 1992.Pharmacology of smokeless tobacco use: nicotine addiction and nicotine related health consequences, Smoking and tobacco control *monograph No 2 NIH publication No. 93-3461:217-244.*

Bishop C.E. and W.D. 1958. Toussaint. Introduction to Agricultural Economic Analysis. New York: John Wiley and Sons, 258 pp.

Boehlje Michael D., and Vernon Eidman. Farm Management, New York: John Wiley and Sons, 1984.

Boylan B.B., V. Brandt, J. Muehlbauer, *M.* Auslander and R. Finger. 1992. Green tobacco Sickness in tobacco harvesters. *MMWR.* **44** : 237-240.

Bradford, Lawrence A., and Glenn L. Johnson. Farm Management Analysis. New York: John Wiley & Sons, 1953, 438 pp.

Bray E.A. 1993. Molecular responses to water deficit. *Plant Physiology*, 3: 61-66.

Brookes J.E. 1937. Tobacco: Its History Illustrated by The Books, Manuscripts and Drawings and Engravings in the Library of George *Arents Jr* ., **1**: 1507-1615. New York.

Brookes J.E. 1952. *The Mighty Leaf*: Tobacco through the Centuries. Boston: Little, Brown.

Burk L. G. and H. E. Heggestad. 1966. The Genus *Nicotiana*: A Source of Resistance to Diseases of Cultivated Tobacco. *Economic Botany*, **20:** 76-88.

Buse, Rueben C. and Daniel W. Bromley. Applied Economics: 1975. ResoUl"Ce Allocation in Rural America. Ames, Iowa: Iowa State University Press, 623 pp.

Butler L.G. 1995. In Allelopathy : Organism, Processes and application: in K. Inderjit, M.M.

Butler L.G. 1995. *In Allelopathy*: Organism, Processes and application: in K. Inderjit, M.M. Dakshini , F.A. Einhelling (Ed.) ACS Symposium Series, **582:** 158-186.

Calatayud A. R. 1944. Accion de la nicotina sobre los movimientos del intestino delgado aplicacion terapeutica, *Med. J*, **11**: 399-406.

Casavant Kenneth L., and Craig Infanger. 1984. Economics and Agricultural Management, Reston, Va. Reston Publishing, 290 pp.

Castle Emery N., Manning H. Becker and A. Gene Nelson. 1987. Farm Business Management, 3rd ed. New York: Macmillan, 413 pp.

Ceccarelli S., S. Grando & A. Impiglia. 1998. Choice of selection strategy in breeding barley for stress environments. *Euphytica* 103: 307-318.

Chaloupka, Frank J. and Warner, Kenneth E (1999), *The Economics of Smoking*, National Bureau of Economic Research Working Paper 7047, March, available at: http://www.nber.org/papers/w7047.

Chaplin J. F. and T. J. Mann. 1961. Interspecific Hybridization, Gene Transfer and Chromosomal Substitution in *Nicotiana*. North Carolina Agricultural Experiment Station *Technical Bulletin* No. 145.

Chari M.S. 1993. *Quality of tobacco in India and its export*. Proc. *Nat. Symp*. Dec-7-8, 1992. Central Tobacco Research Institute, Rajahmundry.

Clausen R. E., and Goodspeed, T. H. 1925. Interspecific hybridization in *Nicotiana*. A tetraploid glutinosa-Tabacum hybrid, an experimental verification of WINGE'S hypothesis. *Genetics*., **10**: 278-284.

Clayton E. E . 1958. The Genetic and Breeding progress in Tobacco during last 50 years. *Agronomy J*., **50**: 352-356.

Clayton E.E. and Mc Murtrey JE Jr. 1958. Tobacco Disease and their control. USDA, *Farmer Bull*. 2023 (Revised).

Clegg M. T. 1990. Molecular diversity in plant population. In plant population genetics, breeding and genetic resource. (ed. A. H. D. Brown, M. T. Clegg, A. L. Kahler and B. S. Weir), pp 98–115 in Sinauer Associates, Sunderland, MA.

CMDR (Centre for Multidisciplinary Development Research). 2000. Economics of shifting from tobacco cultivation, 1997-2000. CMDR, Dharward, Karnataka. Study sponsored by Research in Tobacco Control, IDRC, Canada.

CMIE (Centre for Monitoring the Indian Economy). 1999. National Income Statistics, November 1999. Economic Intelligence Service, CMIE.

Collins, David J. and Lapsley, Helen M. (1997), *The Economic Impact of Tobacco Smoking in Pacific Islands*, Pacific Tobacco and Health Project, Adventist Development and Relief Agency Australia.

CTRI (Central Tobacco Research Institute). 1999. Status Report on Research Programme on Alternative Crops to Tobacco. CTRI, Indian Council of Agricultural Research, Rajahmundry, Andhra Pradesh.

D Alessandro A., N.L. Benowitz, G. Muzi, M.D. Eisner, S. filiberto, P. Fantozzi, L. Montanari and Abbriti G. 2001. Systemic nicotine exposure in tobacco harvesters. *Arch Environ Health.*, **6**:257-263.

Dahl Dale C. and Jerome w. Hammond. 1977. Market and Price Analysis: The Agricultural Industries. New York: McGraw-Hill, 323 pp.

Das S. K., P. Das and S. K. Roy. 2006. *Handbook of Agriculture* ICAR New Delhi.

Decaux F. 1961. Lutilisation du tabac dans la médecine d'autrefois. *Acta Phytotherapeutica*, **8**: 66-71.

Del Piano L., M. Abet, C. Sorrenlino, F. Acanfora, E. Cozzolino, and F. Acanfora. 2000. Genetic variability in *Nicotiana tabacum* and *Nicotiana species* as revealed by RAPD procedure. *Beit. Tabak. forsch. Int.*, **19:** 1–15.

Denev B. Denva, R. Batchvarova .2007. The bio synthetic origin of germination stimulants for Orobacche ramose (L. in tobacco and Arabidopsis. *Biotechnol. & Biotechnol.* EQ.21/2007/1.

DES (Directorate of Economics and Statistics, Ministry of Agriculture). 2000. Cost of Cultivation of Principal Crops in India.

Dickson S.A. Panacea or Precious Bane. 1954. Tobacco in 16th Century Literature. New York Public Library New York

Doctor P.B., V.N. Gokani, P.K. Kulkarni, J.R. Parikh and H.N. Saiyed. 2004. Determination of nicotine and cotinine in tobacco harvesters urine by solid phase extraction and liquid chromatography. *Journal of Chromatography ..* 802:323-328.

Duft Kenneth D. 1979. Principles of Management in Agribusiness. Reston, Va.: Reston, 470 pp.

Dwivedi S.S.L., R. Singh and A.K. Pandey. 1984. Response of chewing tobacco to nitrogenous fertilizers and organic manures. *Tob. Res.* **10**: 118-22.

Einhellig. F.A. 1995. Mechanism of action of allelochemicals in allelopathy. pp. 96-116.In K. Inderjit, M.M. Dakshini, and F.A. Einhellig (ed.). *Allelopathy*: organisms, Processes, and Applications, American Chemical Society, Washington, D.C.

Epp Donald and John W. Malone Jr. 1981. Introduction to Agricultural Economics. New York: Macmillan, 352 pp.

ERC Statistics International Plc. 1998. India. pp. 1-38, in World Cigarette Report, 1998. Evolution trotz Artbestandigkeit. Zeitschr. indukt. Abstamm. U. Vererb., **8**: 325-332.

Evans J.W. and A. D. Imms. 2005. Outline of agricultural Entomology, Asiatic Publishing House, Delhi, India.

Fan S., Hazell, P., Haque, T. 2000. Targeting public investments by agro-ecological zone to achieve growth and poverty alleviation goals in rural India', Food Policy **25:** 411-428.

FAO 1984. Land, food and people. FAO Economic and social development series 30. Food and Agriculture Organization of the United Nations, Rome.

Feinhandler S.J., Fleming H.C. and Monahan J.M. 1979. Pre-Columbian tobaccos in the Pacific. *Economic Botany.*, **33**: 213-26.

Fleming G. W. A. 1840. Tobacco in dropsy, *Bolshoi Med. & Sci. J.*, **22**: 110-11.

Galindo J.C.G., Macías, F.A., García-Díaz, M.D. and Jorrín, J. 2004. Chemistry of Host-Parasite Interactions. pp125-148.In: Allelopathy: Chemistry and Mode of Action of Allelochemicals. Macías, F.A .Galindo, JCG, Molinillo, JMG, Cutler, HG (Ed.). CRC Press, Boca Ratón, FL.

Garner W.W. 1951.*The production of Tobacco.* McGraw-Hill Book Co., Inc., New York.

Gartner C. F. V., 1849. Versuche and Beobachtungen uber die Bastarderzeugung in Pflanzen Gesner C. Epistolarum medicinalum. *Zurich,* **15**: 77.

Gehlbach S. H., W. A. Williams, L.D. Perry and J.S. Woodall. 1974. Green tobacco sickness: An illness of tobacco harvesters. *JAMA.*, **229**:1880-1883.

Gehlbach S. H., W.A. Williams and L.D. Perry. 1975. Nicotine absorption by workers harvesting green tobacco, *Lancet.*, **1**:478-480.

Gehlbach S.H., W.A. Williams and J.I. Freeman. 1979. Protective clothing as a means of reducing nicotine absorption in tobacco harvesters. *Int. Arch Environ Health.*, **34**: 111-121.

Ghosh S.K. , H. N. Saiyed, V. N. Gokani, and M.U. Thakker . 1986. Occupational health problems amnong workers handling Virginia tobacco. *Int Arch Environ Health.*, **58**:47-52.

Ghosh S.K. , J. R. Paridh, V.N. Gokani, M.N. Rao, S.K .Kashyap and S.K. Chatterjee. 1980. Studies on occupation health problems in agricultural tobacco workers, *J Soc Occup Med.*, **29**:113-117.

Ghosh S.K. , J. R. Paridh, V.N. Gokani, S.K. Kashyap and S.K. Chatterjee. 1979. Studies on occupational health problems during agricultureal operation of Indian tobacco workers: A preliminary survey report. *J Soc Occup Med.*, 21-47.

Ghosh S.K., V.N. Gokani, J.R. Parikh, P.B. Doctor, S.K. Kashyap and B.B. Chatterjee. 1987. Protection against green symptoms from tobacco in Indian harvesters-A preliminary intervention study. *Int. Arch Environ Health.*, **42**:121-124.

Goodspeed T. H. 1954. The genus *Nicotiana. Chronica Botanica,* Weltham, MA.

Goodspeed T. H., 1924. Some chromosome numbers in *Nicotiana. Amer. Nat.,* **58**: 381-382.

Goodwin John W. 1982. Agricultural Economics, 2nd ed. Reston, VA. Reston, 400 pp.

Gopalachari N.C. 1994. *Tobacco.* Indian Council of Agricultural Research, New Delhi. India.

Gopioath D. M. 1968. Genetic Improvement of tobacco. *Ind. J. Genet & Pl. Brd.,* **28 A**: 101-106.

Gopioath D. M., Ramana Rao, V. V., Subrahmanyam, M. and Narayana, C. L. 1967. A study of diallel crosses between varieties of *Nicotiana tabacum* L. for yield components. *Euphytica,* **15**: 171-178.

Graves A.B., van Duijn C.M. and Chandra V. 1991. Alcohol and tobacco consumption as risk factors for Alzheimer's disease: a collaborative re-analysis of case-control studies. *Int J Epidemiol;***20**:(suppl 2): 548-57.

Gupta O.P. 2003. *Weed Management, Principle and Practice* (2nd edt.) Agrobios (India) Publishing House, Jodhpur.

Gwartney James D. and Richard Stroup. 1983. Economics: Private and Public Choice, 3rd ed. New York: Academic Press, 792 pp.

Halcrow Harold G. 1980. Economics of Agriculture. New York: McGraw-Hill, 383 pp.

Halcrow Harold G. Food Policy Jr. 1977. America. New York: McGraw-Hill, 564 pp.

Haloph A. 1989. Data Base of Salt Tolerant Plants of the World, by James A. Aronson, Office of Arid Lands Studies (The University of Arizona, 845 North Park Ave., Tucson, AZ 85719).

Hayes H.K. and R.J. Garber . 1927. *Breeding crop plants.* Mc Graw -Hill Book Comp Inc, New Yark.

Hervey E. A. 1869. Tobacco as a hypnotic in a case of chronic illness, *Am. J. Med. Sci.,* **57**: 99-100.

Heyne Paul and Thomas Johnson. 1976. Toward Economic Understanding. Chicago. Science Research Associates, 816 pp.

Heyne Paul T. 1983. The Economic Way of Thinking, 4th ed. Chicago: Science Research Associates, 459 pp.

Heyne Paul and Thomas Johnson. 1976. Toward Economic Understanding. Chicago. Science Research Associates, 816pp.

Heyne Paul T. 1983. The Economic Way of Thinking, 4th ed. Chicago. Science Research Associates, 459pp.

Hipke M. 1993.Green tobacco sickness, *Southern Med J.*, 6:989-992.

Hooper W. D. 1885. A case of irritation of the ganglion of Remak, or the inhibitory nerve of the heart, curcd by smoking tobacco, *Virgil Med. Monthly.*, **12**: 473-75.

Hunngi G. and Krishna R.K. 1998. *Science of field crop Production.* Oxford and IBH Publishing Co. Pvt. Ltd, New Delhi.

ICAR 1997. *Hand Book of Agriculture.* Indian Council of Agricultural Research, New Delhi.

Ingram James C. 1978. International Economic Problems, 3rd ed. New York: John Wiley and Sons, 174 pp.

IPCC 2001. Climate change 2001. The science of climate change. Contribution of working group I to the second assessment report of the intergovernmental Panel on Climate Change. Adaptation to climate change in the context of sustainable development and equity. Cambridge University Press, Cambridge, U.K.

Jain P.C. and M.C. Bhargava . 2007. *Entomology Novel Approach.* New India Publishing Agency, New Delhi.

Jaisoorya N. M. 1942. A preliminary report on the treatment of angiospastic conditions with ergot and Nicotine, *J. Med. Ass.*, **11**: 3861.

James W.C. , P.S. Teng and F.W. Nutter .1990. Estimated loss of crops from plant pathogens. pp 15-50. *In CRC Hand book of Pest Management*, Vol.1 (Edt.) D. Pimentel, Boca Ratoon, CRC Press, Floida.

Jausion Spillman, Karcher and Jaurent 1940. La reparation 'des dermoepidermites et des ulceralions cutanees torpides par la gelose nicotinee, Bot. Soc. fro derm. syph., **47**: 131-3.

Jha Prabhat and Chaloupka, Frank (eds) (forthcoming), *Tobacco Control Policies in Developing Countries*, Oxford University Press.

Jha Prabhat and Chaloupka, Frank 1999, *Curbing the Epidemic. Governments and the Economics of Tobacco Control*, The World Bank, Washington D.C.

Johnson W. D. 1860. Snake bites, New Orleals Med. & Sci. J., July, also in *J. Maleria Med.*, **2**: 354.

Jones W. D. 1891. The use of tobacco in intestinal obstruction, *Med. News.*, **59**: 629.

Kadam B. S., and B.T. Sajnani. 1953. Survey of hookah and chewing tobacco in Darbhanga district of Bihar. *Indian Tob.*, **3**: 133-40.

Kadlec John E. Fann 1985. Management, Englewood Cliffs, N.J.: Prentice-Hall.

Kasuga M., Q. Liu, S. Miura, K. Yamaguchi-Shinozaki and K. Shinozaki, 1999. Improving plant drought salt and freezing tolerance by gene transfer of a single stress inducible transcription factor. *Nature Biotechnology,* 17: 287-291.

Keller-Stanislawski-Keller B., S. Caspary, P.G. Merz, R. Baonn, M. Wolff and N. Reitbrock. 1994. Transdermal nicotine substitution: Pharmacokinetics of nicotine and cotinine. *Int J Clin Pharmacol Therp Toxicol.*, **3**: 31:417-421.

Kiremath G.K. 2000. Economics of bidi vs competing crops. Paper presented at the Tobacco Seminar. 29-30 November 2000, CMDR, Dharwad, Karnataka.

Kohls Richard L. and Joseph N. Uhl. 1980. Marketing of Agricultural Products, 5th ed. New York: Macmillan,612 pp.

Koiwai A, Suzuki F, Matsuzaki T, Kawsashima N. 1983. The fatty acid composition of seeds and leaves of Nicotiana species. *Phytochemistry.,* **22**: 1409-12.

Lakshminarayana R. 1996. *Tobacco.* pp 401-415.In Paroda RS and Chadha KL. (Ed) 50 Years of Crop Science Research in India- ICAR, New Delhi: Vedams.

Lee J.H., A. Hebel and F. Schoffl. 1995. Derepression of the activity of genetically engineered heat shock factor causes constitutive synthesis of heat shock protein and increase thermotolerance in transgenic Arabidopsis. *Plant Journal,* 8: 603-612.

Lee, Warren F., Michael Boehlje, Aaron G. Nelson, and William G. Murray. 1980.Agricultural Finance, 7th ed. Ames, Iowa: Iowa State University Press, 437 pp.

Leftwich Richard H. and Ross D. Eckert. 1982. The Price System and Reso'U1'Ce Allocation, 8th ed. Hinsdale, Ill.: The Dryden Press, 500 pp.

Ljungdahl H. 1922. Zur Zytologie der Gattung Papaver. *Svensk. Bot. Tids.,* **16**: 103-114.

Lock R.H. 1909. A preliminary survey of species crosses in the genus *Nicotiana* from the Mendelian standpoint. *Ann. Roy. Bot. Gar. Peradeniya,* **4**: 195-227.

Mahapatra A.S. Abdus Sattar and S. Amranath . 2008. Thermal requirements of chewing robacco (*Nicotiana tobaccum* L.) under North Bihar condition. *Environmental & Ecology,* **26** (2A): 877-879.

Manuel J. Reigosa, Nuria Pedrol, Luís González.2006. Allelopathy: A Physiological Process with Ecological Implications, Springer Publication 637 p.

Mathur Y. K. and K.D. Uppadhya . 1991. *A textbook of Entomology.* The Goal Publ. Co., Baraut, India.

Mavi H.S. 1986. *Introduction to agrometeorology*. The Oxford IBH Publ. Co., New Delhi, India.

Mc Bride, J.S., D.G. Altman, M. Klein and W. White. 1998. Green tobacco sickness. *Tobacco Control.*, 7:294-298.

Mc Cants C.B., and W.G. Woltz. 1967. Growth and mineral nutrition of tobacco. *Advances in Agronomy.*, **19**: 211-265.

Mitchell E. 1850. Strangulated inguinal hernia reduced by an infusion of tobacco, swallowed by mistake, *Am. J. Med: Sei.*, **9**: 548.

Moll H. 1926. The treatment of post-encephalitic Parkinsonism by nicotine, *Brit. Med. J*, **10** :79-81.

Mullett C.F. 1940. Tobacco as a drug in earlier English medicine. *Ann Med Hist.*, **2**: 110-23.

Murthy G.S., B.T. Sajnani and K.C. Dhyani. 1961. Improvement of hookah and chewing tobacco in North Bihar. *Indian J. Genet.*, **21** : 201-5.

Murthy A.G.K. and N.C. Gopalchari. 1982. Chemical quality characteristics of chewing tobacco (*N. tobacum* L) grown in Bihar State. *Tob. Res.*, **8** : 178-181.

Murty B. R., Murty, G. S. and Pavate. M. V. 1962. Studies on the quantitative inheritance in *Nicotiana tabacum* L. II. Components of genetic variation for flowering time, leaf number grade performance and leaf burn · *Zuchter.*, **32**: 361-368.

NAAS 2002. *Agriculture – interface*: value added farm products policy paper16.

Narayan R. K. 1987. Nuclear DNA changes, genome differentiation and evolution in *Nicotiana* (*Solanaceae*). *Plant Syst. Evol.*, **157**: 161–180.

NCAER (National Council of Applied Economic Research). 1994. Export potential of the tobacco sector. Study, including sample survey carried out through the Rapid Rural Appraisal technique, sponsored by the Tobacco Institute of India, New Delhi.

Newhouse P.A. and J.R Hughes. 1991.The role of nicotine and nicotinic mechanisms in neuropsychiatric disease. *Br J Addict.*, **86**: 521-216.

Nilelas Homberg and Leif, Bulow. 1998. Improving stress tolerance in plants by gene transfer-review. *Trends in Plant Sciences* 3: 61-66.

NIOSH 1993. Update: NIOSH issues warning to tobacco harvesters. USDHHD (NIOSH) Publications. No.93-115.

NSSO (National Sample Survey Organization). 1996. Survey on household consumer expenditure. SARVEKSHANA, No. 69 (October-December). Department of Statistics, Ministry of Planning and Programme Implementation, Government of India.

NSSO. 1998. A note on consumption of tobacco in India. NSS 50th Round (1993-94). Sarvekshana, **21**(3). (No. 74, January-March).

Panchamukhi, P.R. 2000. Economics of shifting from tobacco: an action-cum-basic research study. In: CMDR, 2000, q.v.

Pandey A. K. and R. Sinha. 1984. Relative performance of different combinations of organic manures and inorganic fertilizers in chewing tobacco. *Tob. Res.*, **10** : 113-18.

Patel K.A. and M. S. Swaminathan 1967. Mutation Breeding in Tobacco: *Tobacco Science.*, **5**: 67-69.

Penson John, Rulon Pope, and Michael Cook. 1986. Introduction to Agricultural Economics, Englewood Cliffs, N.J.: Prentice-Hall, 556 pp.

Peterson Willis L. 1983. Principles of Economics: Micm, 5th ed. Homewood, Ill.: Richard D. Irwin, 359 pp.

Peto R, A. D. Lopez, J. Boreham , M. Thun, C. Heath Jr, and R. Doll. 1996. Mortality from smoking worldwide. *Br Med Bull.*, **52**: 12-21.

Poehlman J.M. and D. Borthakar. 1970. *Breeding Asian Field Crops*. Oxford & IBH Publishing Co., New Delhi.

Posselt W Reimann L. 1828. Chemische Untersuchungen des Tabaks und Darstellung des eigenhumlichen wirksamen Principes dieser Pflanze. Geigers Magazin der Pharmazie., **24** : 138-61.

Pradhan S. 2002. *Agricultural Entomology and pest control*. Indian Council of Agricultural Research, New Delhi.

Prasad B., R. Sinha and J. Prasad. 1986. How to get targeted yield of tobacco in calcareous soil. *Tob. News.*, 9 : 15-17.

Prasad B. and R. Sinha. 1987. Fertilizer recommendations for economic yield of chewing tobacco brand in soil test levels in calcareous soil. *Tob. Res.*, **13**: 27-31.

Prasad B., R. Singh and J. Prasad. 1984. Fertilizer requirements for different yield targets of chewing tobacco based on soil test values in a calcareous soil of Bihar. *Tob. Res.*, **10**: 126-81.

Prasad D. 2005. *Crop protection Management Strategies*. Daya Publishing House.

Prasannasimharoa G.S.B and G. Sriramamurthy. 1968. Heterosis and combining ability in tobacco (*N. tobacum* L). *Indian J. Agirc Sci.*, **88** : 221-9.

Prasannasinghrao G.S.B. 1973. HP 6-20, a new high yielding tobacco hybrid selection for North Bihar. *Indian Tobacco Bull.*, 5: 7-8.

Prasannasinghrao, G.S.B. and B.T. Sajnani. 1970. Improvement of chewing tobacco (*N. tobacum L*) by hybridization for North Bihar. *Indian J. Applied Agric. Sci.*, **2**: 41-7.

Prasannasinharao, G.S.B. 1977a. Effect of different level of nitrogen spacing and topping on the quantitative and qualitative improvement of chewing tobacco *(N. tobaccum L). Indian Tob. J.,* **9** : 5-7.

Prasannasinharao, G.S.B. 1977b. Varietal improvement in yield and quality of chewing tobacco. *Tob. Res.,* **3**: 56-58.

Prasannasinharao G.S.B. and B.T. Sajnani. 1971. Improvement of chewing tobacco (*N. tobacum L*) for North Bihar through hybridization in North Bihar. *Indian Tob. Bull,.* **1**: 4-7.

Pujadas-Salvá, A.J. and Velasco, L .2000. Comparative studies on *Orobanche cernua* L. and *Orobanche cumana* Wallr. (Orobanchaceae) in the Iberian Peninsula. *Botanical Journal of the Linnean Society.,* **134**: 513-527.

Quandt S.A., T. A. Arury, J.S. Preisser, D. Northon and C. M. Austin. 2000. Migrations of farm workers and green tobacco sisckness: New issues for an understudied disease. *Am J. Industr Med.,***37**: 307-315.

Raghavaiah C.V. and Subbrao. 1986. Herbicidal Control of weed in monsoon growth barley tobacco. *Indian J. Agron.,* **31**:331-334.

Rangaswami G. 1996. Diseases of Crop plants in India Printice-Hall of India Pvt. Ltd. New Delhi, India.

Ranney H. D. 1848. Tobacco antispasmodic, *Boston Med. & Sci.* J., **37**: 182-183.

Rao K. A. and K. V. Krishnan.murthy 1963. Studies on Multiple Polyploids in *Nicotiana. Genitica.,* **34**: 66-78.

Rao V.S. 1986. *Principle of Weed Science.* Oxford and IBH Publishing Co. Pvt. Ltd, New Delhi.

Rath G.K. and Chaudhry K. 2000. Cost of tobacco-related diseases. Paper presented at the WHO International Conference on Global Tobacco Control Law: Towards a WHO Framework Convention on Tobacco Control. 7-9 January 2000, New Delhi.

Reddy K. Srinath and Prakash C. Gupta. 2004. *Report on Tobacco Control in India.* Ministry of Health and Family Welfare, Gov India.

Reddy P.R.S., K. D. Singh and K.C Jain Achievements in tobacco research .2006. Central Tobacco Research Institute, Rajahmundry.

Reddy S.R. 2000. *Principle of crop production.* Kalyani Publishers, Ludhiana, India.

Reddy S. R. 2004. *Agronomy of field crops.* Kalyani Publishers, Ludhiana, India.

Reddy T. Yalmanda. and G.H. Sankra, Reddi. 2002. *Principle of Agronomy.* Kalyani Publishers, Ludhiana, India.

Reese J. J. 1871. Antagonism of poisons. 2. Antagonism of morphia and atropia. 3. Antagonism of strychnia and prussic acid, tobacco, morphia, etc., *Am. J. Med. Sci.,* **61**: 373-85.

Ren N. and M. P Timko. 2001. AFLP analyis of genetic polymorphism and evolutionary relationships among cultivated and wild *Nicotiana* species. *Genome.*, **44**:559–571.

Rice E.L. 1984. Allelopathy. 2nd ed. Orlando, Florida, USA, Academic Press.

Richardson, B. W. **1867. Half**-yearly report on toxicology, forensic medicine, and hygiene. *Brit. & For. Med.-Chir. Rev.*, 39: 243-245.

Rizvi S. J. H., and V, Rizvi (edt.) 1992. *Allelopathy* - basic and applied aspect - Chapman and Hall Publisher, London.

Rohlf F. J. 1998. NTSYS/PC, numerical taxonomy and multivariate analysis system. version 2.02, Exter Software, Setauket, New York.

Rothman S. and J. H. Mc Creary, 1949. Pustular psoriasis, healed after tobacco juice soaks, *Arch. Den, & Syph.*, **60**: 896-7.

Roy Ewell P., Floyd L. Corty and Gene D. Sullivan. 1981. Economics: Applications to Agriculture and Agribusiness, 3rd ed. Danville, Ill.: The Interstate Printers and Publishers, 455 pp.

Roy P.K., A. Kumar and R.G. Pandey. 1988. Moisture regions of light textured calcareous soil under varying preplan ting irrigations of tobacco crops. *Tob. Res.*, **14** : 36-42.

Saha L. R. 2002. *Hand book of Plant Disease.* Kalyani Publishers, Ludhiana, India.

Sajnani B.T. 1957. Influence of topping and spacing on yield and quality of hookah and chewing tobacco (*N. tobacum L*) in North Bihar. *Indian Tob.*, **7** : 165-76.

Sajnani B.T. and K.C. Dhyani. 1959a. Effect of rabbing and shading of nursery on growth and number of transplantable seedling and yield of transplanted crop of hookah and chewing tobacco (*N. tobacum L*). *Indian Tob,.* **9** : 60-64.

Sajnani B.T. and K.C. Dhyani. 1959b. Effect of different seed rate in the nursery on growth and number of transplanttable seedlings and yields of transplanted crop of hookah and chewing tobacco. *Indian Tob,.* **9** : 121-126.

Sajnani B.T. and K.C. Dhyani. 1959c. Fertilization of hookah and chewing tobacco *(N. tobacum* L) at varying level of NPK. *Indian Tob.*, **9** : 157-64.

Sajnani B.T. and K.C. Dhyani. 1962a. Quality in relation the leaf position on the stalk and the proportion of leaf to stalk in hookah and chewing tobacco (N.*tobac*um L) of North Bihar. *Indian Tob.,* **12** : 53-7.

Sajnani B.T., K.C. Dhyani and A.N. Singh. 1963a. Effect of piercing of main stem after topping on different varieties of hookah and chewing tobacco (*N. tobacum L*). *Indian Tob.*, **13** : 22-30.

Sajnani B.T., K.C. Dhyani and A.N. Singh 1961. Suppression of suckers in hookah and chewing tobacco (*N. tobacum L*) by some vegetable oils. *Indian Tob.*, **11**: 96-102.

Sajnani B.T., K.C. Dhyani and A.N. Singh. 1962b. Effect of mixed cropping with shallow rooted crop on the yield and quality of hookah and chewing tobacco. *Indian Tob.*, **12**: 131-140.

Sajnani B.T., K.C. Dhyani and A.N. Singh. 1963b. Suppresion of suckers in hookah and chewing tobacco (*N. tobacum L*) by chemicals. *Indian Tob.*, **13** : 61-7.

Sajnani B.T., K.C. Dhyani and A.N. Singh. 1964a. Effect of placement of fertilizers in plough furrow and broadcasting versus present method of cultivation. *Indian Tob.*, **14**: 11-14.

Sajnani B.T., K.C. Dhyani and A.N. Singh. 1964b. Effect of different organic and inorganic manures on the yield and quality of hookah and chewing tobacco (*N. tobacum L*). *Indian Tob.*, **14** : 66-70.

Sajnani B.T., K.C. Dhyani and A.N. Singh. 1965. Improved method of growing hookah and chewing tobacco in Bihar for higher yield and better quality. *Indian Tob.*, **15** : 71-6.

Sakamura T. 1918. Kurze Mitteilung uber die Chromosomenzahlen und die Verwandtschaftsverhaltnisse der Triticum-Arten. *Bot. Mag. Tokyo.*, **32**: 151-154.

Sarala K. and R. V. S. Rao 2008. Genetic diversity in Indian FCV and burley tobacco cultivars *J. Genet.*, **87**:2:159-163.

Saxena R. C. and R.C. Srivastava. 2007. *Entomology at glance.* Agrotech Publishing Academy, Udaipur, India.

Schoffl F., R.P. and A. Reindl. 1998. Regulation of heat shock response. *Plant Physiology*, 117: 1135-1145.

Setchell W.A. 1912 Studies in *Nicotiana.* I. Univ. of California Publ. *Botany* **5**: 1-86.

Sevenson C.K. 1987. Clinical pharmacokinetics of nicotine. *Clin. Pharmacokinte.*, **12**: 30-40.

Shamel A. D. 1910. *Tobacco Breeding. American Breeders Association Report.*, **6**: 268-275.

Shepherd Geoffrey S. 1966. Agricultural Price Analysis, 5th ed. Ames, Iowa: Iowa State University Press, 328 pp.

Shepherd, Geoffrey S., Gene A. Futrell and J. Robert Strain. 1976. Marketing Farm Products-Economic Analysis. Ames, Iowa: Iowa State University Press, 485 pp.

Silvette H, Larson PS, Haag HB. 1958. Medical uses of tobacco: past and present. *Virginia Med Monthly.*, **85**: 472-84.

Singh A.N. and G.S.B. Prasannasinharao. 1970. A note on a safeguarding the tobacco nursery form earth worm damage in Bihar. *Indian J. Applied Agric. Sci.,* **1**:1-2.

Singh K. D., S. N. Tripathi, M. Someswararao and R. Subbarao. 1993. Nutritional requirements of chewing, hookah, cigar, wrapper, filler and cheroot tobacco. *In plant nutrition erect on production and quality of tobacco* Ed. *Dev. Et al.* pp 125-140.

Singh K. D., S.N. Tripathi and A. K. Pandey. 1999a. Effect of sucker control methods on yield, quality and economics of chewing tobacco under North Bihar Condition. *Tob. Res.,* **25** : 18-22.

Singh K. D., V. Krishnamurthy and A. K. Pandey. 2000. Studies on soil fertility management in tobacco based sequential cropping system under irrigated and rainfed conditions of Bihar. *Tob. Res.,* **26**: 28-36.

Singh K.D. 2002. Analysis of Indian Situation. *Survey of Indian Agricultural.* The Hindu.

Singh K.D., S.N. Tripathi, R.G. Pandey and A. K. Pandey. 1994. Influence of different methods of inercropping on yield, quality and economics of chewing tobacco (*N. tobacum* L). *Tob. Res.* **20**: 97-101.

Singh AK Vandana, Tyagi Deep Chand and RV Singh. 2005. Indigenous Knowledge and Ethnobotany associated with saffron in Kashmir. *Indian J. Plant Genet. Resour.,* **18** 191-195.

Singh C. 1985. *Modern Technology for raising field crops* Oxford & IBH Publishing Co. New Delhi.

Singh C.P. Singh and R Singh. 2003. *Modern techniques of raising field crops.* Oxford & IBH Publishing Co. Pvt. Ltd., New Delhi.

Singh K. D. and S. Amarnath. 1998. *Management practices for chewing and hookah tobacco in Bihar* .CTRI, Rajahmundry, India.

Singh K. D., A. K. Pandey and J. A. V. Prasadrao. 2004. Influence of different proportion of urea, Ammonium Sulphate and Castor Cake on yield and physical leaf quality characteristics of chewing tobacco in Bihar. *Tob. Res.,* **30**: 11-5.

Singh K. D., and A.K. Pandey.2000. Studies on economically viable ration of oil seed cakes and organic mixtures in combination with fertilizer N in chewing tobacco under non-saline calcareous soils of Bihar. *Tob. Res.,* **26:** 44-51.

Singh K. D., J.A.V. Prasadrao and A.K. Pandey. 2002. Effect of continuous application of NPK fertilizers on chewing tobacco and soil fertility status in sandy loam calcareous soils of Bihar. *Tob. Res.,* **28**: 119-128.

Singh K.D. A. N. Singh and P. S. Upadhyay. 1981. A note on the affect of irrigation and nitrogen levels on yield and quality of chewing tobacco. *Tob. Res.*, **7**: 108-112.

Singh K.D. and R. Sinha. 1994. Parallel multiple cropping of sugarcane and tobacco. *Indian J. Agron.*, **30** : 334-38.

Singh K.D. R. Sinha and P.S. Upadhyay. 1974. Studies on individual and combined effect of NPK on chewing tobacco (*N. tobacum* L) in non-saline calcareous soil of *North Bihar. Indian Tob. Bull.*, **6**: 3-6.

Singh K.D., R. Sinha, A.N. Singh and P. S. Upadhyay. 1976. Comparative study of different cropping sequences with tobacco for small holdings in North Bihar. *Indian J. Agric. Sci.*, **46**: 141-146.

Singh K.D., S.N. Tripathi and A. K. Pandey. 1995. Response of chewing tobacco varieties to plant spacing and nitrogen levels under nonsaline calcareous soil of North Bihar. *Ann. Agric. Res.*, **16** : 330-5.

Singh K.D., S.N. Tripathi and A.K. Pandey. 1999b. Effect of level and ratio of organic and inorganic nitrogen on yield physical leaf quality parameters and economics of chewing tobacco. *Tob. Res.*, **25:** 9-17.

Singh K.D., S.N. Tripathi and A.K. Pandey. 1998. Studies on feasibility and economic viability of tobacco based intercropping system in Bihar. *Tob. Res.*, 24: 70-76.

Singh R.S. 1998. *Plant Disease*. Oxford and IBH Publishing Co. Pvt. Ltd, New Delhi India.

Singh R.V, Deep Chand, Vandna Tyagi, Nidhi Verma, A.K. Singh, S.P. Singh and Surender Singh. 2004. *Plant Germplasm Reporter*. (2003) Vol. 3, No.4, 220p. NBPGR, New Delhi.

Singh R.V., Deep Chand, Vandana Tyagi, A.K. Singh, S..P Singh and Surender Singh.2004. Important Crop Germplasm Introduced into India during 2004. *Indian J. Plant Genet. Resour.*, **17** : 196-205 .

Singh R.V., Deep Chand, Vandana Tyagi, A.K. Singh, S..P Singh and Surender Singh. 2005. *Plant Germplasm Reporter* (2005) Vol. 5 No.1 124 p. NBPGR, New Delhi.

Singh R.V., Deep Chand, Vandana Tyagi, A.K. Singh, S.P. Singh and Surender Singh. 2003. Important Crop Germplasm Introduced into India during 2003 *Indian J.Plant Genet. Resour.*, **16** : 214-221.

Singh R.V., Deep Chand, Vandana Tyagi, A.K. Singh, S.P. Singh and Surender Singh. 2004. *Plant Germplasm Reporte* (2004) Vol. 4 No.3 176 p. NBPGR, New Delhi.

Singh R.V., Deep Chand, Vandana Tyagi, A.K. Singh, S.P. Singh and Surender Singh. 2005. *Plant Germplasm Reporte* (2005) Vol. 5 No.2 267 p. BPGR, New Delhi.

Singh R.V., Deep Chand, Vandana Tyagi, A.K., Singh, S.P., Singh and Surender Singh, 2004. *Plant Germplasm Reporter* (2004) Vol. 4 No. 4, 102 p. NBPGR, New Delhi.

Singh R.V., Deep Chand, Vandana Tyagi, A.K., Singh, S.P., Singh and Surender Singh, 2005. *Plant Germplasm Reporter* (2005) Vol. 5 No. 2, 267p. NBPGR, New Delhi.

Singh R.V., Deep Chand, Vandana Tyagi, A.K. Singh, SP Singh, Surender Singh and P.C. Binda. 2005. Priorities for Introduction of Fruit Crops in India. *Indian J.Plant Genet. Resour* **18** : 67-68.

Singh R.V., Deep Chand, Vandana Tyagi, A.K. Singh., S.P. Singh and Surender Singh. 2004. Important Crop Germplasm Introduced into India during 2004 *Indian J.Plant Genet. Resour.*, **17** : 196-205.

Singh R.V., Deep Chand, Vandana Tyagi, Nidhi Verma, A.K. Singh, S.P. Singh and Surender Singh .2004. *Plant Germplasm Reporter* (2004) Vol. 4 No.1 248 p. NBPGR, New Delhi.

Singh R.V., Deep Chand, Vandna Tyagi, Nidhi Verma, A.K. Singh, SP Singh and Surender Singh. 2004. *Plant Germplasm Reporter.* (2003) Vol. 3, No.3, 203p. NBPGR, New Delhi.

Singh R.V., I.P. Singh, Deep Chand, Vandana Tyagi, A.K. Singh, S P Singh and Surender Singh. 2005. *Plant Germplasm Reporter.* (2005) Vol. 5 No.4 131 p. NBPGR, New Delhi.

Singh R.V., IP Singh, Deep Chand, Vandana Tyagi, A.K. Singh, S. P. Singh and Surender Singh. 2006. *Plant Germplasm Reporte.* (2006) Vol. 6 No.2 324 p. NBPGR, New Delhi.

Singh R.V., IP Singh, Deep Chand, Vandana Tyagi, AK Singh, S P Singh and Surender Singh 2006. *Plant Germplasm Reporter.* (2006) Vol. 6 No.1 222 p. NBPGR, New Delhi.

Singh R.V., IP Singh, Deep Chand, Vandana Tyagi, AK Singh, SP Singh and Surender Singh .2005. *Plant Germplasm Reporter.* Vol. 5 No.3 (2005) NBPGR, New Delhi. 402 p.

Singh R.V., Deep Chand, Vandana Tyagi, Nidhi Verma, AK Singh, SP Singh and Surender Singh 2004. *Plant Germplasm Reporter* (2004) ol. 4 No.2 110 p. NBPGR, New Delhi.

Singh R.V., Deep Chand, Vandna Tyagi, Nidhi Verma, AK Singh, SP Singh and Surender Singh .2003. *Plant Germplasm Reporter* (2003) Vol. 3, No.2, 142p. NBPGR, New Delhi.

Singh S.S. 1993. *Crop management under irrigated and rainfed conditions*. Kalyani Publishers Ludhiana.

Single, E.; Collins, D.; Harwood, H.; Lapsley, H.; and Maynard, A. (1995), *International Guidelines for Estimating the Costs of Substance Abuse*, Canadian Centre on Substance Abuse, available online at: http://www.ccsa.ca).

Sinha R. and A. N. Singh. 1977. Influence of weedicides on occurrence and control of weeds in chewing tobacco (*N. tobacum* L). *Tob. Res.*, **3** :73-76.

Sinha R. and G.S.B. Prasannasinghrao. 1976. Varietal improvements in yields and quality of chewing tobacco (*N. tobacum* L) for North Bihar. *Tob. Res*, **2**: 148-53.

Sinha R., R. G. Pandey and S.S.L. Dwivedi. 1978. Studies on crop performance and economics of chewing tobacco under different source and timing of nitrogen application. *Indian J. Agron.*, 23 : 261-266.

Sinha R. S.S.L. Dwivedi, A.K. Pandey and B. Kumar. 1985a. Critical stages of chewing tobacco for irrigation. *Indian Tob.* J., **16**: 6-8.

Sinha R. and M. Subrahmanyam. 1976. Stability of performance of chewing tobacco (*N. tobacum L*) Varities in North Bihar. *Tob. Res.*, **2**: 159-61.

Sinha R. and B. Prasad. 1986. Fertilizer scheduling for the targeted yield of crops under tobacco-maize cropping pattern of North Bihar. *Indian Tob.* J. **18** : 10-4.

Sinha R., A. N. P. N and Keolyar and A. K. Pandey. 1982a. Plant spacing in hookah tobacco under irrigated and rainfed condition of North Bihar. *Tob. Res.*, **8** : 61-3.

Sinha R., K. D. Singh, A. N. Singh and P. S. Upadhyay. 1975. Effect of different crop rotations on the yield and quality of chewing tobacco. *Indian J. Agric. Sci.*, **48** : 128-31.

Sinha R., K.D. Singh, S.S.L. Dwivedi, A.K. Pandey and RN. Mishra. 1984e. Companion cropping of sugarcane and tobacco. *Maharstra Sugar*. **9** : 53-59.

Sinha R., P. S. Upadhyay, A. N. Singh and K. D. Singh. 1976b. Effect of method of applications and levels of nitrogen and phosphorus on yield and quality of chewing tobacco. *Indian J. Agric.*, **21** : 221-225.

Sinha R., R.C. Jha, S.S.L. Dwivedi and A. K. Pandey. 1985c. Effect of preceding irrigation and cultural practices on growth yield and quality chewing tobacco ratoon topped off at different levels. *Tob. Res.*, **11** : 172-74.

Sinha R., R.G. Pandey and S.S.L. Dwivedi.1976a. Effect of continuous application of organic manure and inorganic fertilizer with and without irrigation on yield and quality of chewing tobacco (*N. tobacum* L) in sandy loam calcareous soils of North Bihar. *Tob. Res.*, **2** : 94-108.

Sinha R., S. Amarnath and G.S.B. Prasannasinharao and A.K. Pandey. 1984d. Promisingh varieties of chewing tobacco (*N. tobacum* L) for yield and quality in North Bihar. *Indian Tob.* J., **16** : 5-8.

Sinha R., S. Amarnath. G.S.B. Prasannasinharao and A. K. Pandey. 1979b. Performance of early maturing hybrid selections of chewing tobacco for North Bihar. *Tob. Res.,* **5** : 109-110.

Sinha R., S. S.L. Dwivedi and A.K. Pandey. 1979a. Effect of source and time of nitrogen application on chewing tobacco and the succeeding maize in North Bihar. *Tob. Res.,* **5** : 70-76.

Sinha R., S.S.L. Dwivedi and A. K. Pandey . 1985b. Effect of time of harvesting of yield and quality of chewing tobacco affected by rain at maturity.. *Tob. Res.,* **11** : 6-11.

Sinha R., S.S.L. Dwivedi and A. K. Pandey. 1982b. Effect of slow release N-fertilizers and nitrification inhibitors on chewing tobacco and on succeeding maize. *Tob. Res.,* **8** : 107-8.

Sinha R., S.S.L. Dwivedi and A. K. Pandey. 1984a. Comparative performance of chewing tobacco (N. tobacum L) in various cropping patterns under rainfed condition of North Bihar. *Indian Tob. J.,* **29**: 61-7.

Sinha R., S.S.L. Dwivedi and A. K. Pandey. 1985d. Effect of seedling age on growth yield and quality of chewing tobacco (*N. tobacum* L). *Tob. Res.,* **11** : 138-42.

Sinha R., S.S.L. Dwivedi and A. K. Pandey. 1985e. Effect of method of harvesting and curing on yield and quality of chewingh tobacco. *Tob. Res.* **11**: 28-34.

Sinha R., S.S.L. Dwivedi and A. K. Pandey. 1985f. Performance of chewing tobacco under varying doses and method of compost application. *Tob. Res.,* **11** : 71-75.

Sinha R., S.S.L. Dwivedi, A. K. Pandey and R. C. Jha. 1984c. Growth and yield pattern of hookah tobacco (*N. rustica* L) under different lavels of planting and topping levels. *Tob. Res.,* **10** : 123-31.

Sinha R., SSL. Dwivedi, A.K. Pandey and R. G. Pandey. 1984b. Planting time, spacing and nitrogen requirements of chewing tobacco (*N tobacum* L) varieties Gandak Bahar and Sona in Bihar. *Indian Tob.* J., **16**: 9-13.

Sinha,R., R.N. Mishra, A. K. Pandey and S.S.L. Dwivedi. 1982. Intercropping of tobacco and sugarcane in North Bihar. *Tob. Res.,* **8** : 119-31.

Sinnha R., A.K. Pandey and S.S.L. Dwivedi. 1986. Effect of *azotobactor* on the growth and yield of chewing tobacco under different fertility levels. *Tob. Res.,* **12** : 56-63.

Sinnha R., A.K. Pandey and S.S.L. Dwivedi. 1991. Effect of soil v/s foliar application of nitrogen on yield and economics of chewing tobacco raised in Bihar. *Tob. Res.,* **17** : 17-20.

Sjo, John. 1976. Economics for Agricultumlists: A Beginning Text in Agricultural Economics. Columbus, Ohio: Grid, 232 pp.

Smith H. H. 1941. Polyploidy in *Nicotiana. American Naturalist.,* **75**: 307-309.

Smith T. E. and E. E. Clayton. 1948. Inheritance of Resistance to bacterial Wilt in Tobacco. *J. of Agricultural Research.,* **76**: 27-32.

Snodgrass, Milton M. and L.T. Wallace. 1980. Agriculture, Economics, and Resource Management, 2nd ed. Englewood Cliffs, N.J.: Prentice-Hall, 546 pp.

Somervail A. 1838. On the use of tobacco in certain cases, *Am. J. Med. Sci.,* **23**: 518.

Sriramamurthy G. and M. Umamaheswararao. 1969. Effect of molding on growth and yields of tobacco crop in relation to soil temperature and moisture. *Indian J. Agric. Sci.,* **39**: 742-748.

Sriramamurthy G. and G.S.B. Prasannasinharao. 1968. Further studies on heterosis and combining ability in dialed crosses in tobacco (*N. tobacum* L). *Indian J. Agric. Sci.,* **38** : 864-69.

Srivastava R. P. and A. S. Bhaskar. 1988c. Biochemical changes in leaf pigments and chemical constituents during curing of chewing tobacco. *Tob. Res.,* **14**: 79-82.

Srivastava R. P. 1984. Can smoking be made less hazardous? *Tob News.,* **7**: 4-5.

Srivastava R. P. and A. S. Bhaskar. 1990. Dry matter accumulation and nutrient up take by chewing tobacco (*N. tobacum* L) at different stages of growth in calcareous soils of Bihar. *Tob. Res.,* **16** : 19-23.

Srivastava R. P. and A. S. Bhaskar. 1987. Influence of viral diseases on chemical composition and quality parameters of chewing tobacco (*N. tobacum* L). *Tob. Res.,* **13**: 153-57.

Srivastava R. P. and A. S. Bhaskar. 1988a. Physical and chemical changes occurring during formulation and aging of chewing tobacco (*N. tobacum* L). *Indian Tob. J.,* **14** :12-14.

Srivastava R. P. and A. S. Bhaskar. 1988b. The effect of method of harvest and curing on yield and quality characteristic of chewing tobacco. *Tob. Res.,* **14**: 32-37.

Stewart G.G. 1967. A history of the medicinal use of tobacco 1492-1860. *Med Hist.,* **11**: 228-68.

Stewart J.S. 1909. 'The Production of a New Strain of Tobacco and its Development. *American Breeders Association Report.,* **5**: 291-298.

Stille 1874. *Tabacum.* Tobacco, *Therapelltics & Materia Medica.,* **2**: 359-385.

Subramanian S. 1999. *All about weed control.* Kalyani Publishers, Ludhiana, India.

Thompson, J. H. 1842. Tobacco in hysteria, *Am. J. Med. Sci.,* **3**: 497-8.

Tomek William G. and Kenneth L. Robinson. 1981. Agricultural Product Prices, 2nd ed. Ithaca, N.Y.: Cornell University Press, 367 pp.

Tripathi N., K. D. Singh, A.K. Pandey and G.S.V. 1995a. Subrahamanyam. Relative efficacy of prilled urea and large granule urea at varying levels of nitrogen on leaf growth, yield and physical quality parameters of chewing tobacco (*N. tobacum* L). *Ann. Agric. Res.*, **16**: 298-302.

Tripathi S. N., A. K. Pandey, K. D. Singh, R.G. Pandey and A. K. Pandey. 1994. Influence of organic and inorganic nitrogen levels on yield and quality of chewing tobacco. *Tob. Res.*, **20**: 118-23.

Tripathi S. N., A. K. Pandey, K.D. Singh and S. Amarnath. 1992. Influence of plant spacing, nitrogen and topping levels on yield, quality and economics of chewing tobacco (*N. tobacum* L) variety PT-76. *Tob. Res.*, **18** : 39-44.

Tripathi S. N., K. D. Singh and A. K. Pandey. 1995b. Influence of crop rotations on yield, quality and economics of chewing tobacco. *Tob. Res.*, **21** : 23-7.

Tripathi S. N., K. D. Singh and A. K. Pandey. 2000. Effect of irrigation, nitrogen and time of application on yield and quality of chewing tobacco of North Bihar. *Tob. Res.*, **26**: 1-7.

Tweeten, Luther. 1979. Foundations of Farm Policy, 2nd ed. Lincoln, Neb.: University of Nebraska Press, 559 pp.

Tyagi Vandana, A.K. Singh, Nidhi Verma, Deep Chand, R.V. Singh and B.S. Dhillon. 2006. Plant Introduction in India during Pre-and Post-CBD periods- An analysis *Indian J.Plant Genet Resour.*, **19** : 436-441.

Vaidya S.G., Vaidya, J.S., and Naik, U.K. 1999. Results of a national study in India: sport sponsorship by cigarettes companies influences the adolescent children's mind and helps initiate smoking. *Journal of the Indian Medical Association*, **97**: 354-359.

Verma D.K. 1999. *Applied Entomology*. Mittal Publication, New Delhi India .

Volkov R.A. 1999. Elimination and rearrangement of parental rDNA in the allotetraploid *Nicotiana tabacum*. *Mol. Biol. Evol.*, **16**: 311–320.

Weaks W.A. (1852-53) Poisoning by totacco, *B. Med. & Sci. J.*, **47**: 461-2.

WHO (World Health Organization). 1997. *Tobacco or health: a global status report*. Geneva: WHO.

Wilcox, Walter W., Willard W. Cochrane, and Robert W. Herdt. 1974. Economics of American Agriculture, 3rd ed. Englewood Cliffs, N.J.: Prentice-Hall, 504 pp.

Willey R.W. 1979. Intercropping – Its importance and research needs. Part1. Competition and yield advantages. Part.2. Agronomy and research approaches. *Field Crop abstract.*, **32**:1-10 and 73-85.

Willey R.W. 1985. Evaluation and presence of intercropping advantages. *Exptl. Agric.*, **21**:119-133.

Willey R.W. 1990. Resource use in intercropping system. *Agric. Water Management.*, **17**:25-31.

Willey R.W., M. R, Rao and M. Natrajan 1981. Traditional cropping systems with pigeon pea and their improvement. pp 11-25. In proceeding of the Internatoional workshop on Pigeon peas. 15-19 Dec 1980. ICRISAT, Hyderabad, India, Vol. 1.

Wills Walter J. 1979. An Introduction to Agribusiness Management, 2nd ed. Danville, Ill. The Interstate Printers and Publishers, 167 pp.

Witcombe John R, Daljit, S. Virk and John Farrington.1998. *Seeds of Choice*. Oxford & IBH Publishing Co. Pvt. Ltd., New Delhi.

Xu M., M. Zheng and G. Liu. 1998. Genetic diversity in cultivars of *Nicotiana tabacum* revealed by RAPD polymorphism. *J. Agric. Biotech.*, **6**: 282–284.

Yadav R.L. 1989. Long term effects of manures and fertilizers on sugarcane based cropping system. *Fertil. News.*, **34**:53-59.

Yi H., Y., R.C. Rufty and E. A. Wernsman. 1998. Mapping root-knot nematode resistance gene (RK) in tobacco with RAPD markers. *Plant Dis.*, **82**: 1319–1322.

Yokota T., S akai H., Okuno K., Yoneyama K., Takeuchi Y. 1998 *Photochemistry.*, **49** : 17 1-173.

Zandstra H.G.1977. *Cropping system research for the Asian rice farmer*. In Proc. Symp. Cropping System Research and Development for Asian Rice Farmer. 21-24 Sept.1976. IRRI, Los Banos, Philippines. pp.11-30.

Zhang H.Y., Liu X. Z., C. S. He and C. M. Zheng. 2005. Random amplified DNA polymorphism of *Nicotiana tabacum* L. *Cultivars. Biol. Plant.*, **49**: 605–607.

□□□

About to Author

Dr. Anil Kumar Singh is presently working in the Division of Land and Water Management (DLWM) at ICAR, Research Complex for Eastern Region, Patna, as a Senior Scientist Agronomy. He did his graduation (1995) and doctorate (2000) from CS Azad University of Agriculture & Technology, Kanpur and post graduation (1997) from GB Pant University of Agriculture and Technology, Pantnagar. He was a brilliant scholar during of his time as he is recipient of prestigious **ICAR- Junior Research Fellowship (JRF)** for pursuing his master degree. He has passed ICAR, National Eligibility Test (NET) with two subject's *viz.*, Agronomy and Economic Botany. He further proves his academic brilliance when he selected scientist (ARS) with **Economic Botany** being a man of **Agronomy** a major branch of agriculture. Dr. Singh started his career as **Scientist (Economic Botany)** and joined NBPGR Regional Station Srinagar (J&K) in 2000 and worked as a Officer In-charge of the centre. He joined as **Senior Scientist Agronomy** (2006) to the then Crop Research Programme, Pusa, Bihar where he serves as Head In-charge with full determination till its merger with IARI, Regional Station Pusa. Dr Singh has been awarded **Scientist of the Year Award -2009,** for understanding and recognition in the field of Agronomy on the occasion of first Indian Scientist and Farmers Congress held on 3-4 October, 2009, by Society for Recent Development in Agriculture, SVP, Univ. of Agril. & Tech. Meerut (UP). Society for upliftment of rural economy has confers his **Rural Scientist of Biennium, 2008-09,** for his technical contribution in the rainfed farming in the tribal belt of Sonebhadra and Mirzapur District (UP). During this period he has published 20 research articles of repute and 43 scientific and technical popular articles. Dr. Singh is presently handling five research project of International, National and Regional importance.